T0318932

Small-Format Aerial Photography

Principles, Techniques and Geoscience Applications

Small-Format Aerial Photography

Principles, Techniques and Geoscience Applications

James S. Aber
Earth Science Department
Emporia State University
Kansas, United States

Irene Marzolff
Department of Physical Geography
Johann Wolfgang Goethe University
Frankfurt am Main, Germany

Johannes B. Ries
Physical Geography
University of Trier
Trier, Germany

AMSTERDAM • BOSTON • HEIDELBERG • LONDON • NEW YORK • OXFORD
PARIS • SAN DIEGO • SAN FRANCISCO • SINGAPORE • SYDNEY • TOKYO

Elsevier
Radarweg 29, PO Box 211, 1000 AE Amsterdam, The Netherlands
Linacre House, Jordan Hill, Oxford OX2 8DP, UK

First edition 2010

British Library Cataloguing in Publication Data
A catalogue record for this book is available from the British Library

Library of Congress Cataloging-in-Publication Data
A catalog record for this book is available from the Library of Congress

ISBN: 978-0-444-63823-6

Printed and bound in The Netherlands

10 11 12 10 9 8 7 6 5 4 3 2 1

Table of Contents

Dedication

James Aber dedicates his contributions for this book to Susan W. Aber, wife and close colleague, who has assisted with kite and blimp aerial photography throughout the United States and Europe. Her kite flying and photographic skills resulted in many of the images presented in this book, identified as SWA in figure captions.

Preface

Photography has the remarkable power to impress into memory a distillation of a particular segment of time.

L. Schwarm (Schwarm and Adams, 2003)

Why small-format aerial photography? This question is often posed to us by people who work on the ground as well as those who analyze conventional aerial photographs and satellite images. Why indeed?

The authors did not start their small-format aerial photography (SFAP) careers as dedicated kite flyers, hot-air blimp developers, UAV fans, or do-it-yourself gadget builders. We have become aerial photographers out of necessity, because we needed to assess landscapes, forms, and distribution patterns in detail and document their changes through time. We required a feasible, cost-effective method that would adapt to the size of the features, the transitory nature of their occurrence, and the speed of their development, and which would yield the best possible spatial and temporal resolutions for the respective research questions. Suitable conventional photographs and satellite images were either inappropriate or unavailable.

Self-made aerial photographs offer the researcher a maximum of flexibility in fieldwork. Within the technical limits of the camera and platform, the photographer may determine not only place and time but also viewing angle, image coverage, and exposure settings. While imagery acquired from external sources may or may not show the study site at the required scale, time and angle, such tailor-made photos show exactly those sites and features we seek.

Small-format aerial photography also represents a way to enter a realm of airspace that is normally difficult or impossible to access via more traditional and accepted means, namely the ultra-low height range, just a few hundred meters above the ground. This height range is restricted for conventional manned aircraft in many countries and is too high to reach via ladders, booms, or towers in most situations. It is possible, in fact, that ultra-low SFAP is the least utilized means for observing the Earth from above. In this respect, SFAP is a way to bridge the gap in scale and resolution between ground observations and imagery acquired from conventional manned aircraft and satellite sensors.

SFAP enables researchers as well as other professionals and the interested public to get their own picture of the world. Large-scale aerial photographs in many cases help to state more precisely the scientific question, to improve the understanding of processes and to deepen the knowledge of our study sites. This enables us to monitor local changes at the spatial and temporal scales at which they occur and to assess their roles and importance in a constantly changing world. In some cases, we might even learn something altogether new.

We began our SFAP efforts in the age of film cameras. Since then, digital photography has revolutionized both field and laboratory methods. It is possible now to examine the results of an aerial survey within a few minutes after landing or even during the flight. Thus, the survey may be continued and further images can be taken if necessary. Likewise, digital image-processing techniques allow many more methods for laboratory analysis and interpretation of the aerial photographs.

The authors would like to share with the reader their experiences gained during many successful operations with different platforms, but also by some failures due to technical deficiencies and human errors. We feel confident that SFAP also would help other scientists and all those interested in pursuing their own questions and applications, and we hope this book may lead to greater knowledge, application, elaboration, and appreciation of small-format aerial photography, particularly in the geosciences, as well as many other human endeavors.

This book is divided into three major portions. Part I covers introductory material, including history, scope and definitions, basic principles, photogrammetry, lighting and atmospheric conditions, and photographic composition (Chapters 1–5). In part II, SFAP techniques are elaborated for cameras, mounts, platforms, field methods, airphoto interpretation, and image analysis (Chapters 6–11). Case studies are presented in part III with an emphasis on geoscience and environmental applications (Chapters 12–19). Many of these applied examples are drawn from the authors' own field work in the United States, Europe, and Africa.

Acknowledgements

This book represents contributions from many individuals and organizations that have encouraged and supported the authors and helped us to pursue small-format aerial photography. Among those who have played significant roles, we thank our colleagues, collaborators and advisors: Kiira Aaviksoo, Ali Aït Hssaine, Klaus-Dieter Albert, DeWayne Backhus, Andrzej Ber, Michael Breuer, Karl-Ludwig Busemeyer, Cornelius Claussen, Don Distler, Lukas Distler, Debra Eberts, Maite Echeverría Arnedo, Jack Estes, Wolfgang Feller, Anja Fengler, Joachim Feuchter, Melinda Flohr, Darek Gałązka, Rafael Giménez, Marco Giardino, Maria Górska-Zabielska, Ralph Hansen, Jürgen Heckes, Günter Hell, Juraj Janočko, Jill Johnston, Paul Johnston, Martin Kähler, Volli Kalm, Edgar Karofeld, Stephan Kiefer, Brooks Leffler, David Leiker, Kham Lulla, Holger Lykke-Andersen, Marco Koch, Michael Niesen, Matt Nowak, Firooza Pavri, Robert Penner, Jean Poesen, Juan de la Riva Fernández, Michael Runzer, Tilmann Sauer, Marcia Schulmeister, Manuel Seeger, Rod Sobieski, Hans-Peter Thamm, Cheryl Unruh, Friedrich Weber, Heribert Willger, Joachim Wolff, Jürgen Wunderlich, and Ryszard Zabielski.

Many undergraduate and graduate students from Emporia State University, Johann Wolfgang Goethe University Frankfurt am Main, University of Trier, Technical University of Košice, and the University of Tartu have provided ample assistance with field projects. In particular, we wish to acknowledge the help of S. Acosta, C. Boyd, V. Butzen, W. Fister, L. Freeman, B. Fricovsky, C. Geißler, B. Graves, K. Harrell, T. Iserloh, W. Jacobson, A. Kalisch, A. Kimmel, K. Kimmel, T. Korenman, B. Landis, M. Landis, M. Lewicki, J. Liira, S. Lowe, N. Lux, G. Manders, K. Möllits, J. Mueller, K. Muru, L. Owens, S. Plegnière, A. Rager, A. Remke, M. Roche, G. Rock, S. Salley, S. Veatch, M. Vilbaste, A. Wachsmuth, J. Wallace, R. Wengel, E. Wilson-Agin, S. Wirtz, B. Zabriskie, and J. Zupancic. We gratefully acknowledge V. Butzen's help with translations for this book and J. Hackenbruch's assistance with editing and proofreading.

Financial support was provided by Kansas NASA EPSCoR, KansasView, Emporia State University, Northeastern Oklahoma A&M College, Kemper Foundation, the U.S. National Research Council, the U.S. Bureau of Reclamation, the Nature Conservancy of Kansas, National Scholarship Programme of the Slovak Republic, Technical University of Košice, University of Tartu, University of Aarhus, Polish Geological Institute, Deutsche Forschungsgemeinschaft, Vereinigung der Freunde und Förderer der Johann Wolfgang Goethe-Universität Frankfurt am Main, Stiftung zur Förderung der Internationalen Wissenschaftlichen Beziehungen der Johann Wolfgang Goethe-Universität Frankfurt am Main, Landesgraduiertenförderung Baden-Württemberg, Forschungsfonds der Universität Trier, and the Prof. Dr. Frithjof Voss Stiftung—Stiftung für Geographie.

Finally James Aber wishes to thank his sons, Jeremy and Jay, who helped with kite and blimp aerial photography at many sites in the United States and Estonia. Special thanks are offered also from Irene Marzolff and Johannes Ries to Jürgen Heckes (Deutsches Bergbaumuseum Bochum), whose unforgettable *Druckbetankung* crash course started our SFAP career these 15 years ago and got us hooked for good.

February 2010
James S. Aber
Irene Marzolff
Johannes B. Ries

Chapter 1

Introduction to Small-Format Aerial Photography

Small is beautiful.

E. Schumacher 1973, quoted by Mack (2007)

1.1. OVERVIEW

People have acquired aerial photographs ever since the means have existed to lift cameras above the Earth's surface, beginning in the mid-19th century. Human desire to see the Earth "as the birds do" is strong for many practical and aesthetic reasons. From rather limited use in the 19th century, the scope and technical means of aerial photography expanded throughout the 20th century. The technique is now utilized for all manners of earth-resource applications from small and simple to large and sophisticated.

Aerial photographs are taken normally from manned airplanes or helicopters, but many other platforms may be used, including balloons, tethered blimps, drones, gliders, rockets, model airplanes, kites, and even birds (Tielkes, 2003). Recent innovations for cameras and platforms have led to new scientific, commercial, and artistic possibilities for acquiring dramatic aerial photographs (Fig. 1-1).

The emphasis of this book is small-format aerial photography (SFAP) utilizing 35- and 70-mm film cameras as well as compact digital and video cameras. In general terms, such cameras are typically designed for hand-held use, in other words of such size and weight that amateur or professional photographers normally hold the camera in one or both hands while taking pictures. Such cameras may be employed from manned or unmanned platforms ranging in height from just 10s of meters above the ground to 100s of kilometers into space. Platforms may be as simple as a fiberglass rod to lift up a point-and-shoot camera, as purpose-designed as a remotely controlled blimp for vertical image acquisition, or as complex as the International Space Station with its

FIGURE 1-1 Vertical view of abandoned agricultural land dissected by erosion channels near Freila, Province of Granada (Spain) during a photographic survey taken with a hot-air blimp (left of center) at low flying heights. The blimp is navigated by tether lines from the ground; camera functions are remotely controlled. Its picture was taken with a compact digital camera in continuous shooting mode from an autopiloted model airplane following *Google Earth*-digitized flight lines at ~200 m height. The takeoff pad at right is 12 × 8 m in size. Photo by C. Claussen, M. Niesen, and JBR, September 2008.

FIGURE 1-2 Artist's rendition of the International Space Station following installation of its nadir-viewing optical-quality window in 2001. Arrow (^) indicates position of nadir window. Image adapted from Johnson Space Center Office of Earth Sciences (Image JSC2001e00360) <http://visibleearth.nasa.gov/>.

specially designed nadir window dedicated to Earth observation (Fig. 1-2).

SFAP became a distinct niche within remote sensing during the 1990s and has been employed in recent years for documenting all manners of natural and human resources (Warner et al., 1996; Bauer et al., 1997). The field is ripe with experimentation and innovation of equipment and techniques applied to diverse situations. In the past, most aerial photography was conducted from manned platforms, as the presence of a human photographer looking through the camera viewfinder was thought to be essential for acquiring useful imagery. For example, Henrard developed an aerial camera in the 1930s, and he photographed Paris from small aircrafts for the next four decades compiling a remarkable aerial survey of the city (Cohen, 2006).

This is still true for many missions and applications today. Perhaps the most famous modern aerial artist-photographer, Y. Arthus-Bertrand, produced his *Earth from above* masterpiece by simply flying in a helicopter using hand-held cameras (Arthus-Bertrand, 2002). Likewise, G. Gerster has spent a lifetime acquiring superb photographs of archaeological ruins and natural landscapes throughout the world from the open door of a small airplane or helicopter (Gerster, 2004).

The most widely available and commonly utilized manned platform nowadays is the conventional fixed-wing small airplane, employed by many small-format aerial photographers (Caulfield, 1987). Among several recent examples, archaeological sites were documented for many years by O. Braasch in Germany (Braasch and Planck, 2005), and by Eriksen and Olesen in northwestern Denmark (2002). In central Europe, Markowski (1993) adopted this approach for aerial views of Polish castles. Bárta and Barta (2006), a father and son team, produced stunning pictures of landscapes, villages and urban scenes in Slovakia.

In the United States, Hamblin (2004) focused on panoramic images of geologic scenery in Utah, and D. Maisel has sought out provocative images of strip mines, dry lake beds, and other unusual landscape patterns in the western

FIGURE 1-3 Closeup vertical view of elephant seals on the beach at Point Piedras, California, United States. These juvenile seals are ~2 to 2.5 m long, and most are sleeping on a bank of seaweed. People were not allowed to approach the seals on the ground, but the seals were not aware of the photographic activity overhead. The spatial detail depicted in such images is extraordinary; individual pebbles are clearly visible on the beach. Kite aerial photograph taken with a compact digital camera. Photo by SWA and JSA, November 2006.

United States (Gambino, 2008). In one of the most unusual manned vehicles, C. Feil pilots a small autogyro for landscape photography in New York and New England of the United States (Feil and Rose, 2005). An ultralight aircraft is utilized for archaeological and landscape scenes in the southwestern United States by A. Heisey (Heisey and Kawano, 2001; Heisey, 2007).

Unmanned, tethered or remotely flown platforms are coming into increasingly widespread use nowadays. This book highlights such unmanned systems for low-height SFAP, including kites, blimps, and drones. Representative recent kite aerial photography, for example, includes C. Wilson's (2006) beautiful views of Wisconsin in the United States, E. Tielkes's (2003) work in Africa, and N. Chorier's magnificent pictures of India (Chorier and Mehta, 2007). Such imagery has large scale and exceptionally high spatial resolution that depict ground features in surprising detail from vantage points difficult to achieve by other means (Fig. 1-3). These photographic views bridge the gap between ground observations and conventional airphotos or satellite images.

1.2. BRIEF HISTORY

Since ancient times, people have yearned to see the landscape as the birds do, and artists have depicted scenes of the Earth as they imagined from above. Early maps of major cities often were presented as bird's-eye views, showing streets, buildings, and indeed people from a perspective that only could be imagined by the artist. Good examples can be found in Frans Hogenberg's *Civitates Orbis Terrarum* (Cologne, 1572–1617). Seventeenth-century artists such as Wenceslaus Hollar engraved remarkable urban panoramas that showed cities from an oblique bird's-eye view. The visual impact of such images is remarkably close to that of aerial photographs.

George Catlin was another leading practitioner of aerial vantages in the early 1800s (Fig. 1-4). Catlin adopted a documentary style of painting in order to represent natural and human conditions in realistic terms. By sheer force of their imagination and their technical mastery, Catlin and previous artists simulated the act of taking images from the air. It was not until the mid-1800s, however, that two innovations combined, namely manned flight and photochemical imagery, to make true aerial photography possible. Since then, photography and flight have developed in myriad ways leading to many manned and unmanned methods for observing the Earth from above (Fig. 1-5).

1.2.1. Nineteenth Century

Louis-Jacques-Mandé Daguerre invented photography based on silver-coated copper plates in the 1830s, and this process was published by the French government in 1839

FIGURE 1-4 Bird's-eye view of Niagara Falls, Canada, and the United States. George Catlin, 1827, gouache, ~45 × 39 cm. Adapted from Dippie et al. (2002, p. 36).

(Romer, 2007). The earliest known attempt to take aerial photographs was made by Colonel Aimé Laussedat of the French Army Corps of Engineers (Wolf and Dewitt, 2000). In 1849, he experimented with kites and balloons, but was unsuccessful. The first documented aerial photograph was taken from a balloon in 1858 by Gaspard Félix Tournachon, later known as "Nadar" (Colwell, 1997). He ascended in a tethered balloon to a height of several hundred meters and photographed the village of Petit Bicêtre, France. Later that same year, Laussedat again tried to use a glass-plate camera lifted by several kites (Colwell, 1997), but it is uncertain if he was successful. The oldest surviving airphoto was taken by S.A. King and J.W. Black from a balloon in 1860 over Boston, Massachusetts (Jensen, 2007).

Hydrogen-filled balloons were utilized for observations of enemy positions during the American Civil War (1861–1865); photographs reputedly were taken, although none have survived (Jensen, 2007). Meanwhile, Tournachon continued his experiments with balloons and aerial photography in France with limited success. In 1887, a German forester obtained airphotos from a balloon for the purpose of identifying and measuring stands of forest trees (Colwell, 1997).

Already in the 1850s, stereophotography was practiced, and new types of glass led to modern anastigmatic camera lenses by 1890 (Zahorcak, 2007). Experimental color photography was conducted by F.E. Ives in the 1890s (Romer, 2007).

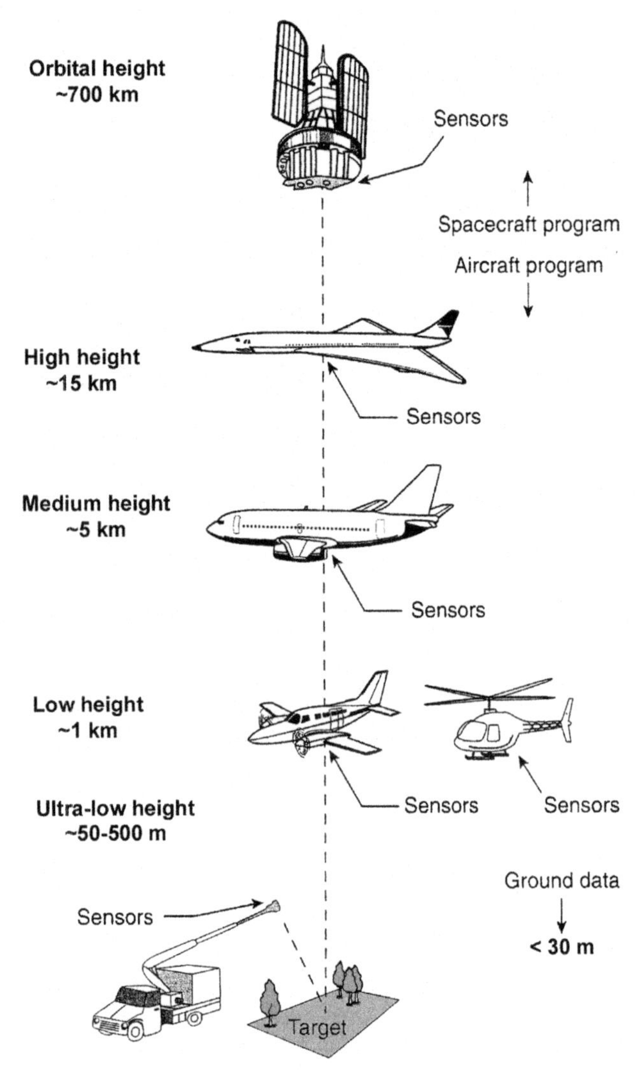

FIGURE 1-5 Schematic illustration of the "multi" approach for remote sensing of the Earth's surface from above. Multiple types of platforms and instruments operating at multiple heights. SFAP as emphasized in this book deals primarily with the ultra-low height range of observations. Not to scale, adapted from Avery and Berlin (1992, fig. 1-1).

Considerable debate and uncertainty surround the question of who was first to take aerial photographs from a kite. By some accounts, the first person was the British meteorologist E.D. Archibald, as early as 1882 (Colwell, 1997). He is credited with taking kite aerial photographs in 1887 by using a small explosive charge to release the camera shutter (Hart, 1982). At about the same time, the Tissandier brothers, Gaston and Albert, also conducted kite and balloon aerial photography in France (Cohen, 2006). Others maintain that kite aerial photography was invented in France in 1888 by A. Batut, who built a lightweight camera using a 9 × 12-cm glass plate for the photographic emulsion (Beauffort and Dusariez, 1995). The camera was attached to the wooden frame of a diamond-shaped kite and was triggered by a burning fuse. Later he built a panoramic system that included six cameras in a hexagonal arrangement for 360° views (Tielkes, 2003).

In 1890, Batut published the first book on kite aerial photography entitled *La photographie aérienne par cerf-volant*—aerial photography by kite (Batut, 1890; translated and reprinted in Beauffort and Dusariez, 1995). In that same year, another Frenchman, Emile Wenz, began practicing kite aerial photography. Batut and Wenz developed a close working relationship that lasted many years. They quickly gave up the technique of attaching the camera directly to the kite frame in favor of suspension from the line some 10s of meters below the kite. The activities of Batut and Wenz gained considerable attention in the press, and the method moved across the Atlantic. The first kite aerial photographs in the United States were taken in 1895 (Beauffort and Dusariez, 1995). Thereafter the practice of taking photographs from kites advanced rapidly with many technological innovations.

1.2.2. Twentieth Century

The early 20th century may be considered the golden age of kite aerial photography. At the beginning of the century, kites were the most widely available means for lifting a camera into the sky. Aerial photographs had been taken from balloons since the mid-1800s, but balloon aerial photography was a costly and highly dangerous undertaking and so was not widely practiced. Meanwhile powered flight in airplanes had just begun, but it also was quite a risky way to take aerial photographs. Kites were the "democratic means" for obtaining pictures from above the ground. In the first decade of the 20th century, kite aerial photography was a utilitarian method for scientific surveys, military applications, and general viewing of the Earth's surface. Its reliability and superiority over other methods were well known (Beauffort and Dusariez, 1995).

In the United States, G.R. Lawrence (1869–1938) became a photographic innovator in the 1890s, using the slogan "The Hitherto Impossible in Photography is Our Specialty" (U.S. Library of Congress, 2007). He built his own large, panoramic cameras that he mounted on towers or ladders. He tried ascending in balloons, but had a near fatal accident when he fell more than 60 m. Thereafter, he took remarkable aerial photographs with kites. His best-known photograph was the panoramic view of *San Francisco in Ruins* taken in May 1906 a few weeks after a devastating earthquake and fire had destroyed much of the city (Fig. 1-6).

Some controversy has surrounded Lawrence's camera rig, which he called a "captive airship." Some have interpreted this to mean he used a balloon (Beauffort and Dusariez, 1995). However, strong historical documentation exists for kites as the lifting means (Baker, 1994, 1997). Lawrence utilized a train with up to 17 Conyne kites that

FIGURE 1-6 Panoramic kite aerial Photograph of San Francisco by George R. Lawrence (1906). Caption on the image reads: photograph of San Francisco in ruins from Lawrence "captive airship" 2000 feet above San Francisco Bay overlooking waterfront. Sunset over Golden Gate. Image adapted from the collection of panoramic photographs, U.S. Library of Congress, Digital ID: pan 6a34514.

was flown from a naval ship in San Francisco Bay. The panoramic camera took photographs with a wide field of view around 160°. The remarkable quality of this photograph was due to a series of mishaps that delayed the picture until late in the day, when the combination of clouds and low sun position provided dramatic lighting of the scene.

On the same trip to California, Lawrence photographed many other locations in a similar manner, including Pacific Grove (Fig. 1-7). On the centennial of this event, we attempted to recreate Lawrence's panoramic view using modern kite aerial photography techniques. With a single, large rokkaku kite, we lifted a small digital camera rig from a position near Point Pinos. Because the city has grown substantially since Lawrence's time, we had to move outward (along the shore) to capture the cityscape. Nonetheless, we achieved a similar height, direction, and field of view (Fig. 1-8). At the top of his fame and fortune in kite aerial photography, Lawrence left the field in 1910 and pursued a career in aviation design.

The most daring method of this era was manned kite aerial photography undertaken by S.F. Cody in the early 1900s (Robinson, 2003b). An American who immigrated to England, Cody made a fortune with his "wild west" show. Cody and his sons began flying kites in the 1890s. They

FIGURE 1-7 Panoramic kite aerial Photograph of Pacific Grove and Monterey Bay, California by G.R. Lawrence (1906). View from near Point Pinos looking toward the southeast at scene center. Image adapted from the collection of panoramic photographs, U.S. Library of Congress, Digital ID: pan 6a34645.

FIGURE 1-8 Panoramic kite aerial photograph of Pacific Grove and Monterey Bay, California. Two wide-angle images were stitched together to create this picture. Photo by JSA and SWA, October 2006.

quickly progressed to larger kites and multiple-kite trains designed to lift a human. Cody experimented with kites and eventually developed a "bat" design, which is a double-celled Hargrave box kite with extended wings. He patented this "Cody kite" in 1901, and it was the basis of his ingenious man-lifting system. He eventually succeeded to interest the British military in the man-lifting kites, and a demonstration was conducted at Whale Island, Portsmouth, England in 1903. At first one son ascended 60 m and took photographs. Then Cody went up to 120 m, and finally another son rode up to 240 m and photographed naval ships in the harbor (Robinson, 2003b). Further trials were undertaken in 1904–1905, and Cody achieved a record height of 800 m for manned kite flight. However, few others followed Cody into manned kite flying, because of the cost and enormous risk involved.

During the period 1910–1939, René Desclée became the pre-eminent European kite aerial photographer of his day (Beauffort and Dusariez, 1995). His main subjects were the city of Tournai (France) and its cathedral. Over a period of three decades he produced more than 100 superb aerial photographs, among the best kite aerial photography portfolios prior to World War II. Desclée's career marked the end of kite aerial photography's golden age. Rapid progress in military and commercial photography from airplanes reduced kites to a marginal role (Hart, 1982), and kite aerial photography nearly became a lost art during the mid-20th century.

The first photograph from a powered flight was taken by L.P. Bonvillain in an airplane piloted by W. Wright in 1908 (Jensen, 2007). They shot motion-picture film over Camp d'Auvours near Le Mans, France. The original film is lost, but one still frame was published that same year in a French magazine. Aerial photography from manned airplanes gained prominence for military reconnaissance during World War I. Aerial cameras and photographic methods were developed rapidly, and stereo-imagery came into common usage. A typical mission consisted of a pilot and photographer who flew behind enemy lines at relatively slow speed (Fig. 1-9). Tens of 1000s of aerial photographs were acquired by Allied and German forces, and the intelligence gained from these images had decisive importance for military operations (Colwell, 1997).

The first near-infrared and near-ultraviolet photographs were published by R.W. Wood in 1910 (Finney, 2007). Practical black-and-white infrared film was perfected and made available commercially in the late 1930s, and early types of color film were developed.

During the 1920s and 1930s, civilian and commercial use of aerial photography expanded for cartography, engineering, forestry, soil studies, and other applications. In his landmark paper on the potential of aerial photography for such applications, and especially for studies of what he termed landscape ecology, the German geographer Carl Troll (1939) highlighted the capability of aerial photographs for viewing the landscape as a spatial and visual entity and strongly advocated their use in scientific studies. Many branches of the United States government employed aerial photography routinely during the 1930s, including the Agricultural Adjustment Administration, Forest Service, Geological Survey, and Navy, as well as regional and local agencies such as the Tennessee Valley Authority and Chicago Planning Commission (Colwell, 1997).

The advent of World War II once again stimulated rapid research, testing, and development of improved capabilities for aerial photography. Cameras, lenses, films and film handling, and camera mounting systems developed quickly for acquiring higher and faster aerial photography. Large-format aerial mapping cameras were designed for 9-inch (23 cm) format film (Malin and Light, 2007). The

FIGURE 1-9 Restored 1917 Curtiss JN-4D "Jenny" displayed at the Glenn H. Curtiss Museum in Hammondsport, New York, United States. Commonly used for aerial photography during World War I and in the 1920s. The photographer in the second seat has a clear vertical view behind the lower wing. Photo by JSA, August 2005.

most important innovation was color-infrared photography intended for camouflage detection. The global extent of this war led to ever-increasing types of terrain, vegetation, urban and rural settlement, military installations, and other exotic features to confuse photo interpreters. From Finland to the South Pacific, all major combatants utilized aerial photography extensively to prosecute their military campaigns on the ground and at sea. In the end, the forces with the best airphoto reconnaissance and photointerpretation proved victorious in the war, a lesson that was taken quite seriously during the subsequent Cold War (Colwell, 1997).

The art and science of aerial photography benefited substantially immediately after World War II in the United States and other countries involved in the war, as military photographers and photo interpreters returned to civilian life (Colwell, 1997), and surplus photographic equipment was sold off (Fig. 1-10). Many of these individuals had been drawn from professions in which aerial photography held great promise for further development, and it is not surprising that aerial photography expanded significantly in the post-war years for non-military commercial, governmental, and scientific applications. Meanwhile, as the Cold War heated up, military aerial photography moved to yet higher and faster platforms, such as the manned U-2 and SR-71 U.S. aircraft. Unmanned, rocket-launched satellite photographic systems, such as Corona (U.S.) and Zenit (Soviet), were operated from orbital altitudes during the 1960s and 1970s (Jensen, 2007).

Non-military uses of aerial photography continued to expand apace. As an example, the U.S. *Skylab* missions in the early 1970s demonstrated the potential for manned, space-based, small-format photography of the Earth (Fig. 1-11). *Skylab* 4 was most successful; about 2000 photographs were obtained of more than 850 features and phenomena (Wilmarth et al., 1977). Such photographs by astronauts and early satellite images provided dramatic pictures that inspired a new appreciation of the Earth's beauty as seen from above. The lessons learned during *Skylab* missions formed the basis

FIGURE 1-10 Surplus aerial camera for 5-inch (125 mm) format film of the type commonly available following World War II. The handle on top indicates overall size of the camera. Displayed at Moesgård Museum, near Aarhus, Denmark, in connection with a special exhibit on The Past from above: Georg Gerster's aerial photos from all over the world, and Aerial archaeology in Denmark, 9 October 2004–27 February 2005. Photo by JSA, October 2004.

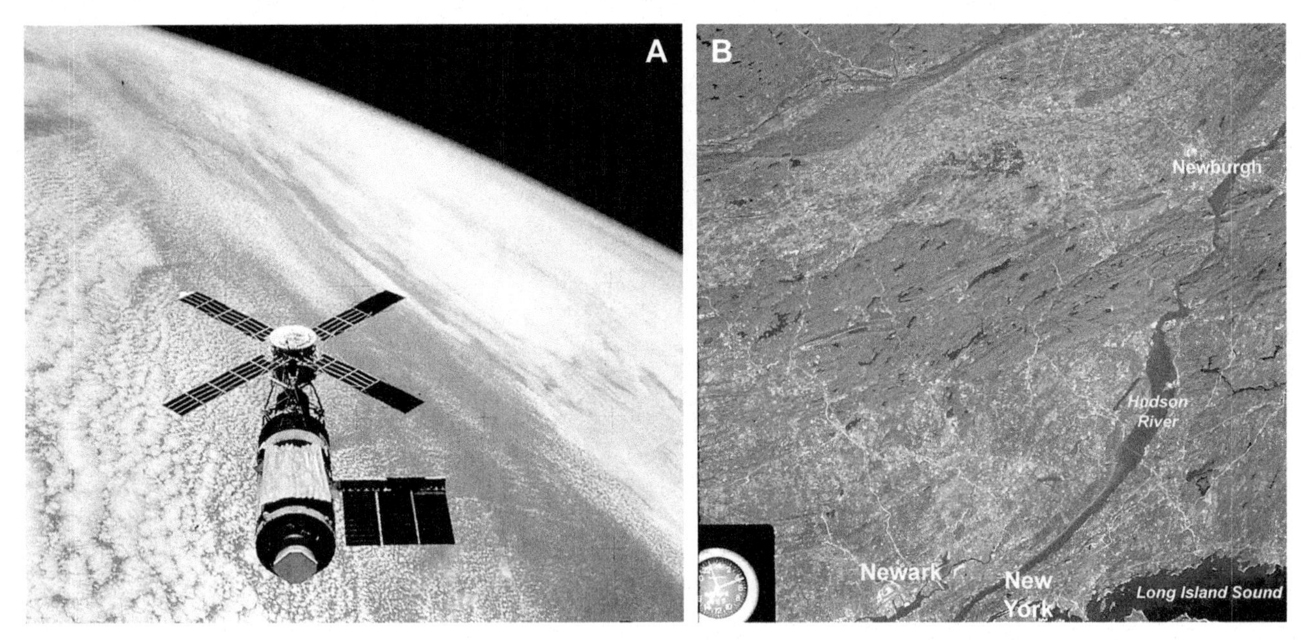

FIGURE 1-11 (A) Photograph of *Skylab* in orbit around the Earth taken from the manned rendezvous module. NASA photo SL4-143-4706, January 1974. (B) Near-vertical view of New York City and surroundings. Color-infrared, 70-mm film, Hasselblad camera; active vegetation appears in red and pink colors. NASA photo SL3-87-299, August 1973. Both images courtesy of K. Lulla, NASA Johnson Space Center.

for the program of U.S. space-shuttle photography in the 1980s and 1990s. These trends culminated early in the 21st century with astronaut photography of the Earth from the International Space Station for scientific and environmental purposes.

Closer to ground, renewed interest in kites began in the United States following World War II. Aeronautical engineering was applied to kites, parachutes, hang gliders, and other flying devices. For example, the *Flexi-Kite* designed and built by F. and G. Rogallo in the late 1940s was the inspiration for many modern kites as well as hang gliders and ultralight aircraft (Robinson, 2003a). The *Sutton Flowform*, a soft airfoil kite, was invented as a byproduct of experiments to create a better parachute during the 1970s (Sutton, 1999). This kite has become a leading choice for lifting camera rigs.

Small-format aerial photography began to make a slow but definite comeback during the 1970s and 1980s, particularly in the United States, Japan, and western Europe. Unmanned purpose-built platforms for off-the-shelf cameras, in particular, were taken up again for archaeology and cultural heritage studies, and also in forestry, agriculture, vegetation studies, and geo-ecology. Since the 1990s, SFAP has become quite widely utilized for diverse applications around the world, from Novaya Zemlya (arctic Russia) to Antarctica. The late 20th century saw rapid development in methods and popularity for unconventional manned flight, including unpowered hot-air balloons, gliders and sailplanes, as well as powered ultralight aircraft of diverse types. All these platforms have been utilized for small-format aerial photography (Fig. 1-12).

Developments in computer hardware and software have encouraged the use of small-format, non-metric photography for applications hitherto reserved to large-format metric cameras, particularly photogrammetric and GIS techniques, and SFAP has expanded from mostly scientific studies into the service sector. Acquisition, enhancement, and communication of aerial images are now possible in ways that were unimagined only a few years ago, and rapid technical advances will facilitate continued innovation and development of SFAP in the near future (Malin and Light, 2007).

1.3. PHOTOGRAPHY AND IMAGERY

The word *photograph* means literally "something written by light," in other words an image created from light. For the first century of its existence, photography referred exclusively to images made using the light-sensitive reaction of silver halide crystals, which undergo a chemical change when exposed to near-ultraviolet, visible, or near-infrared radiation. This photochemical change can be "developed" into a visible picture. All types of film are based on this phenomenon.

Beginning in the mid-20th century, however, new electronic means of creating aerial images came into existence. For example, Landsat I was the first satellite to provide images of the Earth using a remarkable device called the multispectral scanner (MSS). At that time, scanners were viewed with great skepticism by most engineers and scientists for two reasons (Hall, 1992). First, the scanner employed a moving part, an oscillating mirror, which

FIGURE 1-12 Airport at the National Soaring Museum at Harris Hill, near Elmira, New York, United States. Several gliders are visible on and next to the runway. Photo taken with a compact digital camera through the open window of the copilot's seat in an unpowered, two-person glider several 100 m above the ground, an example of an unconventional platform utilized for SFAP. Photo by JSA, July 2005.

was considered unreliable. Second, the scanner was not a full-frame imaging device; it created images from strips. Cartographers were suspicious of the scanner's geometric integrity. But, the scanner did have one important advantage, its multispectral capability for visible and infrared wavelengths.

Within hours of Landsat's launch in 1972, the first MSS images created a sensation with their amazing clarity and synoptic views of the landscape (Williams and Carter, 1976; Lauer et al., 1997). Landsat imagery revolutionized all types of cartographic, environmental, and resource studies of the Earth (Fig. 1-13). Rapid development of electronic scanners followed for both airborne and space-based platforms, and the remote sensing community embraced many types of sensors and imaging systems during the 1980s and 1990s.

As electronic imagery became more common, many restricted use of the term *photograph* to those pictures exposed originally in film and developed via photochemical processing. Thus, aerial imagery was classed as photographic or non-photographic; the latter included all other types of pictures made through electronic means. Traditional film-based photographs are referred to as *analog* images, because each silver halide crystal in the film emulsion records a light level within a continuous range from pure white to pure black. The spatial resolution of a photograph is determined by the size of minute silver halide crystals. In contrast, electronic imagery is typically recorded as *digital* values, for example 0–255 (2^8) from minimum to maximum levels, for each picture element (cell or pixel) in the scene. Spatial resolution is given by pixel size (linear dimension).

In the late 20th century, a basic distinction grew up between analog photographs exposed in film and digital images recorded electronically. Analog photographs generally had superior spatial resolution but limited spectral range—panchromatic, color visible, color infrared, etc. Digital imagery lacked the fine spatial resolution of photographs, but had a much broader spectral range and enhanced multispectral capability. Photographic purists maintained the superiority of analog film and viewed electronic imagery as lesser in quality.

The dichotomy between analog and digital imagery faded quickly in the first decade of the 21st century for several reasons. The advantages of digital image storage, processing, enhancement, analysis, and reproduction are major factors spurring adoption by users at all levels—amateur to professional specialist. Analog airphotos are routinely scanned and converted into digital images nowadays. Digital cameras have achieved equality with film cameras in terms of spatial resolution and geometric fidelity (Malin and Light, 2007). Film photography is rapidly becoming obsolete, in fact, except for certain artistic and technical uses and where the lower cost of film remains attractive. For most people today, nonetheless, the word *photograph* is applied equally to images produced from film

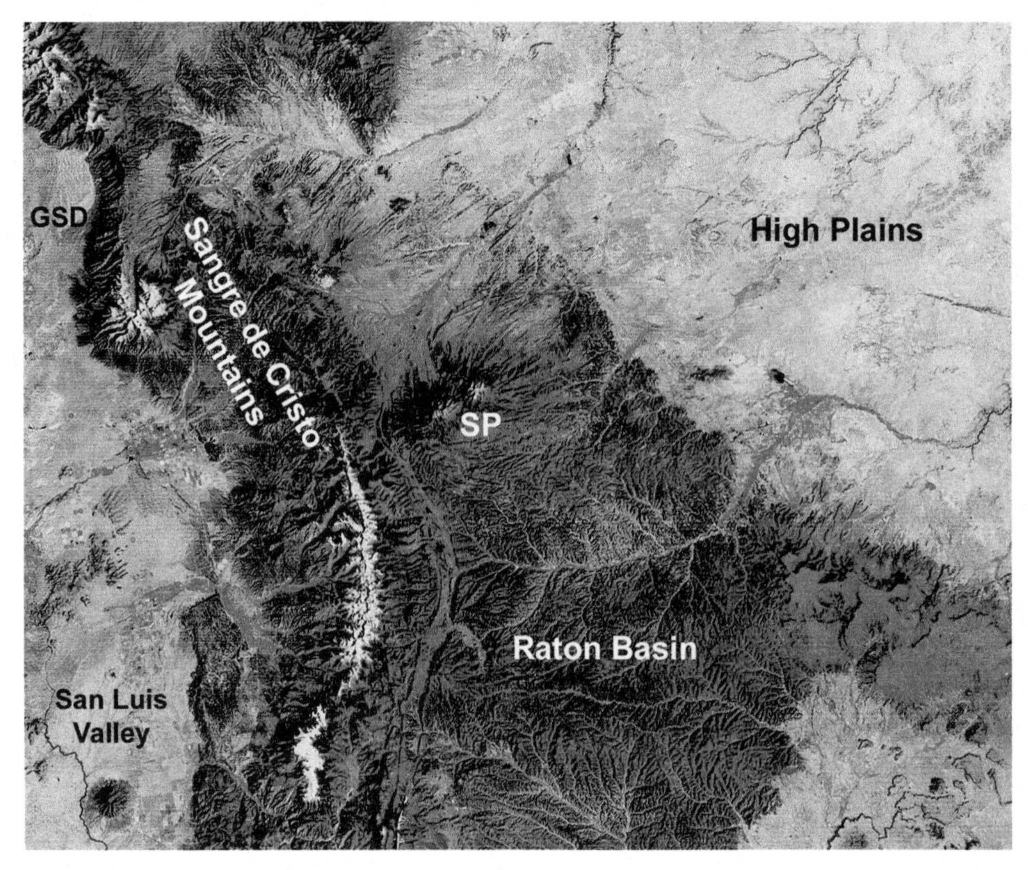

FIGURE 1-13 Early Landsat MSS composite image of the Rocky Mountains and High Plains in south-central Colorado, United States. This false-color composite resembles color-infrared photography; active vegetation appears in red colors. SP = Spanish Peaks, *GSD* = Great Sand Dunes. NASA ERTS E-2 977-16311-457, September 25, 1977, image adapted from the U.S. Geological Survey, EROS Data Center.

or electronic sensors. We follow this liberal use of the term photograph in this book, in which we place primary emphasis on digital photography regardless of how the original image was recorded.

1.4. CONVENTIONAL AERIAL PHOTOGRAPHY

Since World War I, aerial photography has evolved in two directions, larger formats for accurate mapping and cartographic purposes and smaller formats for reconnaissance usage (Warner et al., 1996). The former became standardized with large, geometrically precise cameras designed for resource mapping and military use. The science of photogrammetry was developed for transforming airphotos into accurate cartographic measurements and maps (Wolf and Dewitt, 2000). Standard, analog aerial photography today is based on the following:

- *Large-format film*: panchromatic, color-visible, infrared, or color-infrared film that is 9 inches (23 cm) wide. This format is the largest film in production and common use nowadays.
- *Large cameras*: bulky cameras weighing 100s of kilograms with large film magazines. Film rolls contain several hundred frames. Standard lenses are 6- or 12-inch (152- or 304-mm) focal length.
- *Substantial aircraft*: twin-engine airplanes are utilized to carry the large camera and support equipment necessary for aerial photography. Moderate (3000 m) to high (12,000 m) altitudes are typical for airphoto missions.
- Taking photographs is usually controlled by computer programming in combination with global positioning system (GPS) to acquire nadir (vertical) shots in a predetermined grid pattern that provides complete stereoscopic coverage of the mapping area.

Large-format aerial photography is expensive—$10s to $100s of thousands to acquire airphoto coverage. This cost can be justified for major engineering projects and extensive regional mapping of the type often undertaken by provincial or national governments—soil survey, environmental monitoring, resource evaluation, property assessment and taxation, topographic mapping, and basic cartography.

An example of this approach is summarized here for production of an orthophoto atlas of the Slovak Tatra Mountains (Geodis, 2006). Original vertical photographs were taken on large-format color film from a *Cessna* C402B twin-engine airplane guided by GPS. The flight pattern resulted in images with 80% end and 30% side overlaps to achieve complete stereoscopic coverage of the ground area at a nominal scale of 1:23,000. All photographs were taken on a single day with optimum weather conditions—no cloud cover or ground mist.

The analog photographs were scanned and georectified to create digital orthophotos, in which lens perspective, terrain relief, and curvature of the Earth were eliminated, and the images were recast into the national coordinate system. Individual orthophotos were joined into a mosaic, which was then cut into separate map sheets for publication in the atlas at a scale of 1:15,000. Primary consumers of the orthophoto atlas are tourists, who come to the Tatra Mountains by the thousands each year to enjoy the relatively unspoiled natural environment. The first edition of this atlas was produced in 2003, when most of the lower mountain slope was forested. However, a major windstorm in November 2004 blew down forest trees over broad areas of the mountain front in the portions most frequented by tourists (Fig. 1-14). A second edition of the orthophoto atlas was produced from air-photo coverage in the summer of 2005 to document the extent of this natural event.

Analog aerial photography is mature with many cameras, films, airplanes, and other equipment readily available worldwide. It is the accepted norm by most governmental agencies and commercial enterprises. Large-format digital cameras are relatively new, and several types of optical and sensor systems are in use, such as *Leica* ADS-40 (linear array), *DSS* (one CCD array), or *Z/I DMC* (4 CCD arrays). These cameras operate in the same spectral range as conventional analog film, and spatial resolution and geometric fidelity are comparable. Large-format digital cameras, thus, have achieved technical parity with analog cameras. The main digital drawback at present is higher cost, but that is changing rapidly. Analog cameras will continue to be used extensively for the next few years, but gradually large-format digital cameras should come to dominate the market within the next decade.

1.5. SFAP

Small-format aerial photography is based on lightweight cameras with 35- or 70-mm film format as well as equivalent digital cameras and other electronic imaging devices. For the most part, these are "popular" cameras designed for hand-held or tripod use by amateur and professional photographers. Such cameras lack the geometric fidelity and exceptional spatial resolution of aerial mapping cameras. However, the case for SFAP depends on cost and accessibility.

- *Low cost*: SFAP cameras are relatively inexpensive, few \$100 to several \$1000, compared with large-format aerial cameras at several \$100,000. The cost of SFAP platforms ranges from only a few \$100 for kites to tens of \$1000 for larger and more sophisticated aircraft. These costs put SFAP within the financial means for many individuals and organizations that could otherwise not afford to acquire conventional aerial photography suitable for their needs.

FIGURE 1-14 Overview of the forest blowdown zone on the southern flank of the Tatra Mountains at Tatranská Polianka, Slovakia. Three years after the windstorm, pink-purple common fireweed (*Epilobium angustifolium*) bloomed profusely in the blowdown zone across the middle portion of this scene. Kite aerial photo by SWA and JSA, July 2007.

- *Logistics*: low-height, large-scale imagery is feasible with various manned or unmanned platforms. SFAP may be acquired in situations that would be impractical, risky, or impossible for operating larger aircraft.
- SFAP has high portability, rapid field setup and use, and limited need for highly trained personnel, all of which makes this means for aerial photography logistically possible for many applications.

Low-cost availability of cameras and lifting platforms is a combination that renders SFAP desirable for many people and organizations (Malin and Light, 2007). SFAP is self-made remote sensing—system design, technical implementation, and image analysis may be in the hands of a single person, granting utmost flexibility and specialization.

Manned platforms include single-engine airplanes, helicopters, ultralight aircraft, hot-air balloons, large blimps, and sailplanes. These are necessarily more expensive and require specialized pilot training in contrast to most unmanned platforms, such as balloons, blimps, kites, model airplanes, and drones. Within the field of aerial photography, much innovation is taking place nowadays with all types of platforms and imaging equipment.

As a specialty within remote sensing, SFAP fills a niche of observational scale, resolution, and height between the ground and conventional aerial photography or satellite imagery—a range that is particularly valuable for detailed site investigations of environmental conditions at the Earth's surface. SFAP is employed in various applications ranging from geoscience to wildlife habitat, archaeology, crime-scene investigation, and real-estate development.

Within the past decade, commercial satellite imagery of the Earth has achieved 1-m, panchromatic, spatial resolution, and resolution less than half a meter may come soon (Tatem et al., 2008). Such resolution may be possible in principle; however, satellite systems must look through atmospheric haze 100s of kilometers thick, which degrades image quality. Operating close to the surface, SFAP provides sub-decimeter, multispectral, spatial resolution with usually insignificant atmospheric effects.

As an example, consider mapping vegetation at Kushiro wetland on Hokkaido, northern Japan. Aerial photography is hampered at Kushiro by persistent sea fog derived from cold offshore currents during the summer growing season when vegetation is active. Miyamoto et al. (2004) utilized two tethered helium balloons to acquire vertical airphotos of a study site in Akanuma marsh (Fig. 1-15). A standard *Nikon F-801* camera was utilized with a 28 mm lens, skylight filter, and color negative film. Combined weight of the radio-controlled camera rig and balloon tether was ~3.5 kg. In total, 66 pictures were taken from 120 m height over the study site.

FIGURE 1-15 Ground views of balloons and camera rig used for aerial photography at Kushiro wetland, Japan. Each balloon is 2.4 m in diameter with a helium capacity of 7 m^3. Taken from Miyamoto et al. (2004, fig. 3); used with permission of the authors.

Original photographs were scanned at 600 dots per inch (dpi), which yielded spatial resolution of 15 cm/pixel. Twenty-three images were selected for mosaicking and georeferencing based on ground control points surveyed with GPS. The mosaic was inspected visually; vegetation types were identified with the help of ground observations, and polygons were digitized on-screen to delineate each vegetation class. On this basis, a detailed map of vegetation was prepared (Fig. 1-16). These results were considered superior to an earlier attempt using high-cost Ikonos satellite imagery, which was obscured by typical fog. The balloon system allowed the investigators to take quick advantage of brief fog-free conditions to acquire useful imagery. This example demonstrates the spatial, temporal, and cost advantages of SFAP to succeed in a situation where other remote-sensing techniques had not proven capable.

1.6. SUMMARY

For more than 150 years, aerial photography has provided the means to see the Earth "as the birds do." During its first half-century of development, aerial photography was little used because of high cost and risk. Perhaps the most impressive early pictures were the panoramic photographs taken from kites by G.R. Lawrence in the first decade of the 20th century. With the introduction of powered flight, aerial photography expanded tremendously throughout the 1900s based on many technological inventions for various imaging devices plus the airborne and space-based platforms to carry those devices. These

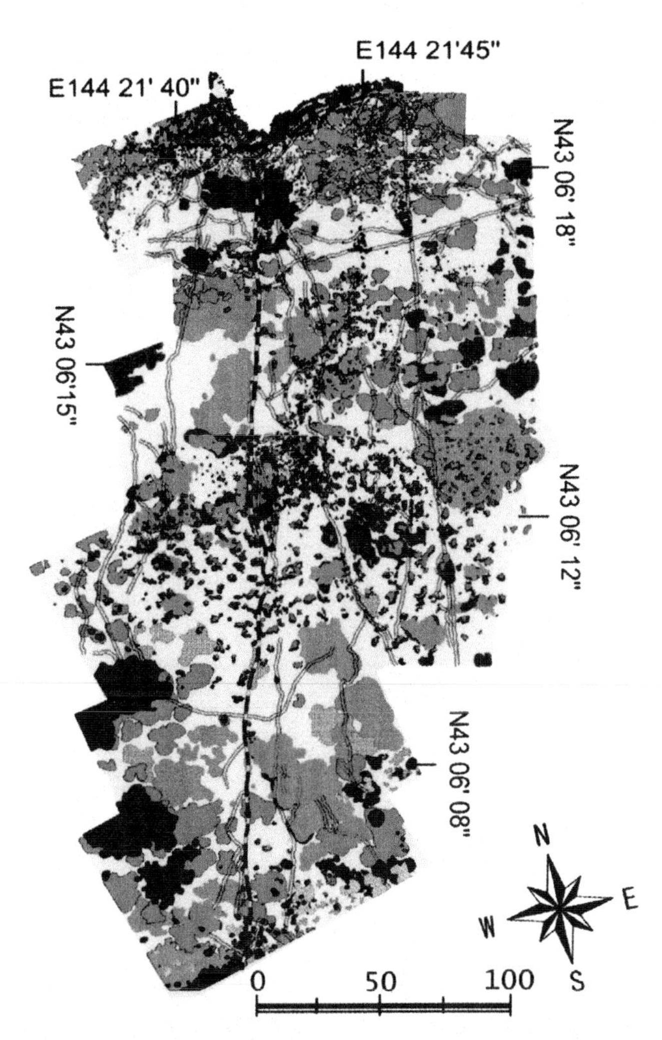

FIGURE 1-16 Final vegetation map of Akanuma marsh study site in Kushiro wetland, Japan. The colors represent 10 main classes and 27 subclasses of vegetation; scale in meters. Adapted from Miyamoto et al. (2004, fig. 6a); used with permission of the authors.

innovations were accelerated by military needs, particularly during World Wars I and II as well as the Cold War. Since World War I, aerial photography evolved in two directions—larger formats for accurate mapping and smaller formats for reconnaissance usage. During the last 30 years, technical advances in electronic devices and desktop computing have encouraged the use of small-format aerial photography with increasingly sophisticated analysis methods. Various types of electronic sensors and digital imagery progressively have taken the place of analog film photography in recent decades.

This book emphasizes small-format aerial photography based on lightweight and inexpensive 35- or 70-mm film format and digital cameras operated from platforms at relatively low height (<500 m). Photographs acquired by such means possess large scale and exceptionally high spatial resolution that portray ground features in surprising detail. Within the field of remote sensing, SFAP has established a niche that bridges the scale and resolution gap between ground observations and conventional large-format airphotos or satellite images.

Both manned and unmanned platforms are utilized for SFAP. The former includes fixed-wing airplanes, helicopters, autogyros, gliders and sailplanes, hot-air balloons, and large blimps. Unmanned platforms are model airplanes and helicopters, balloons, blimps, kites, and drones. The advantages of SFAP are based primarily on lower cost and greater accessibility compared with conventional large-format aerial photography or other means of remote sensing. The combination of inexpensive cameras and lifting platforms renders SFAP desirable for many people and organizations and is particularly valuable for detailed site investigations of environmental conditions and human influence at the Earth's surface.

Chapter 2

Basic Principles of Small-Format Aerial Photography

The sky is where I work, and the land and people are what I see ... my aerial vantage point is a privileged one.

A. Heisey (Heisey and Kawano 2001)

2.1. REMOTE SENSING

Small-format aerial photography is a type of remote sensing, which is the science and art of gathering information about an object from a distance. In other words, measurements or observations are taken without making direct physical contact with the object in question. The electromagnetic spectrum is the energy that carries information through the atmosphere from the Earth's surface to the small-format camera. Most film and digital cameras are capable of operating in the spectral range that includes near-ultraviolet, visible, and near-infrared radiation, which is the spectrum emphasized in this book. For the most part, this electromagnetic radiation represents reflected solar energy – in other words natural sunlight that illuminates the scene, reflects from surface objects, and exposes the film or electronic detector of the camera. There are several aspects in which actual SFAP deviates from ideal remote sensing (adapted from Lillesand and Kiefer, 1994).

2.1.1. Ideal Remote Sensing

- *Sunlight*: constant solar energy over all wavelengths at known output, irrespective of time and place.
- *Neutral atmosphere*: totally transparent atmosphere that would neither absorb nor scatter solar radiation.
- *Unique spectral signatures*: each object would have a distinctive and known spectral response everywhere and at all times.
- *Super sensor*: camera that would be highly sensitive through all wavelengths of interest and would be economical and practical to operate.
- *Real-time data handling*: system that would allow instant downloading, processing, and presentation of images.
- *Multiple data users*: images would be useful to scientists, managers, and others from all disciplines and applications.

2.1.2. Actual SFAP

- *Sunlight*: varies with time and place in ways that cannot be fully predicted. Calibration is sometimes possible, but the exact nature of available solar energy is usually not known.
- *Atmosphere*: varies according to latitude, season, time of day, local weather, etc. Selective absorption and scattering are the rule at most times and places.
- *Spectral signatures*: all objects have theoretically unique signatures, but in practice these may change and cannot always be distinguished; many objects appear the same.
- *Real cameras*: no existing small-format camera system can operate practically in all wavelengths of interest. Each camera is limited by its optical, film, or electronic nature to certain wavelengths. Likewise, certain cameras are limited by their high cost.
- *Data handling*: many cameras now generate far more imagery than can be handled instantly by either visual inspection or computer analysis.
- *Multiple users*: no single combination of imagery and analysis satisfies all users. Many users are not familiar with subjects outside their immediate disciplines and thus cannot appreciate the full potential or limitations of small-format aerial photography.

Small-format aerial photography, like other types of remote sensing, is a compromise between the ideal and what is logistically feasible and financially affordable for a given project. In this regard, the relatively low cost, high spatial resolution, and field portability of SFAP offer some advantages not possible with other means of aerial remote sensing.

SFAP normally exploits the so-called visible atmospheric window consisting of wavelengths from ~0.3 to 1.5 μm long (Fig. 2-1). On a cloud-free day, this range of wavelengths passes through the atmosphere with little scattering or absorption by gas molecules, aerosols, or fine

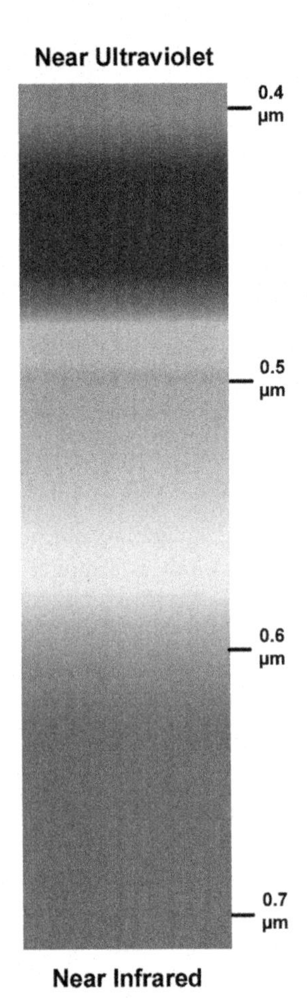

FIGURE 2-1 Spectrum of visible light in micrometers (μm) wavelength. All visible colors are made up of three primary colors: blue (0.4–0.5 μm), green (0.5–0.6 μm), and red (0.6–0.7 μm). Near-ultraviolet is ~0.3 to 0.4 μm, and near-infrared is ~0.7 to 1.5 μm wavelengths.

dust. Given the low-height operation for most SFAP below 300 m or even <100 m, images are acquired in which the reflected radiation has suffered minimal degradation from atmospheric scattering or absorption. This is an important consideration in terms of spectral signatures of objects depicted in SFAP images.

2.2. COMMON ASPECTS OF SFAP

Among different types of remote sensing, small-format aerial photography undoubtedly has the greatest variety in terms of aerial platforms and camera systems. Some basic aspects are common to all approaches, nonetheless, regardless of the type of platform, camera, or purpose for SFAP. These common aspects are introduced here and elaborated in more detail in subsequent chapters.

2.2.1. Image Vantage

Aerial photographs can be taken in three vantages relative to the Earth's surface, as determined by the tilt of the camera lens relative to the horizon (Fig. 2-2). The amount of tilt is called depression angle.

- *High-oblique vantage*: side view, horizon is visible, depression angle typically <20°.
- *Low-oblique vantage*: side view, horizon is not visible, depression angle typically 20°–87°.
- *Vertical vantage*: view straight down, also called nadir, depression angle >87°–90°.

Vertical images are generally preferred for mapping and measurement purposes, as explained below, because the geometry of vertical images can be calculated. However, such vertical views may be difficult for many people to interpret unless they are quite familiar with the site and objects shown in the image. Oblique shots, on the other hand, provide overviews of sites and their surroundings that are quite easy for most people to recognize visually and understand readily (e.g. Ham and Curtis, 1960). Yet, oblique photographs have substantial distortions in scene geometry that render accurate measurements difficult or impossible.

2.2.2. Photographic Scale and Resolution

The scale of a vertical aerial photograph can be calculated simply in two ways. The scale (S) depends on the average height above the ground (H_g) and the lens focal length (f) of the camera. In either case, the units of measurement must be the same.

$$S = \text{photo distance}(d)/\text{ground distance}(D) \quad \text{(Equation 2-1)}$$

$$S = f/H_g \quad \text{(Equation 2-2)}$$

In cases where objects of known size appear in the vertical photograph, the first method may be utilized for scale calculation (Fig. 2-3). If no objects of known size are visible in the photograph and the flying height above ground is known, the second method is employed. Scale is usually expressed as a fraction or ratio, such as 1/1000 or 1:1000, meaning one linear unit of measurement on the photograph equals 1000 units on the ground. In rugged terrain, however, photo scale varies because of large height differences within the photograph. Likewise oblique photos also display large variations in scale.

Scale is a fundamental property of routine aerial photographs, and is especially important for vertical airphotos used for measurements and photogrammetric purposes. Interpretability of aerial photographs is often determined by photo scale. For analog aerial photographs, the original film scale is the most commonly used characteristic for describing the amount of detail identifiable in the image. The actual photographic resolution, determining the size of the smallest identifiable feature within an image, is

FIGURE 2-2 Three views of Gammelsogn Kirke (old parish church), near Ringkøbing, western Denmark. (A) High-oblique view showing the horizon. (B) Low-oblique view in which the horizon is not visible. (C) Near-vertical view. The church dates from the 1170s when the Roman nave and choir were built; the tower and entry house were added later. Kite aerial photographs by JSA, SWA, and IM, September 2005.

measured as the number of resolvable lines per inch (or mm) and depends on film emulsion and image contrast.

For digital images, the original scale on the image sensor is not of much interest, as the scale of a digital image is easily changed when viewing it on a display device and becomes a property of this device. However, the original image resolution does not change with varying display scale, and the size of the smallest visible object depends directly on the size of the sensor cells or pixels in the electronic detector. In the case of digital imagery, ground sample distance (*GSD*) is, therefore, more appropriate as a measure for image scale (Comer et al., 1998).

Consider a digital camera with a charged-couple device (CCD): collection *GSD* is related to the size of each pixel element within the detector array. Using the scale calculations noted above, *GSD* can be determined as follows:

$$GSDs = (\text{pixel element size}) \times H_g/f \quad \text{(Equation 2-3)}$$

However, a single pixel usually cannot be identified as a unique object by itself. For visual identification of distinct objects, generally a group of 4–9 pixels is the minimum necessary (Comer et al., 1998). This leads to a general rule of thumb (Hall, 1997).

- Positive recognition of objects in aerial photographs requires a *GSD* 3–5 times smaller than the object size.

Digital images as well as small-format analog photographs are rarely, if ever, displayed at the original camera scale, which would be much too small for normal visual examination. Most usually, digital images are enlarged substantially for display on a computer monitor, in which the dot pitch controls the image size and scale, assuming that one image pixel is displayed for each monitor dot. In this case, the display scale is a ratio of the collection *GSD* to the monitor dot pitch.

$$\text{Display scale} = (\text{monitor dot pitch})/GSD \quad \text{(Equation 2-4)}$$

As an example, take a digital vertical photograph acquired at a height of 100 m using a camera with a 35-mm lens focal length and CCD pixel element size of 0.009 mm.

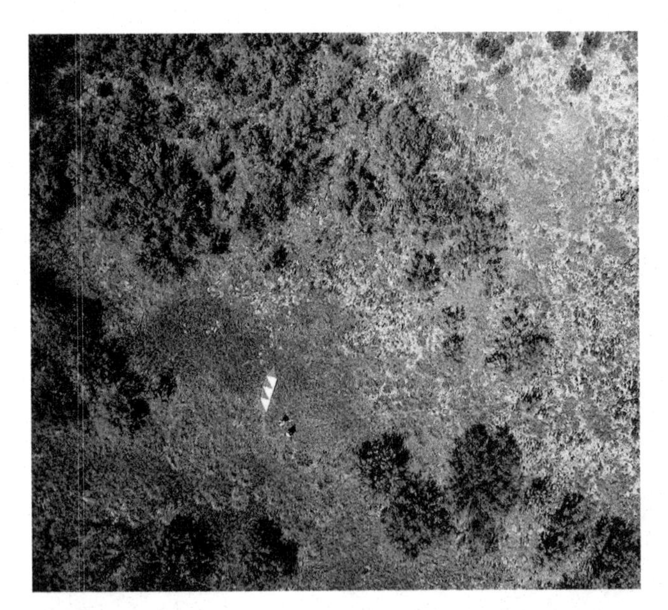

FIGURE 2-3 Biological study site near Pueblo, Colorado, United States. The north arrow is 4 m long by 1 m wide; it provides both a scale bar and directional indicator for this vertical kite aerial photograph. Note two people standing next to the survey arrow. Photo date August 2003 (Aber et al., 2005).

FIGURE 2-4 Enlargement of the arrow and people in the previous figure. This image contains no more spatial or spectral information than the previous image; each pixel represents exactly the same ground area and color as before. All the details visible in the enlarged image are present in the original image.

Collection *GSD* would be $(0.000009 \times 100) \div 0.035 = 0.026$ m ($\sim$2.5 cm or 1 inch). Now, displaying this image at full size on a monitor with dot pitch of 0.26 mm, the display scale would be $0.00026 \div 0.026 = 0.01$ (or 1:100). A similar calculation can be done for printed digital images. The nominal pixel size for standard printing at 300 dpi (dots per inch) is about 0.085 mm. In this case, printed scale would be $0.000085 \div 0.026 = 0.00327$ (or about 1:300). Displaying or printing the image at smaller scales would mean losing some of its information content when viewing it on the screen or printout.

This example demonstrates that display and printed scales are usually many times greater than original digital image scale, because the display/print pixels are many times larger than the electronic detector pixel elements. The larger scales employed for display and printing of digital images do not imply more information or better interpretability, however, compared to the "raw" image data (Fig. 2-4). A digital number is simply a color value for a single pixel, regardless of the size at which the pixel is displayed.

So far, we have used the term resolution, which is employed in different ways in remote sensing, for describing the spatial aspects of small-format aerial photography. Other aspects of resolution include spectral and temporal dimensions, which may be equally or even more important than is spatial resolution for identification of certain objects. For example, color-infrared photography was developed originally for camouflage detection and is widely employed now for vegetation, soil, and water studies (Finney, 2007). The combination of green, red, and near-infrared radiation reveals objects that may appear similar in visible light (Fig. 2-5).

Temporal resolution refers to how objects change through time on scales ranging from diurnal to decadal. Deciduous vegetation, as an example, is strongly seasonal in character, and this situation may be exploited for identification of plant types (Fig. 2-6). Finally, the term radiometric resolution refers to the number of digital levels, also called precision, that the sensor uses for recording different intensities of radiation—usually 0–255 or 2^8 per image band for SFAP cameras.

2.2.3. Relief Displacement

The camera lens operates much like the human eye, both of which produce single-point perspective views of the scene. This perspective produces increasing relief displacement of objects nearing the edge of view. This is most noticeable in vertical airphotos, because tall objects appear to lean away from the photo center. Conversely, low objects are displaced toward the center. Relief displacement is minimal near the photo center and becomes extreme at the edge. This allows for a side view of tall objects near the edge of a vertical photograph, particularly for wide-angle fields of view (Fig. 2-7). The height of a tall vertical object may be calculated from its relief displacement, and the height of tall objects also may be determined from measurements of shadows.

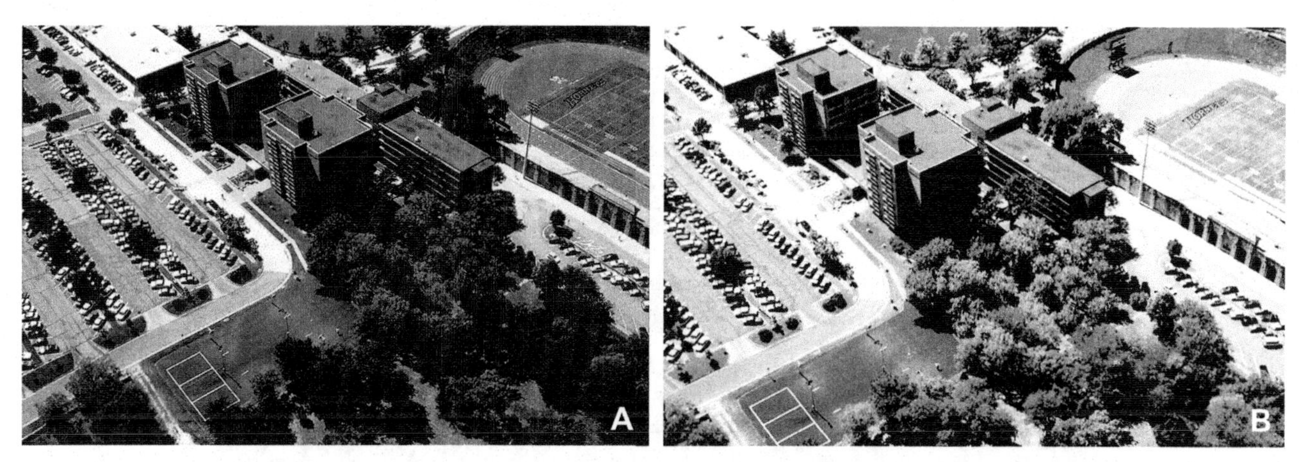

FIGURE 2-5 Color-visible (A) and color-infrared (B) kite aerial photographs of the campus of Emporia State University, Kansas, United States. A portion of the football field, dormitory buildings, parking lots, automobiles, grass, and deciduous trees. Photosynthetically active vegetation is depicted in red and pink colors in the infrared image. Note variations in tree appearance in the color-infrared version. Taken from Aber, Aber and Leffler (2001, fig. 3).

FIGURE 2-6 Vertical kite aerial photograph in visible light showing water pools and vegetated hummocks in the central portion of Männikjärve Bog, Estonia. Moss species display distinctive green, gold, and red autumn colors along with pale green dwarf pine trees on hummocks. These dramatic colors are not displayed at other times of the year. Field of view ~60 m across. Based on Aber et al. (2002, fig. 2).

2.2.4. Stereoscopic Images

Humans see in three dimensions, in other words, we have depth perception, because our eyes provide overlapping fields of view from slightly different vantage points. The amount of depth perception in humans is limited to about 400 m distance, however, because of the relatively close spacing of our eyes, i.e., only 6–7 cm apart (Drury, 1987). Stereoscopic photography has been practiced since the middle 19th century to provide 3-D imagery (Osterman, 2007; Fig. 2-8). Aerial stereo photographs may be taken from widely separated positions (Fig. 2-9). Such overlapping pictures typically are viewed through a stereoscope in order to produce exaggerated depth perception (Fig. 2-10). Vertical stereophotos are important for visual photointerpretation and are the basis for many photogrammetric techniques (Ogleby, 2007).

FIGURE 2-7 Wide-angle, vertical view over conifer forest at Kojšovská hol'a in southeastern Slovakia. Note how trees appear to lean away from photo center toward outer edges of the scene. Kite aerial photo by SWA and JSA, August 2007.

FIGURE 2-8 Antique stereoscopic viewer and stereophotos of Yosemite Valley, California, United States. These devices were popular in the late 19th and early 20th centuries.

2.3. PHOTOGRAPHIC STORAGE

SFAP is more than just snapshots; usually the images are intended for long-term storage and reproduction years and even decades after they were acquired. However, neither film nor digital photography is everlasting; all photographic media are subject to long-term decay (Rosenthaler, 2007). Thus, proper storage of the images becomes a significant issue for most SFAP projects. Geographic information typically consists of two kinds of data. First is the primary dataset composed of location information and attribute data about individual features. Second is so-called metadata, which includes such information about the geographic dataset as its grid system, map projection, units of measurement, date of creation, camera model, lens focal length, and history of processing.

Aerial photographs are one type of geographic information. The original image itself is the primary dataset. Metadata should contain information about location, date of image acquisition, type of camera/lens, altitude, and other relevant facts. For analog (film, print) photographs, such information could be written directly on the image, so there is no chance the image could be separated from its metadata (see Fig. 1-6). A more common approach is to place metadata on the margins, back, or frame of the photographic medium (Fig. 2-11). Still this approach is often omitted or incomplete for SFAP, and years later nobody can remember the where, when, or what aspects for a photograph. Some metadata (EXIF header) is built into image files collected using common digital cameras. The image file contains

FIGURE 2-9 Pair of oblique stereophotos showing a residential scene in Emporia, Kansas, United States. The pictures were taken simultaneously with two cameras spaced ~1 m apart. Note slight left–right offset in views; compare vehicles in lower left corner and house in lower right corner of each photograph. Kite aerial photographs by JSA and SWA, December 1998.

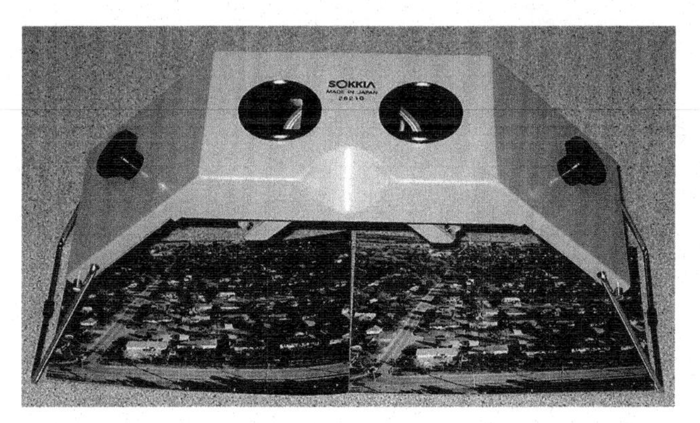

FIGURE 2-10 *Sokkia* mirror stereoscope (MS16). This model is the ideal size for viewing 4 × 6 inch (10 × 15 cm) prints made from 35-mm film.

FIGURE 2-11 Example of 35-mm color film mounted in a plastic frame (slide). Metadata written on the frame include location, date of acquisition, and picture number. Kite aerial photograph by JSA and SWA, Estonia, August 2000.

metadata, under file properties, such as image dimensions, file size in bytes, date and time the image was taken, type of data compression, and camera model.

Both analog photographs and digital images are stored in some type of physical medium, the properties of which determine how long the image is likely to survive. The potential longevity of panchromatic (b/w) film and prints is on the order of one century to 150 years; color film and prints have only about half these lifetimes or less (Jensen, 2005; Rosenthaler, 2007). In contrast, magnetic media, such as disks and tapes, last only one or two decades, before they degrade under the Earth's magnetic field. Optical disks (CD, DVD) are thought to survive for a century or more, although they have not been in service long enough to really know their longevity in practice.

Another, perhaps more serious, long-term issue for storage of digital datasets concerns the computer hardware and software necessary to read, transfer, and display digital files stored in a particular medium. Medium types, file storage formats, and computer operating systems have changed rapidly since digital data became commonplace (Fig. 2-12). So, although the magnetic tape or optical disk may survive intact, an operational reading device and software to interpret the file format may no longer exist in many cases. The computer industry has done little, unfortunately, to solve the problem of long-term digital archival storage.

Burge (2007) emphasized the importance of data migration periodically as new digital media are established and old media become obsolete; such migration is necessary about every five years (Rosenthaler, 2007). Large

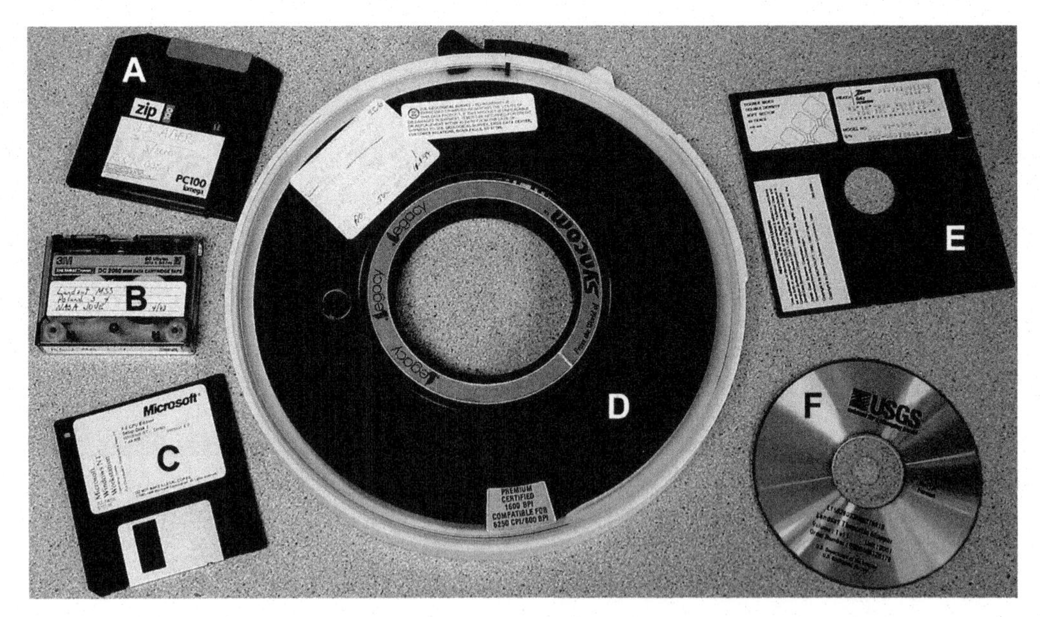

FIGURE 2-12 Collection of various removable storage media for digital image data in the late 20th century. (A) Zip disk, (B) compact cartridge tape, (C) 3½-inch floppy disk, (D) 9-track reel tape, (E) 5¼-inch floppy disk, and (F) optical disk (CD). Of these devices, only the last (F) is still commonly used in the early 21st century.

companies and governmental agencies may have the capability to transfer and reformat digital files from one medium to another periodically. In one geospatial analysis laboratory, for example, digital image files were updated from original 9-track tapes, to compact cartridge tapes, to zip disks, and finally to optical disks (CD) during the past 15 years. The investment in time and labor to accomplish this was quite significant.

Lawrence's photograph of *San Francisco in Ruins* (see Fig. 1-6) has survived for a century. One could ask what the chances are for modern digital photographs to survive with their metadata for so long (Burge, 2007). Recent history suggests that technical innovation and obsolescence will continue to happen rapidly for digital storage devices and formats. Means of long-term storage for digital imagery is an issue yet to be fully resolved.

2.4. SUMMARY

Small-format aerial photography is based primarily on solar radiation reflected from the Earth's surface in the visible and near-infrared portions of the spectrum. SFAP is a compromise between ideal photographic conditions and the reality of what is possible to accomplish under natural conditions within logistical constraints and financial limitations.

SFAP may be taken in oblique and vertical vantages. The former is appropriate for depicting the relationships of landscape elements over broad areas; the latter is best suited for accurate measurements. Photographic scale (*S*) and ground sample distance (*GSD*) are closely related, but distinct, concepts dealing with spatial aspects of photographs. Resolution relates to spatial, spectral, temporal, and radiometric aspects of photographs.

A single photograph has geometric characteristics similar to the picture sensed through a single human eye, which creates a single-point perspective view. Two overlapping views of the same area result in stereoscopic imagery in which depth perception is apparent. This is the basis for much photointerpretation as well as photogrammetry.

Photographic information consists of both the primary image data as well as metadata about the image. Metadata should include date of image acquisition, location, type of camera and lens, and other relevant information. This metadata should be saved in a manner that is difficult to separate from the image data.

Various media and file formats have been used over the years for storing photographic images and digital datasets. Among the most long-lived are panchromatic film negatives and optical disks; the least stable are magnetic media. Storage of digital datasets is subject to rapid changes in media types and related computer hardware and software necessary for reading the digital files. Continued technical changes and obsolescence are likely to happen in the near future, which raises questions for long-term archival storage of photographs.

Chapter 3

Photogrammetry

Despite all the exciting possibilities for 3D imaging techniques, users of this technology must keep in mind [that] light will continue to travel in straight lines and an understanding of geometry will be just as valid tomorrow as it is today.

John G. Fryer (Fryer et al., 2007)

3.1. INTRODUCTION

Photogrammetry is *the art, science, and technology of obtaining reliable information about physical objects and the environment through processes of recording, measuring, and interpreting photographic images and patterns of recorded radiant electromagnetic energy and other phenomena* (Wolf and Dewitt, 2000; McGlone, 2004). Photogrammetry is nearly as old as photography itself. Since its development approximately 150 years ago, photogrammetry has moved from a purely analog, optical–mechanical technique to analytical methods based on computer-aided solution of mathematical algorithms and finally to digital or softcopy photogrammetry based on digital imagery and computer vision, which is devoid of any opto-mechanical hardware. Photogrammetry is primarily concerned with making precise measurements of three-dimensional objects and terrain features from two-dimensional photographs. Applications include the measuring of coordinates; the quantification of distances, heights, areas, and volumes; the preparation of topographic maps; and the generation of digital elevation models and orthophotographs.

Two general types of photogrammetry exist: aerial (with the camera in the air) and terrestrial (with the camera handheld or on a tripod). Terrestrial photogrammetry dealing with object distances up to ca. 200 m is also termed close-range photogrammetry. Small-format aerial photogrammetry in a way takes place between these two types, combining the aerial vantage point with close object distances and high image detail.

This book is not a photogrammetry textbook and can only scratch the surface of a continuously developing technology that comprises plentiful principles and techniques from quite simple to highly mathematical. In this chapter, an introduction to those concepts and techniques is given that are most likely to be of interest to the reader who plans to use small-format aerial photography, but has little or no previous knowledge of the subject. For a deeper understanding, the reader is referred to the technical literature on photogrammetry, for example, the textbooks by Wolf and Dewitt (2000), Kasser and Egels (2002), Konecny (2003), Luhmann (2003), Kraus (2004), McGlone (2004), Kraus et al., (2007), and Luhmann et al. (2007).

The basic principle behind all photogrammetric measurements is the geometrical–mathematical reconstruction of the paths of rays from the object to the sensor at the moment of exposure. The most fundamental element therefore is the knowledge of the geometric characteristics of a single photograph.

3.2. GEOMETRY OF SINGLE PHOTOGRAPHS

3.2.1. Vertical Photography

Other than a map and similar to the images we perceive with our eyes, a photograph—either analog or digital—is the result of a central projection, also known as single-point perspective. The distances of the central point of convergence—the optical center of the camera lens, or exposure station—to the sensor on one side and the object on the other side determine the most basic property of an image, namely its scale.

Figure 3-1 shows the ideal case of a vertical photograph taken with perfect central perspective over completely flat terrain. The triangles established by a ground distance D—e.g. the distance D_2 between points A and P—and the flying height above ground H_g on the terrain side and by the corresponding photo distance d and the focal length f on the camera side are geometrically similar for any given D and d: the scale S or $1/s$ of the photograph is the same at any point.

$$S = 1/s = f/H_g = d/D \qquad \text{(Equation 3-1)}$$

Once the image scale S is known, this equation can be resolved for D in order to calculate the width or length of the ground area covered by the image by using the image width or length on the film or sensor chip as d.

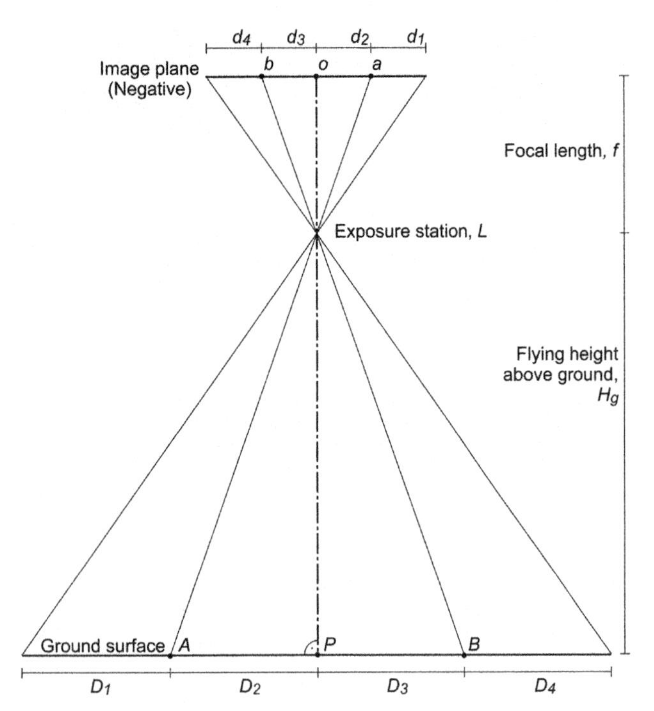

FIGURE 3-1 Vertical photograph taken over completely flat terrain. The optical axis, which intersects the image plane at its center *o*, meets the ground in a right angle at the principal point *P*. Note that the respective distances between all points are the same on the ground and on the image.

Several other important characteristics of the photograph can be derived from the basic relationships described in Figure 3-1. By transforming Equation (3-1), the commonly used equation for H_g can be obtained:

$$H_g = f \times s \qquad \text{(Equation 3-2; compare with Eq. 2-2)}$$

If the image format is square, the ground area *A* covered by the photograph can be derived from squaring Equation 3-1. For SFAP cameras, this usually would not be the case, so the rectangular image format (with d_L as the image length d_W as the image width) must be taken into account:

$$1/s^2 = (d_L \times d_W)/(D_L \times D_W) = a/A \text{ or}$$
$$A = (d_L \times d_W) \times s^2 \qquad \text{(Equation 3-3)}$$

For digital images, the ground sample distance *GSD* (see Chapter 2) determines the spatial resolution or smallest visible detail in the photograph. The exact size of the picture element (sensor cell), which is needed for *GSD* calculation, can be determined from the manufacturer's information on the sensor size in pixels and millimeters (see Eq. 2-3 in Chapter 2).

$$GSD = (\text{picture element size} \times H_g)/f$$
$$\text{(Equation 3-4; compare with Eq. 2-3)}$$

Because the relationship is a direct linear one, any change in H_g or f will change the scale and the image distances by the same factor. For example, doubling the flying height results in an image with half the scale *S* and halved image distances *d*, tripling the focal length will enlarge the scale and image distances by a factor of three. In reality, most aerial photographs—especially by far the most SFAP—deviate from the situation in Figure 3-1 for three reasons:

- The ground is not completely flat, i.e. the distance between image plane and ground varies within the image.
- The photograph is not completely vertical, i.e. the optical axis is not perpendicular to the ground.
- The central projection is imperfect due to lens distortions, i.e. the paths of rays are bent when passing through the lens.

All three situations spoil the similarity of the triangles and result in scale variations and hence geometric distortions of the objects within the image. The last two problems can be minimized with modern survey and manufacturing techniques for professional high-tech survey cameras and mounts, but may be quite severe for the platforms and cameras often used in small-format aerial photography. The first problem, scale variations and geometric distortions caused by varied terrain, does not depend on camera specifications and occurs with any remote sensing images. However, it also normally would be more severe for SFAP than for conventional aerial photography because of the lower flying heights and thus relatively higher relief variations.

Figure 3-2 illustrates the effects of different elevations on the geometry of a vertical photograph. All points lying on the same elevation as the image center have the same scale—points above this horizontal plane are closer to the camera and therefore have a larger scale, points below this horizontal plane are farther away and have a smaller scale. At the same time, the positions of the points in the image are shifted radially away from (for higher points) or toward (for lower points) the image center (*o*)—compare the different positions of *a* and *a*′, and *b* and *b*′, respectively in Figure 3-2. This happens because they appear under another angle than they would if they were in the same horizontal plane as the ground principal point, seemingly increasing or reducing their distance to the point at the image center.

This so-called relief displacement increases with the distance to the image center (see Fig. 2-7). It is inversely proportional to the flying height and the focal length: the displacement is less severe with larger heights and longer focal lengths because in both cases the rays of light are comparatively less inclined. This effect of varying relief displacement can be exploited according to the requirements of image analysis. Images for monoscopic mapping or image mosaicking—demanding minimum distortion by relief displacement—are best taken using a telephoto lens from a greater flying height. However, images utilized for stereoscopic viewing and analysis should have higher relief displacement and stereo-parallaxes and are best attained

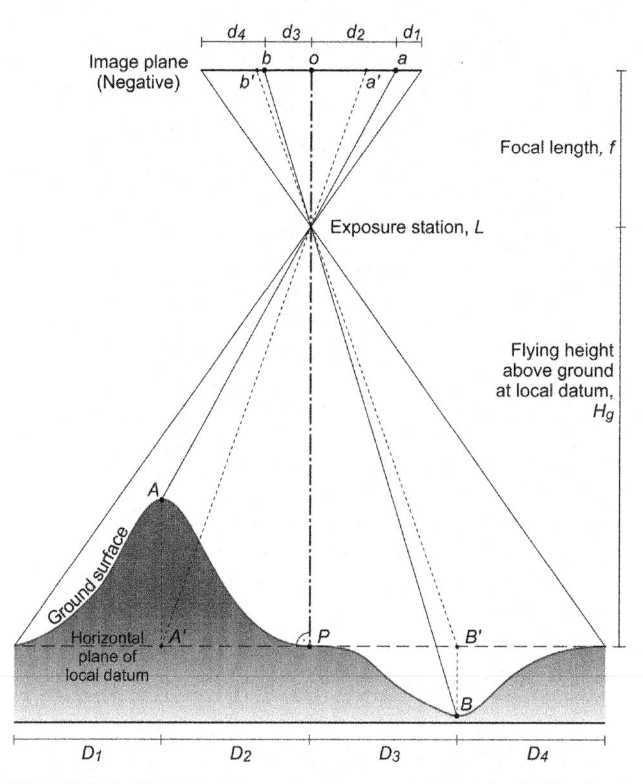

FIGURE 3-2 Vertical photograph taken over variable terrain. The elevation of the principal point *P* determines the horizontal plane of local datum. Points lying on this plane remain undistorted, whereas points above or below are shifted radially with respect to the image center. Note that the horizontal distances D_1–D_4 are the same in the object space but not in the image.

with wide-angle lenses and lower heights (see section on base–height ratio and Fig. 3-11 below).

3.2.2. Tilted Photography

None of the equations given above is valid for oblique photographs with non-vertical optical axes, because the scale varies with the magnitude and angular orientation of the tilt (Fig. 3-3). The magnitude of the tilt is expressed by the nadir angle ν, which is the angle between the optical axis and the vertical line through the perspective center (nadir line) and is the complement of depression angle. Topographic relief introduces additional scale variations and radial distortions relative to the nadir point n—much the same as for true vertical photographs (see Fig. 3-2), where n is identical with the principal point or image center o. Oblique images are useful for providing overviews of an area and they are easier to understand and interpret for most people (see Chapter 2). However, obliqueness undermines the validity of many principles and algorithms used in photogrammetry. Oblique photographs therefore usually are avoided for measurement purposes, and images are classified according to their degree of tilt from vertical (see Fig. 2-2).

- True vertical: $\nu = 0°$, often difficult in practice (especially with typical SFAP platforms).
- Near-vertical or slightly tilted ($\nu < 3°$).
- Low oblique ($\nu \geq 3°$, horizon not visible).
- High oblique (usually $\nu > 70°$, horizon visible).

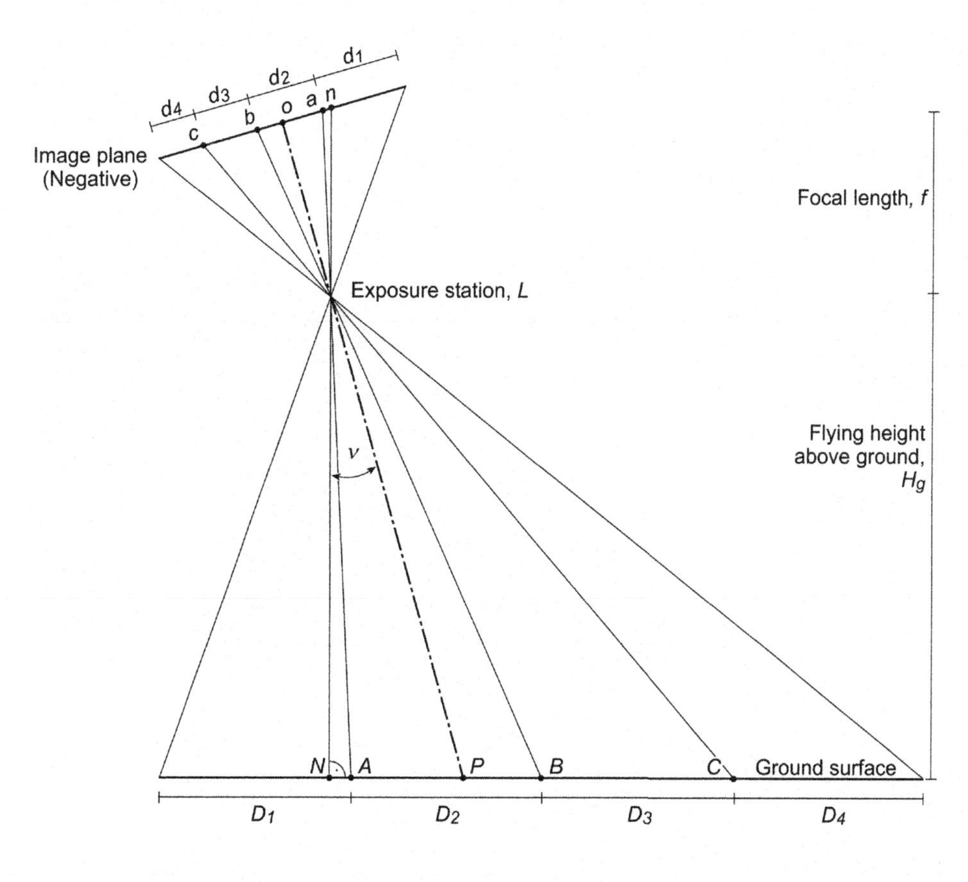

FIGURE 3-3 Low-oblique photograph taken over completely flat terrain. The vertical line through the perspective center *L* intersects the image plane at the photographic nadir point *n* and meets the ground in a right angle at the ground nadir point *N*. The angle between the nadir line and the optical axis is the nadir angle ν. Note that the distances *D* between all points on the ground are the same, but the corresponding distances *d* on the image vary continually.

FIGURE 3-4 Top right corner of an aerial photograph taken with a conventional analog (film) metric camera, showing one of four fiducial marks (see enlarged insert) composed of a dot, cross and circle. The text block on the left indicates lens type, number, and calibrated focal length in mm; the mechanical counter records the sequential image number.

For many practical applications, the errors resulting in simple measurements from slightly tilted images ($\nu < 3°$) can be considered negligible, but this is not the case with oblique images.

3.2.3. Interior Orientation

The geometrical principles and equations mentioned above are sufficient for simple measurements from analog or digital photographs. For the precise calculation of 3D coordinates, however, the paths of rays both within and outside the camera need to be reconstructed mathematically with high precision (see also further below, Fig. 3-12). The necessary parameters for describing the geometry of the optical paths are given by the interior and exterior orientations of the camera.

The interior orientation of an aerial camera comprises the focal length (measured at infinity), the parameters of radial lens distortion, and the position of the so-called principal point in the image coordinate system. The principal point (o) is defined as the intersection of the optical axis with the image plane and falls quite close to the origin of the image coordinate system at the image center. For measurements within the image, this coordinate system has to be permanently established and rigid with respect to the optical axis. For metric analog cameras this is realized with built-in fiducial marks that protrude from the image frame and are exposed onto each photograph (Fig. 3-4). For 35-mm film cameras, fiducials do not normally exist unless the camera has been specially fitted with them (Fig. 3-5). Digital cameras do not need fiducials because a Cartesian image coordinate system is already given by the pixel cell array of the image sensor.

Various camera calibration methods exist for the determination of interior orientation values (Fraser, 1997; Wolf and Dewitt, 2000). While metric cameras are usually calibrated with laboratory methods by the manufacturer, off-the-shelf small-format cameras as used in SFAP do not come with calibration reports. However, they may be calibrated by dedicated companies or institutions or even by the user with test-field calibration or self-calibration methods (see also Chapter 6.6.5).

3.2.4. Exterior Orientation

The exterior orientation includes the position X, Y, Z of the camera in the ground coordinate system and the three rotations of the camera ω, ϕ, κ relative to this system. The elements of exterior orientation can be determined theoretically with modern high-tech GPS/INS systems simultaneous to image acquisition, but this is hardly an option for SFAP owing to the associated costs, weights, and insufficient precisions. The commonly used method of determining exterior orientation is the post-survey reconstruction using ground control points (GCPs) with known X, Y, and Z coordinates. A theoretical minimum of three GCPs is

FIGURE 3-5 35-mm transparency taken with an analog *Pentax* SLR camera fitted with additional fiducial marks (*PhotoModeler* plastic film plane inserts). Archaeological excavations at Tell Chuera settlement mound, northeastern Syria. Kite aerial photograph taken by IM and J. Wunderlich, September 2003.

required for the orientation of a single photograph—in practice, multiple photographs are usually oriented together using least-squares adjustment algorithms (see section on bundle adjustment below) which allows the use of less than three points per image.

The precision and abundance of GCPs are crucial for the precision of the exterior orientation. Standard GPS measurements or control points collected from maps are useful enough for working with small-scale traditional airphotos and satellite images, but SFAP imagery with centimeter resolution and quite large scale requires accordingly precise ground control (Chandler, 1999). Usually this has to be accomplished by pre-marking control points in the coverage area and determining their coordinates using a total station survey (see Chapter 9).

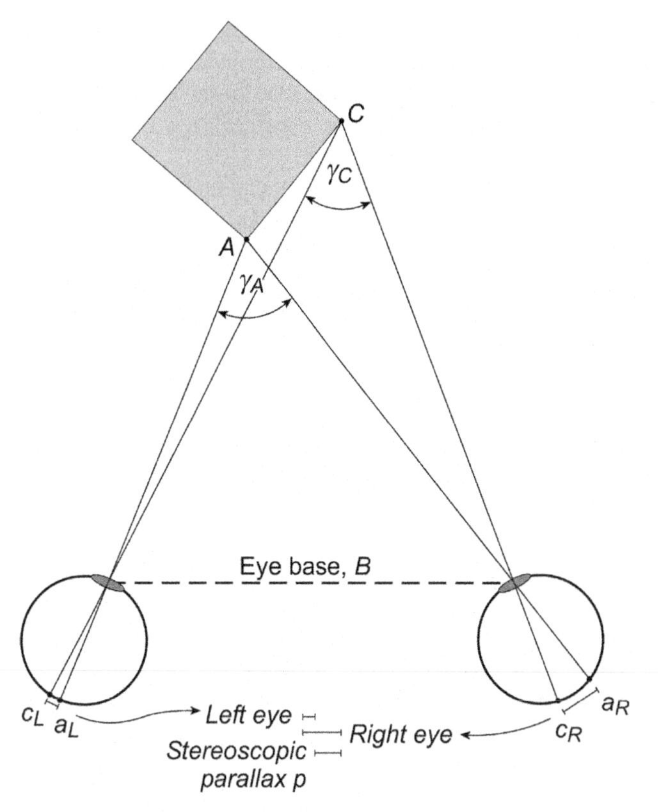

FIGURE 3-6 Stereoscopic parallax for points at different distances in binocular vision. The differences of the angles of convergence γ result in different distances of *A* and *C* projected onto each retina. Their disparity, the differential or stereoscopic parallax, is used by the brain for depth perception. After Albertz (2007, fig. 111, adapted).

3.3. GEOMETRY OF STEREOPHOTOGRAPHS

3.3.1. Principle of Stereoscopic Viewing

Distorting effects of the central perspective are usually undesirable for the analysis of single photographs, but they also have their virtues. Because the magnitude of radial distortion is directly dependent on the terrain's elevation differences, the latter can be determined if the former can be measured. We make use of this fact in daily life with our own two eyes and our capability of stereoscopic viewing. People with normal sight have binocular vision, that is, they perceive two images simultaneously, but from slightly different positions which are separated by the eyes' distance (the eye base *B*, Fig. 3-6).

When the eyes focus on an object, their optical axes converge on that point at an angle (the parallactic angle γ). Objects at different distances appear under different parallactic angles. Because the eye's central perspective causes radial distortion for objects at different distances, quite similar to the effects of mountainous terrain in airphotos, the two images on the retinae are distorted. The amount of displacement parallel to our eye base, however, is not equal in the two images because of the different positions of the eyes relative to the object. This difference between the two displacement measures is the stereoscopic parallax *p*. The stereoscopic parallax and thus 3D perception increase with increasing parallactic angle γ, making it easier to judge differences in distances for closer objects.

Stereoscopic vision also may be created by viewing not the objects themselves but images of the objects, provided they appear under different angles in the images (Fig. 3-7). By viewing the image taken from the left with the left eye and the image taken from the right with the right eye, a virtual stereoscopic model of the image motif appears where the lines of sight from the eyes to the images intersect in the space behind. Although it is possible to

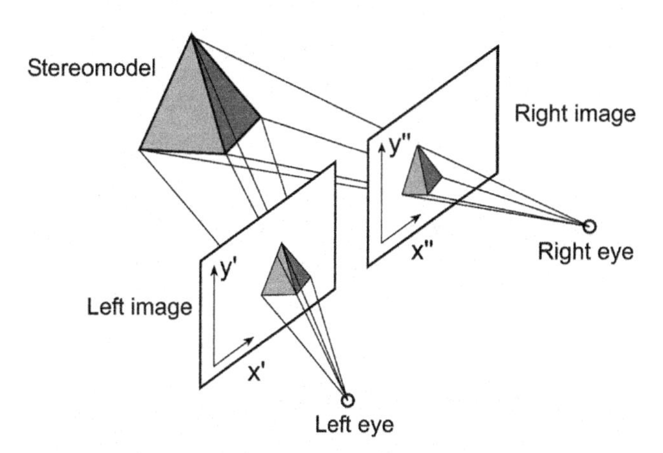

FIGURE 3-7 Stereoscopic viewing of overlapping images showing the same object under different angles. A three-dimensional impression of the object—a stereomodel—appearing in the space behind the images is perceived by the brain. After Albertz (2007, fig. 111, adapted).

achieve this without the aid of optical devices, just by forcing each eye to perceive only one of the images and adjusting the lines of sight accordingly, it is difficult especially for larger photographs and may cause eye strain. Devices like lens or mirror stereoscopes (for analog images), anaglyph lenses (for both analog and digital anaglyphs, i.e. red/blue-images), or electronic shutter lenses (for stereographic cards and computer screens) make stereoviewing much easier and also provide facilities for zooming in and moving within the stereoview (see Chapter 11.5.1).

For any stereoscopic analysis, the images need to be orientated so they reflect their relative position at the moment of exposure. The direction of spacing between the images needs to be parallel to our eye base, because only the x-component of the radial distortion vector is different in the two images, effectuating stereoscopic (or x-) parallax. The amount of distortion in y-direction is the same in both images and needs to be aligned perpendicular to our eye base. This alignment procedure can be as simple as wiggling two photo positives under a lens stereoscope until they merge into a stereomodel (for using SFAP images under a stereoscope see Chapter 11.5.1). Or it may be as laborious as identifying hundreds of tie points in a series of digital images in order to compute the exact relative orientation of a large number of neighboring stereopairs prior to establishing their precise position in space as a prerequisite for the automatic extraction of digital terrain models (see following sections).

3.3.2. Base–height Ratio and Stereoscopic Coverage

From Figure 3-6 it is also evident that the stereoscopic parallax p and thus depth impression can be increased when the eye base B is increased, an option we do not have in real-life viewing, but is possible and desirable in artificial stereoscopic viewing. If the base between the exposure stations in relation to their height (the photographic base–height ratio) is larger than the base of our eyes in relation to the viewing distance (the stereoviewing base–height ratio), the stereomodel appears exaggerated in height by the same factor (Wolf and Dewitt, 2000). The exaggeration factor varies with image overlap and focal length and is typically between two and six times for conventional aerial photographs.

Stereoscopic photographs from professional aerial surveys are acquired in blocks of multiple flightlines in such a way that full stereoscopic coverage of the area is ensured with multiple stereopairs (Fig. 3-8). Each photograph overlaps the next photograph in a line by approximately 60% (forward overlap or endlap), while adjacent lines overlap by 20–30% (sidelap). In mountainous terrain, the overlap can be

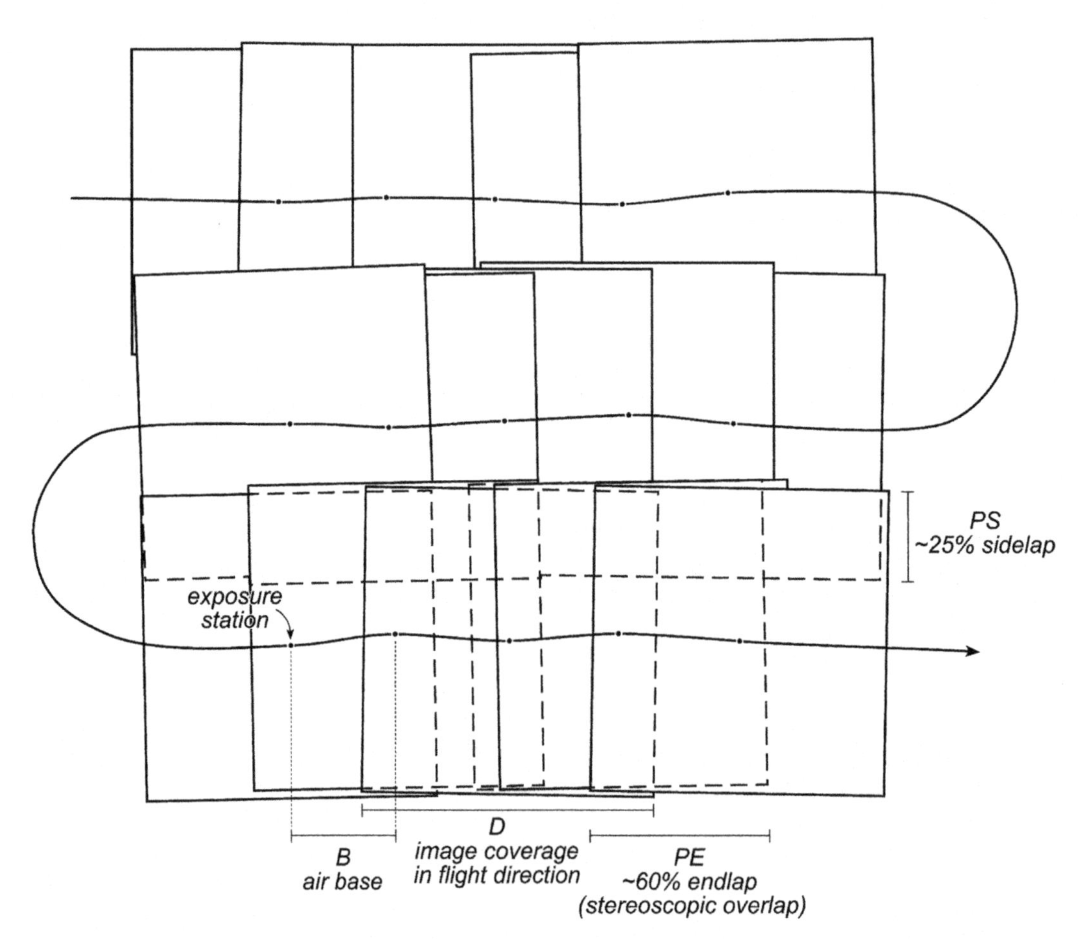

FIGURE 3-8 Block of aerial photographs with 60% endlap in three flightlines with 25% sidelap. This survey design is ideal for ensuring gapless stereoscopic coverage while minimizing image number and redundancy.

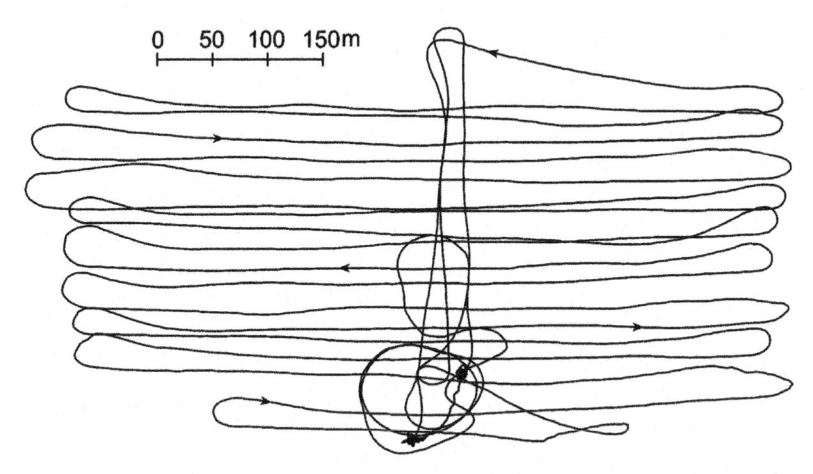

FIGURE 3-9 Flight path of an SFAP survey by autopiloted model airplane conducted at Freila, Province of Granada, Spain (see Fig. 1-1). The two swirly patterns result from the airplane's circling in order to gain and loose height during ascent and descent.

increased in order to avoid gaps of stereoscopic coverage by relief displacement and sight shadowing.

The required air base or distance between exposure stations B is dependent on the dimensions of the image footprint and the desired endlap. If D is the image coverage in direction of the flightline and PE the percent endlap, B calculates as:

$$B = D \times (1 - PE/100) \qquad \text{(Equation 3-5)}$$

The same equation can be used for calculating the required distance between adjacent flightlines for a desired sidelap PS; keep in mind, however, that SFAP cameras most probably feature a rectangular image format ($d_L \times d_W$) and a decision has to be made as to its longitudinal or transversal orientation along the flightline (substituting D in Eq. 3-5 by D_L or D_W, respectively).

For platforms with approximately constant ground velocity V_g, the time interval ΔT between exposures is:

$$\Delta T = B/V_g \qquad \text{(Equation 3-6)}$$

The more precisely a platform is navigable, the easier it is to follow a pre-arranged flightplan with systematic design. Figure 3-9 shows the actual flightpath of a survey taken by the autopiloted model airplane presented in Chapter 8.5.2, which comes quite close to the conventional blocks of aerial photography. However, with most SFAP platforms, it is usually difficult to achieve such a regular flightline pattern, and it is indeed not a prerequisite for stereoscopic analysis. Stereomodels also can be created from images which overlap in a much more irregular pattern as long as all parts of the area are covered by at least two photos each with sufficient base–height ratio (Fig. 3-10). It should be pointed out, however, that some digital photogrammetry systems designed for analyzing standard aerial

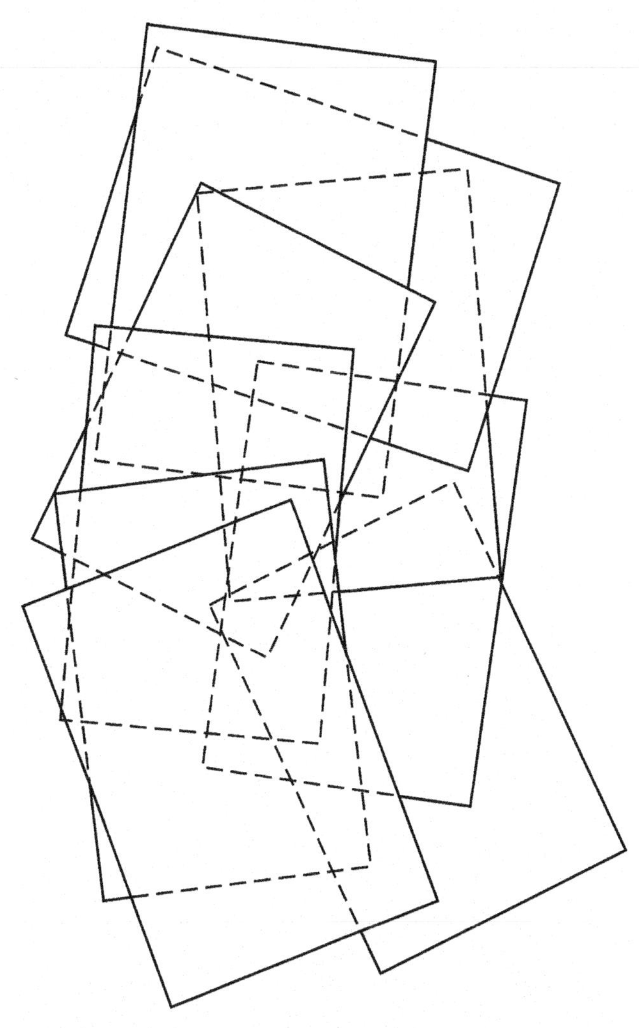

FIGURE 3-10 Block of aerial photographs taken by a tethered hot-air blimp from ~120 m flying height. Although the image alignment and overlaps are irregular and the photographs even have slightly different scales, full stereoscopic coverage of the desired ground area was given.

imagery may have difficulties in dealing with irregular image blocks.

The vertical exaggeration of relief in the stereomodel plays an important role in 3D analysis. However, which base–height ratio is "sufficient" for stereoscopic analysis depends on the desired degree of depth impression and (if applicable) measurement accuracy for the particular relief type. Small height variations of gently undulating terrain are better viewed and measured with an extra-large base–height ratio, while the analysis of high-relief terrain, which already exhibits larger stereo-parallaxes due to radial distortion, tolerates lower base–height ratios. In general, errors in planimetry and height decrease with increasing base–height ratio and increasing stereo-parallaxes (Wolf and Dewitt, 2000; Kraus, 2004; Kraus et al., 2007).

For a stereo survey with a given image scale *S*, wide-angle lenses may be used at lower flying heights than smaller-angle lenses, thus increasing the base–height ratio (Fig. 3-11). In addition, wide-angle lenses cause larger relief displacement and thus larger stereoscopic parallaxes than smaller-angle lenses, even if both are taken with the same base–height ratio. For stereoscopic purposes, images taken with wide-angle lenses from lower flying heights are therefore preferable to images taken with smaller-angle lenses from larger flying heights—quite contrary to the case of monoscopic mapping or mosaicking (see below) where distortion-minimized images are desired.

A typical SFAP problem regarding the stereo-capability of photographs taken from kites, balloons, and drones is the variability of scales and tilting angles that is often unavoidable. From the authors' experience, difficulties with stereoscopic viewing and image matching (see below) may arise if the image scales differ more than 10% or so. Stereoviewing is also hampered by image obliqueness. For unfavorable constellations of nadir angles, tilting may even have the additional effect of extinguishing the stereo-parallaxes regardless of good base–height ratios.

The only solution is to take as many photographs as possible during a survey and carefully choose the best of them for analysis (see Chapter 11.5.). Taking more photographs is always a good way to increase chances for success, regardless of the mission's nature or purpose.

3.3.3. 3D Measurements from Stereomodels

Beyond the simple viewing and interpreting of stereopairs, stereomodels also enable various kinds of 3D measurements using photogrammetric techniques. The 3D ground

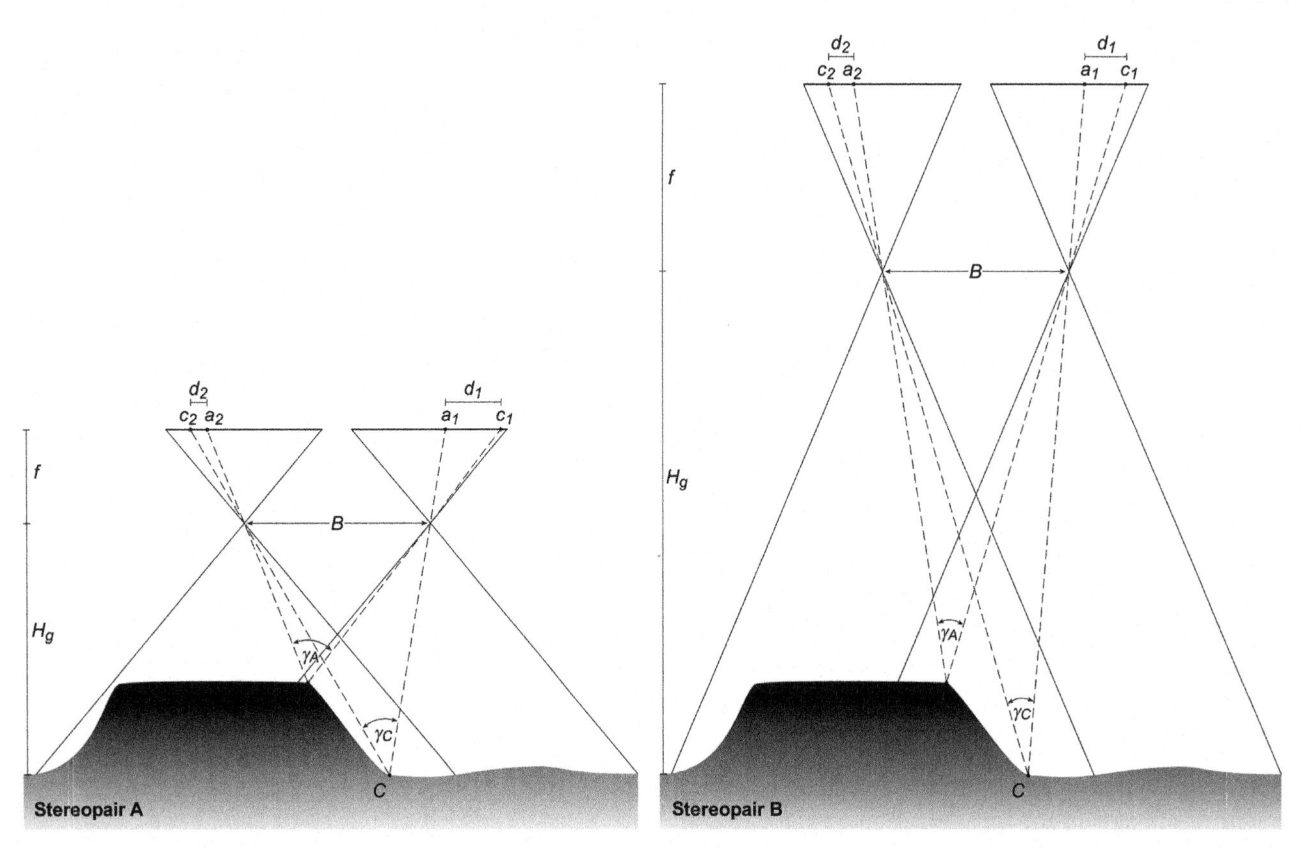

FIGURE 3-11 Two stereopairs with equal image scale and air base *B* taken over the same terrain. Focal length and flying height of stereopair B (normal-angle lens) are twice those of stereopair A (wide-angle lens), resulting in a halved base–height ratio. Note how the parallactic angles γ and the stereo-parallax—the difference between d_1 and d_2—decrease from stereopair A to stereopair B.

coordinate of any given object point can be determined if the corresponding 2D image coordinates in a stereopair and the position of the camera within the ground coordinate system are known. This is illustrated in Figure 3-12, where the position of an object point *A* in the landscape is reconstructed by tracing the rays from the homologous image points a_1 and a_2 back through the lens. With a single image (the left photo in the figure), no unique solution can be found for the position of *A* along the reconstructed ray. By adding a second (stereo) image on which *A* also appears, a second ray intersecting the first can be reconstructed and the position of *A* can be determined. This method is called a space-forward intersection; it is based on the formulation of collinearity equations describing the straight-line relationship between object point, corresponding image point and exposure station.

Deviating from the schematic situation in Figure 3-12, the differences between the focal length and the flying height for aerial surveys are in reality much greater. For SFAP surveys, the focal length is usually well below 10 cm and the flying height somewhere between 20 m and 500 m; for conventional aerial surveys, the focal length is normally between 9 and 16 cm and the flying height somewhere between 2000 m and 5000 m. Therefore, the accuracy of the reconstructed 3D coordinate of *A* is highly dependent on the precision with which a_1, a_2, and the focal lengths are measured. This is why photogrammetry is a science with many decimal places! To make the extrapolation from the (known) interior of the camera into the (unknown) space beyond as exact as possible, both the values of the interior orientation and the exterior orientation (see above) need to be known precisely.

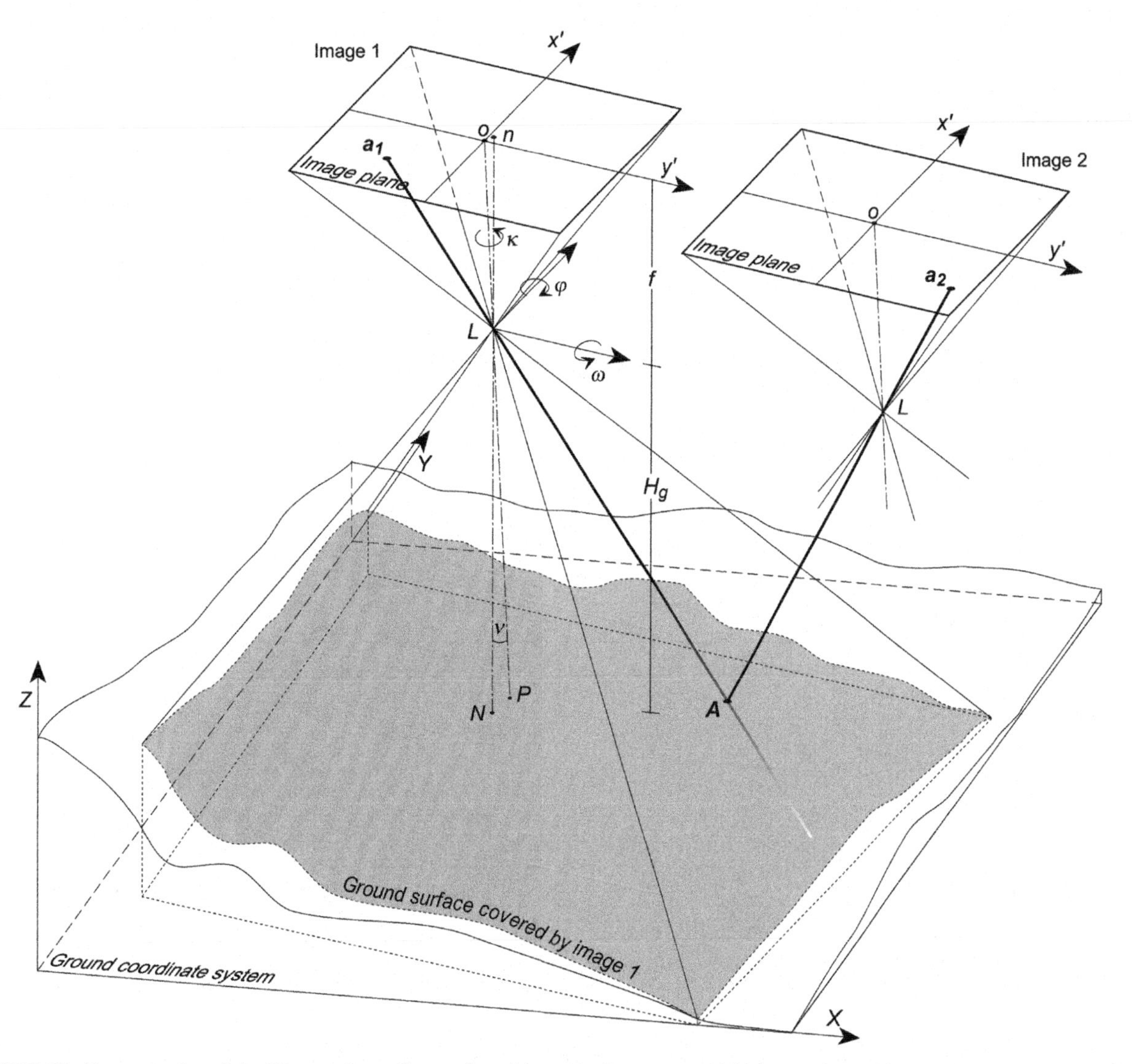

FIGURE 3-12 Reconstruction of the 3D ground coordinates of an object point in a stereomodel by space-forward intersection. From the image coordinates x', y' of the homologous points a_1 and a_2, the rays of light are traced back through the exposure center *L* to the object point *A*. The interior orientation—the geometric relation of *L* to the image plane—and the exterior orientation—the position *X, Y, Z* of *L* and the rotations ω, ϕ, κ of the image plane with respect to the ground coordinate system—must be known for this reconstruction.

As outlined in the previous section, the required exact exterior orientation for SFAP images is not known a priori. However, it can be determined using the same reconstruction method backwards in a space resection. With the known 3D coordinates of three ground control points and their corresponding 2D image coordinates, the position X, Y, Z of the exposure center at the intersection of the three rays and the three rotations of the camera ω, ϕ, κ relative to the ground coordinate system can be calculated. Afterwards, these exterior orientation parameters can be used in the space-forward intersection in order to calculate any other object point coordinate in the area covered by the stereomodel.

3.3.4. Creating Stereomodels with Aerial Triangulation by Bundle-Block Adjustment

In praxis, photogrammetric analysis is mostly done using not one, but several or even many stereopairs for covering larger areas. In order to avoid the individual orientation of each stereomodel with accordingly large numbers of ground control points, multiple overlapping images forming a so-called block (see Fig. 3-8 and 3-9) can be oriented simultaneously with fewer ground control points using aerial triangulation techniques. One of the most commonly used and most rigorous aerial triangulation methods is the bundle adjustment or bundle-block adjustment.

In theory, bundle-block adjustment allows the absolute orientation of an entire block of an unlimited number of photographs using only three GCPs. This requires that the relative orientation of the individual images within the block first be established by additional tie points—image points with unknown ground coordinates which appear on two or more images and serve as connections between them (Fig. 3-13). These tie points can be identified either manually or (in digital photogrammetry) with automatic image matching procedures. The term bundle refers to the bundle of light rays passing from the image points through the perspective center L to the object points. The bundles from all photos are adjusted simultaneously so that the remaining errors at the image points, GCPs and perspective centers are distributed and minimized. Thus, the ground coordinates of the tie points as well as the six exterior orientation parameters of each image can be calculated in a single solution for the entire block. The stereomodels may now serve for further quantitative analysis, i.e. manual or automated measuring and mapping of unknown object point coordinates, heights, distances, areas, or volumes.

Regarding the accuracy of bundle-block adjustment, several equations for accuracy estimations are given in the photogrammetric textbooks (e.g. Kraus, 2004; Kraus et al., 2007). However, they are not easily assignable for the SFAP case, as accuracy is a function not only of scale and flying height, but also of the quality of interior orientation, base–height ratio, etc. Table 3-1 summarizes typical results for bundle-block adjustment carried out with the type of vertical SFAP appearing throughout this book. Nine images, taken with an 8 megapixel DSLR (*Canon EOS 350D*)

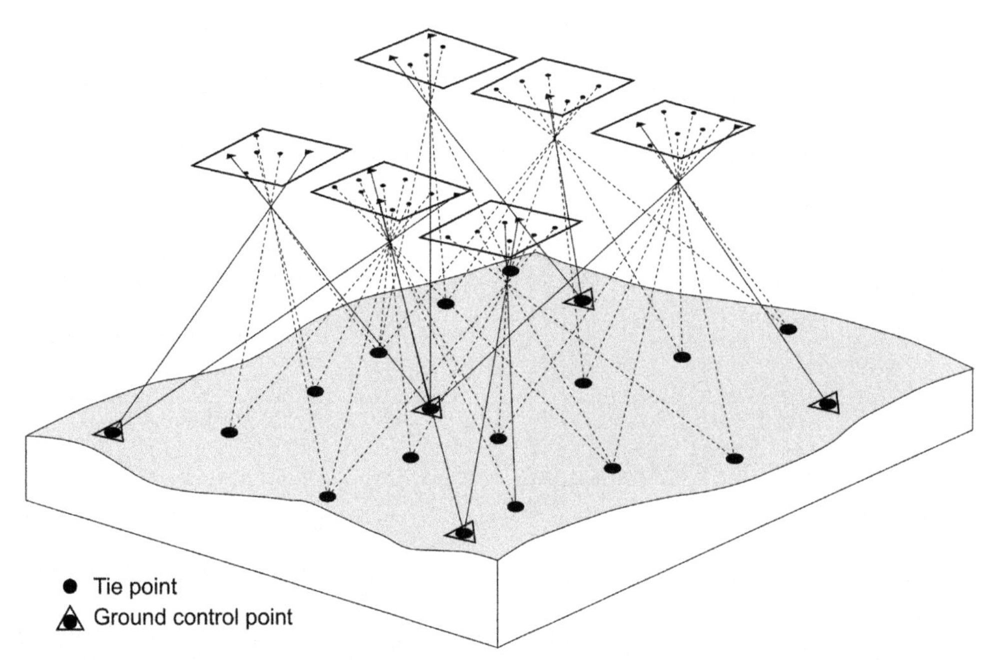

FIGURE 3-13 Principle of bundle-block adjustment; example of a possible SFAP survey. The relative orientation of the images in the block is established by both tie points and GCPs, the absolute orientation of the block within the ground coordinate system is realized using the GCP coordinates. After Kraus (2004, fig. 5.3.1, adapted).

TABLE 3-1 Summary of bundle-block adjustment results for nine digital images taken at Gully Bardenas 1, Province of Navarra, Spain; image *GSDs* 1.9–2.6 cm. Triangulation performed without camera calibration, with self-calibration and with precedent test-field calibration, using varying numbers of control and check points. Photogrammetric processing by IM.

Bundle-block adjustment		No calibration			Self-calibration			Test-field calibration		
GCP number (control/check)		6/45	12/39	26/25	6/45	12/39	26/25	6/45	12/39	26/25
Total image unit-weight RMSE		0.907	1.154	1.227	0.458	0.453	0.462	0.480	0.482	0.480
Control point RMSE	Ground X (cm)	4.10	6.36	4.83	0.21	0.51	0.61	0.22	0.54	0.50
	Ground Y (cm)	2.90	5.09	3.73	0.45	0.41	0.62	0.45	0.55	0.57
	Ground Z (cm)	11.70	14.94	11.65	0.84	0.55	0.92	0.67	0.70	0.63
	Image (pixels)	5.34	5.91	4.16	0.69	0.67	0.62	0.62	0.67	0.58
Check point RMSE	Ground X (cm)	8.68	11.00	14.95	2.05	1.64	2.44	1.55	1.23	1.66
	Ground Y (cm)	14.61	10.65	21.51	3.04	2.33	2.01	2.88	2.36	1.80
	Ground Z (cm)	91.87	46.19	76.74	7.11	5.14	6.66	7.18	5.25	5.05
	Image (pixels)	3.72	2.01	1.05	0.45	0.44	0.40	0.45	0.44	0.41
Mean exterior orientation error	Position (cm)	6.21	5.61	4.78	4.73	6.24	3.48	3.45	2.44	1.94
	Angle (°)	0.04	0.04	0.03	0.03	0.02	0.01	0.02	0.02	0.01

camera at flying heights between 60 and 82 m above ground, were triangulated using a total of 51 GCPs in *Leica Photogrammetry Suite*.

The study site was an agricultural area with an erosion gully, in parts densely vegetated, cutting between two fallow fields (see Fig. 11-1). Of the GCPs, six, 12 or 26 were used as control points, the rest as check points for error assessment. The interior orientation of the camera was not defined (nominal focal length of lens only; "no calibration"), determined by the software in a self-calibrating approach (correction of focal length and principal point position, lens-distortion model applied; "self-calibration"), or determined by previous test-field calibration (see Chapter 6.6.5).

Results show that bundle adjustment without any camera calibration performs by far the poorest. The nonlinear fluctuation of error values with increasing control point number (12 control points are better than 26) indicates that no satisfactory solution can be found for the adjustment. Errors at the independent check points are around 10–20 cm for horizontal and up to 92 cm for vertical position which must be considered intolerable. Both the self-calibration and the precedent test-field calibration show much better results, with the latter only slightly superior to the former. All residuals at the control points are well below 1 cm, and the horizontal errors are within the range of image *GSDs* (1.9–2.6 cm). Z errors are, as to be expected with aerial triangulations, somewhat poorer but decrease just as horizontal errors with the number of control points employed. Altogether, the accuracies achieved in this study benefit from the large number of tie points (900) generated for improved image-block stabilization.

3.4. QUANTITATIVE ANALYSIS OF PHOTOGRAPHS

The interpretation of remote sensing images may involve both qualitative and quantitative aspects. Visual interpretation techniques, which are discussed in Chapter 10, are utilized for identifying and characterizing objects and for interpreting spatial patterns. Most methods of extracting quantitative information from remote sensing images do not work without some amount of image interpretation, but their primary concern is to obtain spatial measurement values.

Photogrammetry comprises a large number of different measurement methods from single or stereo images and from manual to fully automatic, which are extensively described in the literature (e.g. Pfeiffer and Weimann, 1991; Warner et al., 1996; Wolf and Dewitt, 2000; Jensen, 2007; Lillesand et al., 2008). Some approaches applicable for (digital) SFAP that the authors of this book judge the most useful and commonly applied are introduced briefly in the following sections.

3.4.1. Measuring and Mapping from Single Photographs

Individual measurements of lengths and sizes—the width of a brook, the diameter of a tree crown, the distance between vegetation patches—can be taken easily in analog or digital photographs using Equation 3-1, if the image shows an object or distance with known length. This could, for example, be a scale bar as in Figure 2-3 or the (calculable) distance between two ground control points as in Figure 9-8. Adapting Equation 3-1, the required ground measure D can be calculated by comparing its image length d with that of the scale bar.

$$D = d \times D_{\text{scale bar}}/d_{\text{scale bar}} \qquad \text{(Equation 3-7)}$$

Other possibilities exist for measuring object heights in single photographs, but they are restricted to objects rising vertically from the ground. Here, the relief displacement between the top and bottom point of the object, e.g. a building or tree, can be measured and related to the object's distance from the image principal point in order to calculate the object's height (Jensen, 2007; Lillesand et al., 2008). If the exact time, date and place of image acquisition is known, another possibility is to use the shadow cast by the object for calculating its height (Jensen, 2007). Neither method can be used for objects where the top and bottom points are not perpendicularly aligned and both visible. The height of a mountain, sloping river bank, or rock face of uncertain verticality cannot be determined with these methods, making them rather useless for geoscientific applications.

It has to be kept in mind that the accuracy of such measurements from single images depends not only on the measurement precision but also more importantly on the validity of the equation, i.e. closeness of the image acquisition situation to perfect central projection over flat terrain. Even for flat terrain, SFAP images rarely meet these requirements, as the optical axis is rarely exactly vertical and SFAP camera lenses often show considerable distortions (see Chapter 6).

Usually, continuous spatial mapping is more of interest than single measurements when analyzing aerial photographs. In order to be suitable as a base for mapping or monitoring the photographs have to be geometrically corrected and georeferenced. This can be done either by polynomial rectification using a set of ground control points or by orthorectification, as follows.

3.4.1.1. Polynomial Rectification by GCPs

Polynomial equations formed by ground control point coordinates and their corresponding image-point coordinates are used in order to scale, offset, rotate and warp images and fit them into the ground coordinate system (Mather, 2004). The highest exponent of the polynomial equation (the polynomial order) determines the degree of complexity used in this transformation—1st order transformations are linear and can take account of scale, skew, offset and rotation, while 2nd and 3rd order transformations (quadratic and cubic polynomials) also may correct nonlinear distortions. Because the polynomials are computed from the GCP points only and then applied to the entire image, they only produce good results if the GCP locations and distribution adequately represent the geometric distortions of an image. For aerial photographs, which are subject to radial displacement, this is usually difficult to achieve.

Vertical or oblique images of flat terrain (Figs. 3-1 and 3-3) can be quite successfully rectified with 1st or 2nd order polynomials, but the relief distortions present in images of variable terrain are much more difficult to correct (consider the impossibility of fitting a 2nd or 3rd order polynomial to the surface shown in Fig. 3-2). Although higher orders of transformation can be used to correct more complicated terrain distortions, the risk of unwanted edge effects by extrapolation beyond the GCP-covered area increases.

Further similar methods of image rectification may utilize ground control points, i.e. spline transformation or rubber sheeting, which optimize local accuracy at the control points, but all of these methods have in common that the rectification of the image areas between the GCPs is interpolated by the polynomial equation and not a direct function of radial relief displacement. Thus, the rectified image would not be truly and completely distortion-free. While polynomial rectification by GCPs may well be sufficient for low-distortion images or applications with limited demand for accuracy, seriously distorted images and more precise applications require full modelling of the distortion parameters (relief displacement, lens distortion, and image obliqueness) for geocorrection.

3.4.1.2. *Orthorectification or Orthophoto Correction*

The continuously changing distortions caused by relief displacement for images of variable terrain cannot be removed sufficiently with polynomial rectifications, because the surface is too complex to be described by simple mathematical algorithms. A complete and differential rectification of the distorted photograph into a planimetrically correct image with orthographic (= map) projection can be achieved only if the exact amount and direction of displacement for each pixel can be calculated and removed. Following the principles shown in Figures 3-2 and 3-12, this is possible if the interior and exterior orientation of the camera and the terrain heights are known. Orthorectification procedures make use of digital elevation models (DEMs) in relation to which the photographs are oriented in space so that the relief displacement (with the added effect of image obliqueness and lens distortion) of each single pixel can be determined. In the new, orthorectified image file, each pixel is then placed in its correct planimetric position.

Depending on the type and source of the DEM, the elevation values either describe the ground surface (digital terrain model or DTM) or the true surface including all objects rising above the terrain (e.g. woodland, buildings; digital surface model or DSM) (Kasser and Egels, 2002; Jensen, 2007). Thus, it can be differentiated between conventional orthophotos and true orthophotos, where the latter present an image truly geocorrected for all terrain and object elevations. For orthorectifying SFAP images, a DEM with appropriately high resolution is normally not available from external sources—thus, the best solution would be to generate a DEM from the SFAP images themselves first and subsequently use this for orthophoto correction. If the DEM is generated by automatic elevation extraction procedures, it would necessarily be a DSM, resulting in a true orthophoto and thus in the most geometrically accurate image product achievable. A brief overview of DEM generation is given below in this chapter.

If a single rectified photograph or orthophoto does not fully cover the study area, an aerial mosaic may be constructed by stitching the georeferenced images together (controlled mosaic, Wolf and Dewitt, 2000; see Chapter 11.3.3). With semi-controlled mosaics, even areas devoid of ground control points can be bridged by registering the images of the gap areas visually to the georeferenced images around them. There are several image-processing techniques that help to minimize a jigsaw-puzzle appearance of the resulting mosaic. The seamlines between the individual mosaic pieces can be manually or automatically placed in order to be most inconspicuous, and radiometric matching techniques may be used for taking color and brightness differences into account.

The resulting GCP-rectified or orthorectified images or mosaics may now serve a variety of purposes—as photomaps annotated with reference grid and coordinates, for measuring distances and areas of individual objects, as a background for compiling thematic map layers in a GIS environment, or as components of a time series for change monitoring. For further details and examples, see Chapters 11 and 13.

3.4.2. Manual Measuring and Mapping from Stereomodels

In a single image, the location of a point is given by its position x, y. Its corresponding 2D object space coordinate X, Y can be determined by some method of georeferencing as described above. When measuring from stereo images that have been relatively and absolutely oriented in a ground coordinate system, the stereoscopic parallax of a point is measured additionally for deriving its height Z. All stereoscopic measurement devices—analog parallax bars for stereoscopes, analytical stereoplotters, digital photogrammetric workstations—make use of the principle of the floating mark or cursor. Small marks etched onto a glass bar or superimposed as dots or crosses of light are displayed over the left and right images independently. By adjusting their position so that they fuse into a single mark at the exact location of the point to be measured, the x-parallax and thus the height of the point can be determined.

The floating mark is perceived by the operator as resting exactly on the terrain surface, when the left and right marks are placed on corresponding points, i.e. over the same feature in the two images (Fig. 3-14). Most digital photogrammetric workstations now offer the possibility of automatically adjusting the floating cursor height when the mouse is moved over the terrain. This is accomplished by the same digital image correlation techniques that also are used for automatic tie point generation and DEM extraction (see below). This technique saves the operator continually adjusting the x-parallax to changing terrain height; however, its success does significantly depend on local image contrast, texture and pattern.

Apart from individual measurements of 3D point positions, distances, areas, volumes, angles, and slopes, digital photogrammetry systems with stereoviewing facilities allow the direct stereoscopic collection of georectified 3D GIS vector data as point, line, or polygon features. These can then be used in GIS software for further editing, analysis, and visualization as well as combination with orthophoto maps. Such softcopy stereoplotting techniques are basically the fully computerized implementation of traditional stereoplotting methods, where the movements of the floating mark in the analog stereomodel were mechanically transmitted to a tracing pencil on a sheet of paper (mechanical stereoplotters) or converted into coordinates recorded by a linked CAD system (analytical stereoplotters).

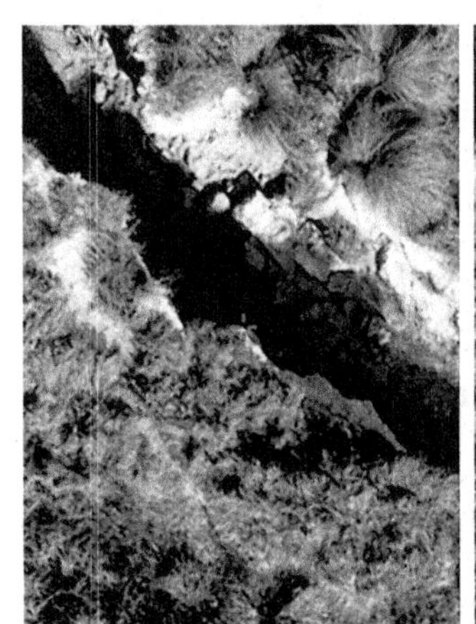

FIGURE 3-14 Floating mark (stereo cursor) positioned at the edge of an erosion channel in a digital stereomodel (subset). Image width ~4.3 m, height of channel wall ~2.3 m. Kite aerial photograph by IM and JBR at the Bardenas Reales, Province of Navarra, Spain, February 2009.

3.4.3. Automatic DEM Extraction from Stereomodels

A digital elevation model (DEM) is a digital representation of terrain heights. The most common forms are a regular grid (usually saved in raster format) or a triangular irregular network (TIN) of triangle facets (vector format). They can be created by manual collection (see preceding section) of height points, breaklines, and contours and subsequent application of interpolation algorithms or Delaunay triangulation, which convert these data to regular grids or TINs (Li, Zhu and Gold, 2005). With the advent of digital photogrammetry, however, it has become possible to extract elevation information automatically from stereomodels using stereo-correlation or image-matching techniques.

A human operator manually mapping topographic features in stereomodels basically accomplishes two tasks: placing the floating mark onto an object and point by merging the left and right image of the object into a stereoscopic view, and interpreting the nature of the object, classifying it, for example as a tree, a street, a house, or a height point. The first task only involves comparing two image subsets for their similarity and also can be performed automatically by computer processing. The second task involves recognizing, distinguishing, and interpreting individual objects, at which human beings are still much superior to computers, although pattern recognition and feature extraction techniques continue to be important research topics in all imaging sciences, including photogrammetry and remote sensing (e.g. Bartels and Wei, 2007; Grün, 2008).

Based on the two main categories of image-matching techniques (area-based matching and feature-based matching), various hybrid methods have been developed (Wolf and Dewitt, 2000). All are concerned with statistically determining the correspondence between two or multiple image areas or features in order to identify homologous image points. A variety of parameters constrains and controls the process of searching and correlating image points. The quality and amount of height points may be affected both by the ground-cover type and by image characteristics such as noise and local contrast. Dark and shadowed regions often show low correlation coefficients owing to increased noise, and homogeneous smooth surfaces may have too little texture for successful image matching.

Figure 3-15 illustrates the results of an investigation into the role of different image types and ground-cover classes. Both analog and digital images of the same scene (taken simultaneously with a 35-mm SLR and 6.3 megapixel DSLR on a double-camera mount) were used for DEM extraction. The analog transparency slides were scanned with 2900 dpi and subsequently resampled to 1800 dpi, all images were processed with *Leica Photogrammetry Suite*. Results show that the digital camera is clearly superior to the film camera in terms of point density; not only are the

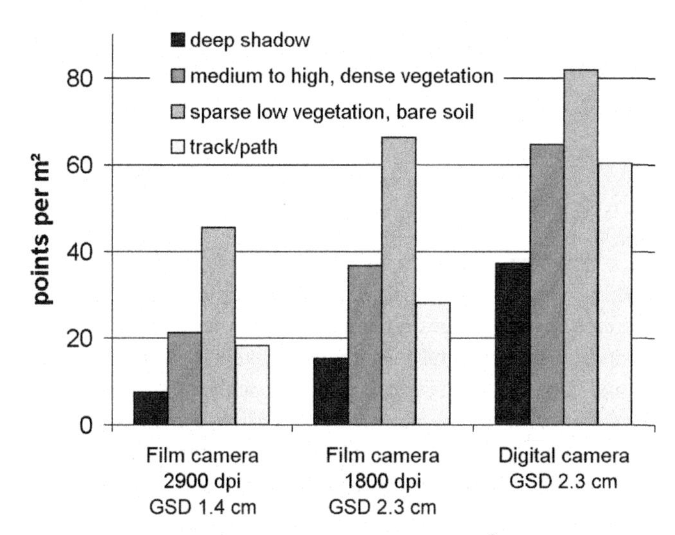

FIGURE 3-15 Autocorrelated height points per m^2 ground area for varying ground-cover types, derived by automatic image matching from analog and digital stereopairs. SFAP taken simultaneously with a double-camera mount from a hot-air blimp. Image processing by IM.

absolute numbers of points higher but also the differences for varying ground-cover types lower. The high-resolution scan of the analog film performs considerably poorer than the resampled lower resolution; this can be attributed to the effect of capturing and averaging out, respectively, the film grain structure. These results correspond to similar findings reported by Gruber et al. (2003) for metric aerial cameras.

Once a set of corresponding points in the overlapping images has been identified, their 3D coordinates are computed by space-forward intersection using the block triangulation results. The primary result is a dataset of more or less irregularly spaced points, which are usually directly processed by interpolation techniques into a raster DEM file or TIN. Typically, the resolution of the DEM is about one magnitude lower than the image *GSD*, because not every pixel can be matched with the current photogrammetry algorithms. However, recent developments are moving toward generating higher-density results in automatic terrain extraction, staying abreast of the increasingly higher resolutions provided by modern sensors (Woodhouse, 2009).

Because the camera records the light reflected from the visible surface of all objects present in a scene, the matching algorithms also extract height points from whatever is closest to the camera and possibly covers and obscures the actual terrain surface. Digital elevation models extracted by automatic matching procedures from stereomodels are therefore inherently digital surface models (DSMs) rather than digital terrain models (DTMs) devoid of aboveground features. However, the accuracy with which objects rising above the terrain are captured, or the degree to which they may irritate the matching algorithms, depends greatly on their properties. Wispy vegetation, vertical walls, moving shadows, vehicles, people, or animals can result in serious mismatches, just as deep shadows and hidden areas (sight shadow) may result in information gaps (Kasser and Egels, 2002).

DEM extraction from SFAP images is especially prone to such errors because of the high spatial resolution and small areal coverage. The relative height errors caused by shrub cover or trees can be quite large compared to the relief undulations, and the areas obscured by shadow, where few matching points would be found, may result in large information gaps with respect to DEM pixel size. Also, vertical objects in SFAP cause comparatively large sight shadowing because of the low flying height, while proportionally large vertical objects rarely exist in small-scale images.

Consider for example the following problems (Fig. 3-16; see also Fig. 11-22) that were identified at various gully monitoring test sites (Marzolff and Poesen, 2009), but

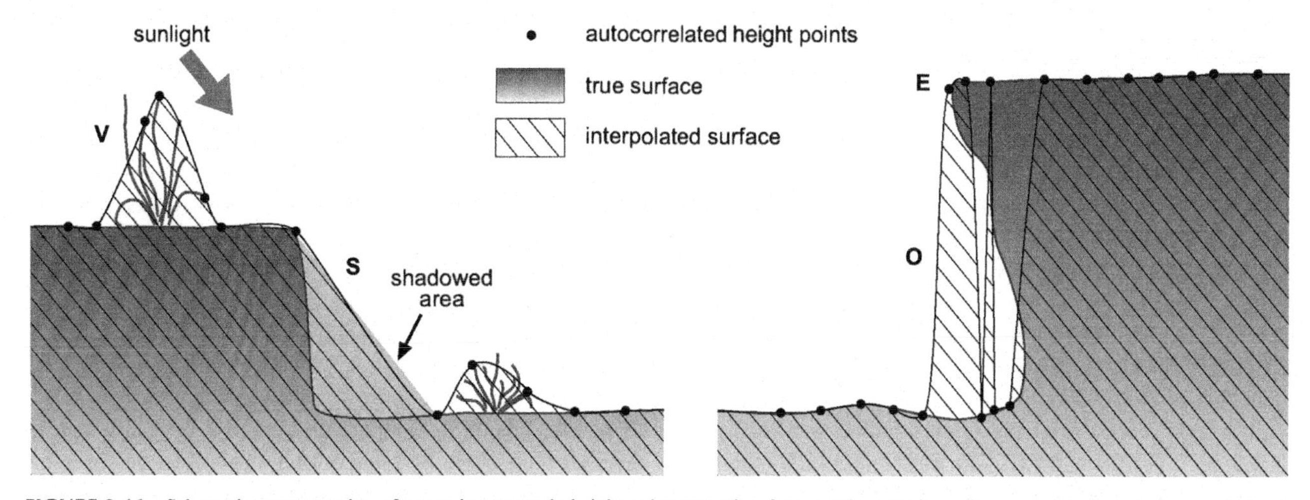

FIGURE 3-16 Schematic representation of errors in automatic height-point extraction for a varying terrain surface caused by morphology and illumination effects. V—vegetation cover; S—shadowed area; E—illuminated edges; and O—overhanging walls. Adapted from Marzolff and Poesen (2009, fig. 5).

would apply similarly for river banks, coastal cliffs, built-up areas and other typical SFAP motifs.

- The image matching process makes no difference between the soil surface and objects covering this surface, e.g. vegetation (V). Although the irregular, discontinuous surface of trees, shrubs, or tall grasses does not result in a high density of height points, correlations are often found for individual branches and for small patches of herbs. The interpolated surface represents fragmentary vegetation canopy rather than soil surface in these cases.
- In sunny survey conditions, steep walls averted from the sun cast a deep shadow (S) onto parts of the ground surface. Little image texture is available for the matching process in these areas, and a lack of matched points may lead to false or less-inclined slopes. The same effect happens where sight shadowing occurs due to the photograph's central perspective (see also Giménez et al., 2009).
- Where walls and edges (E) are directly exposed to the sun, lack of image texture by their direct illumination may lead to a low number of matched points along sharp edges or to mismatches resulting from the large x-parallax caused by the steep drop. This may result in rounded rather than sharply defined edges of gullies, river banks, cliffs, or buildings.
- Natural scarps occurring with gullies, river banks, or cliffs commonly feature overhangs (O) where fluvial or coastal erosion has undercut the steep walls. Because the central perspective of the camera may look beneath this undercut, height points may be sampled here as well. The algorithm for surface interpolation (raster DEM or TIN) would then link neighboring (i.e. as determined from horizontal coordinates) height points by a best-fitting surface and consequently produce artifacts of peaks and pits. This is unavoidable as neither the connectivity of the points with respect to the (unknown) true surface is known nor the possibility of more than one z value per x/y location given in the 2.5D format of TIN or raster DEMs.

In order to correct matching errors caused by vegetation, shadow, excessively steep relief, or overhanging objects, the primary dataset or the finished DEM or TIN usually needs some editing (see Chapter 11.5.3). The statistical errors and accuracies of the resulting DEM or TIN can be judged by various measures if suitable reference data for terrain elevations exist. Several researchers have discussed the difficulties of quality assessment for DEMs derived from small-format stereo photographs, emphasizing the need for distinguishing accuracy, precision, and reliability (e.g. Chandler, 1999; Lane, 2000). The most common method of error assessment, although not truly independent, is to reserve a number of ground control points as check points and compare their z values with those indicated by the DEM for the same location. This requires that sufficient control points have been installed during the survey preparation (see Chapter 9.5).

3.5. SUMMARY

Photogrammetry comprises all techniques concerned with making measurements of real-world objects and terrain features from images. These may be aerial as well as terrestrial images, and they may be taken by film cameras, digital cameras or electronic scanners on tripods, airborne or spaceborne platforms. Applications include the measuring of coordinates, quantification of distances, heights, areas and volumes, preparation of topographic maps, and generation of digital elevation models and orthophotographs.

Understanding the geometry of a single photograph is the basis for all measurement techniques. A number of simple equations may be used for deriving measurements from single vertical photographs taken over flat terrain. However, they do not apply without restriction for oblique imagery or varying relief. Both cause image scale to vary across the scene, and undulating terrain results in radial relief displacement of features rising above or dipping below the horizontal plane at the image center.

The relief displacement caused by varying image depth changes with the position and angle from which the terrain is viewed. This fact is exploited by the human ability for stereoscopic vision, both in real life and in stereophoto analysis. By determining the relative displacement of an object within two images (the stereoscopic parallax), its height may be visually or mathematically estimated. Ideally, stereoscopic airphotos are acquired with regular overlaps along blocks of parallel flightlines, but any set of images taken with similar scales and sufficient overlap for a satisfactory base–height ratio may be used for creating stereomodels.

Quantitative analysis of photographs may be done manually or fully automatically from single or stereo images. When using single images for measuring, mapping, and monitoring, they need to be geometrically corrected and georeferenced, which may be accomplished either by polynomial rectification using a set of ground control points or by orthorectification using a high-resolution digital elevation model.

When using stereomodels for precise measuring and mapping in the 3D object space, the paths of rays both within and outside the camera need to be mathematically reconstructed with high precision, requiring knowledge of the internal camera geometry and external camera position. Using dedicated digital photogrammetry software, entire blocks of overlapping images may be oriented simultaneously with relatively few ground control points using aerial triangulation techniques such as bundle-block adjustment. From the resulting oriented stereomodels, measurements may be taken and 3D objects may be mapped manually in a stereoviewing environment. Alternatively, automatic extraction of height points based on image matching techniques allows creating digital terrain representations in the form of raster or vector elevation models.

Photogrammetric analysis of SFAP must be expected to yield somewhat lower accuracies (relative to image scale) than metric aerial imagery owing the lower quality of the optical components of consumer-grade cameras. Also, effects associated with typical error sources such as vegetation, vertical walls, and shadowing are usually stronger for SFAP because of its shorter object distance and higher resolution. However, with due understanding of error sources and suitable editing actions, excellent results may be achieved in small-format aerial photogrammetry.

Chapter 4

Lighting and Atmospheric Conditions

Light is all important.

(Caulfield 1987, p. 32)

4.1. INTRODUCTION

How an object appears in an aerial photograph depends on the way in which it is illuminated and the position of the camera relative to the object and source of illumination (Fig. 4-1). Natural sunlight is tremendously diverse in its characteristics, which include direction, diffusion, harshness, and color (Zuckerman, 1996). Thus, natural light varies with season, time of day, latitude, altitude, cloud cover, humidity, dust, and other ephemeral conditions.

Light quality may be described as hard or soft (Defibaugh, 2007). The former is typical under clear sky during mid-day hours, while the latter is found in cloudy conditions and early or late in the day. To achieve gentle illumination and warm colors, professional landscape photographers prefer the lighting conditions just after sunrise or before sunset, when a diffuse golden glow fills the sky. However, typical aerial photography is taken during mid-day hours when the sun is relatively high in the sky, in order to provide for full illumination or toplighting of objects as seen from above. Under these conditions, more blue light is available, especially in shadows. For purposes of the following discussion, we assume that most SFAP takes place under clear sky between mid-morning and mid-afternoon hours.

Traditional large-format aerial photography is done in the vertical mode; however, many newer remote sensing systems are designed to operate in a variety of vertical and oblique modes. This introduces much greater flexibility for possible viewing angles relative to a given target. In recent years, considerable theoretical and experimental research has been undertaken in order to understand better the phenomenon of multiview-angle reflectance, which is the variation in reflectivity depending on the location of the sensor in relation to the ground target and sun position (Asner et al., 1998). In addition, multiangular reflectance involves *interaction of light with three-dimensionally structured surfaces into which light partly penetrates* (Lucht, 2004, p. 9).

In principle, multiview-angle reflectance varies in three dimensions around an object. In practice, this situation is often restricted to the solar plane, that is the vertical plane that includes the sun, ground object, and aerial sensor. The bidirectional reflectance distribution function (BRDF) refers to variations of reflectivity with different viewing angles, as demonstrated within the solar plane (Fig. 4-2). Note the position of the sun and differences in reflectivity both toward and away from the sun. Reflection toward the sun is called backscatter, and reflections away from the sun are forward scatter. Of course, considerable radiation is also scattered in the third dimension to either side of the solar plane in both the back and forward directions.

In addition to viewing angle and sun position, many other factors influence lighting and reflectivity typically encountered for small-format aerial photography. These include the nature of objects on the Earth's surface as well as atmospheric conditions. Some of these vary with seasonal regularity; others are temporally irregular or ephemeral in nature. The following sections elaborate various lighting and atmospheric effects for SFAP.

4.2. MULTIVIEW-ANGLE EFFECTS

Multiview-angle effects are everywhere. Consider a newly mowed lawn or harvested crop field. Distinct stripes are visible both on the ground and from the aerial vantage (Fig. 4-3). These stripes reflect passage of the mower back and forth across the field such that the grass or crop stubble is bent at opposed angles for alternate stripes. As this example demonstrates, multiangular reflectance is a basic property of the natural world, and this has many implications for small-format aerial photography. Directional reflectance anisotropy represents the sum of differential reflection by objects that have complicated three-dimensional geometry (Lucht, 2004). Any scene is composed of certain geometrical properties in terms of the sizes, shapes, angles, and positions of reflective elements, such as tree and grass leaves, water surfaces, roof tiles, beach pebbles, glacier ice crystals, and so on. Furthermore, each of these

FIGURE 4-1 Chalk buttes surrounded by short-grass prairie in the Smoky Hill River valley of western Kansas, United States. (A) Low-oblique view showing cluster of buttes with an arch. Individual buttes stand 10–15 m high; road to lower right is ~6 m wide. (B) Vertical view of same butte cluster. Note shadows and light coming through the arch in the butte wall. Shadow of a person is visible just above the arrow (^). Kite aerial photographs (Aber and Aber, 2009, fig. 35).

geometric elements reflects certain wavelengths of light more strongly than others (see Section 4.4 below).

Vegetation has particularly complex reflectance geometry. Consider, for example, a forest. Some sunlight is reflected directly from the leaves in the crown of the canopy, some is reflected from middle levels, and some light penetrates to the forest floor and is reflected upward through the canopy. Volumetric scattering and transmission by leaves send some radiation off in other directions within the canopy, where further scattering, transmission, or absorption may take place.

Another strong influence on reflected radiation is shadow casting. Trees, grass, soil clumps, pebbles, or other irregularities of the surface cast shadows. Each discrete object has an illuminated (bright) side and a shadowed (dark) side. The sizes, shapes, and spatial arrangement of shadows have a strong influence on overall scene brightness, and depend on the angle of viewing in relation to the sun

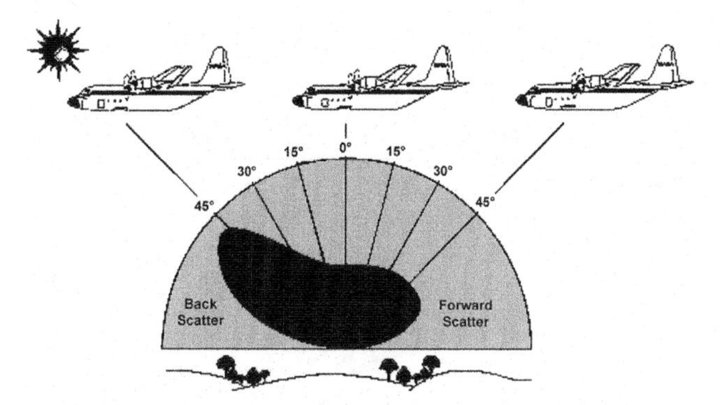

FIGURE 4-2 Diagram of aerial photography and typical BRDF. Amount of reflectivity in the solar plane is indicated by the black oval. Maximum reflectivity occurs directly back toward the sun. Illustration not to scale; adapted from Ranson et al. (1994, fig. 1).

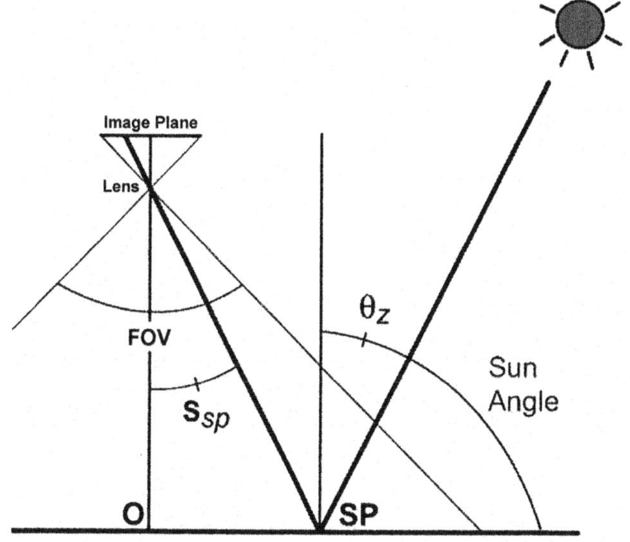

FIGURE 4-4 Geometric elements for specular reflection of sun light from a horizontal surface in a wide-angle vertical photograph. The angle at the focal point (S_{sp}) is equal to the solar zenith angle (θ_z), which is the complement of the sun angle at the specular point (SP). If this angle (S_{sp}) is less than half the angular field of view (FOV), sun glint could appear in the image. Modified from Mount (2005, fig. 2).

position. These factors give rise to two special lighting effects—sun glint and the hot spot, both of which are observed in the solar plane.

Sun glint is direct, specular, forward reflection from an optically smooth (mirror) surface, such as water, glass, or metal, in which the angle of incident sunlight equals the angle of reflection (Fig. 4-4). This phenomenon is also called the sunspot or solar flaring (Teng et al., 1997). The most common reflective materials involve water bodies and man-made metal structures. Water surfaces with ripples or waves may produce a variant called sun glitter, in which each wave surface forms a small reflecting facet (Fig. 4-5). Sun glint is quite common in oblique views taken toward the sun; it is seen less often in vertical views.

Sun glint and glitter are observed in small-format aerial photographs frequently compared to traditional large-format images, because of the wider field of view and rectangular-format cameras often employed for SFAP (Mount, 2005). The occurrence of sun glint in vertical views increases for photographs taken in late spring or early summer when the sun is highest in the sky or in imagery from low latitudes where the sun is always high overhead. In extreme cases, views taken toward the sun may display internal lens reflections, if sunlight strikes the lens directly (Fig. 4-6). This can happen in high-oblique views when the sun is low in the sky.

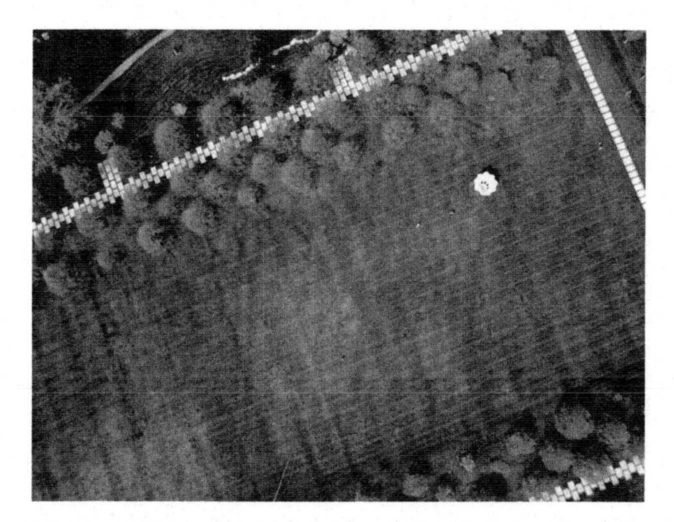

FIGURE 4-3 Vertical photograph of lawn with mowed grass. Note linear mowing pattern. The large "birdie" is an outdoor sculpture on the campus of the Nelson-Atkins Museum of Art, Kansas City, Missouri, United States. Helium blimp aerial photograph by JSA, April 2005.

The hot spot is the position on the ground in direct alignment with the sun and camera. The hot spot is located at the antisolar point, which is the point on the ground opposite the sun in relation to the camera (Lynch and Livingston, 1995; Teng et al., 1997). The hot spot appears brighter than its surroundings, as if that position is overexposed in the image (Fig. 4-7). The hot spot is also known as the opposition effect or shadow point, because the shadow of the platform that carried the camera may appear at the center of the hot spot (Fig. 4-8; Murtha et al., 1997).

The hot spot phenomenon has received a great deal of investigation in recent years, and most agree that its primary cause is a result of shadow hiding (Hapke et al., 1996; Leroy and Bréon, 1996; Lucht, 2004). The absence of visible shadows causes the spot to display substantially brighter than the surroundings in which shadows appear (Fig. 4-9). This effect derives from shadows cast by objects of all sizes, from sand grains on a beach to forest trees. Furthermore, the color of the hot spot differs from that of the surroundings (Lynch and Livingston, 1995). This is because scattered

FIGURE 4-5 Sun glint and glitter. (A) High-oblique view of sun glint from lake surface in left background and from metal roof in right foreground. Lake Kahola, Kansas, March 1997. (B) Low-oblique view of sun glint from smooth water (*); sun glitter from ripple and wave surfaces elsewhere in scene. South Padre Island, Texas, October 2005. (C) Vertical view of sun glitter from rippled stream surface upper left; sun glint from steel track of railroad (^). Palisades State Park, South Dakota, July 1998. Kite aerial photographs by JSA and SWA, United States.

FIGURE 4-6 The streak of bright octogons was created by internal reflections in the lens from direct sunlight. Winter image taken in late afternoon looking toward the sun. Kite aerial photograph by SWA and JSA; Lake Kahola, Kansas, United States, December 2002.

FIGURE 4-7 Hot spot displayed in a harvested agricultural field. Note bright spot next to arrow (>). Kite aerial photograph by D. Gałązka and JSA; Mława, Poland, September 1998.

FIGURE 4-8 Antisolar point marked by shadow of a small helium blimp in lower-right portion of this vertical view of a formal rose garden. Image taken by JSA; Loose Park, Kansas City, Missouri, United States, June 2006.

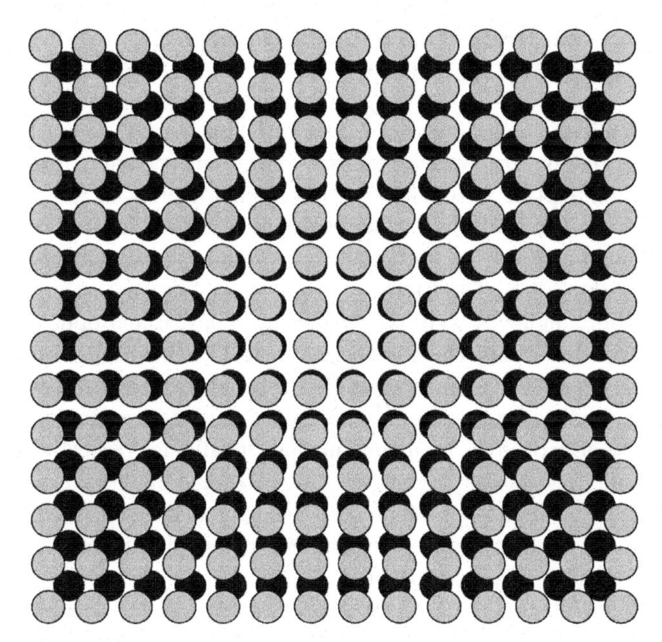

FIGURE 4-9 Schematic illustration showing the vertical view of shadows cast by objects around the antisolar point at scene center. The lack of visible shadows at the antisolar point renders the scene brighter at the hot spot. The position of each shadow is a function of relief displacement (see Chapter 2.2.3). Illustration adapted from Lynch and Livingston (1995, fig. 1-5B).

blue light, which normally illuminates shadowed zones, is absent from the hot spot vicinity. Thus, the hot spot typically appears more yellow (lacking in blue). For example, dark green forest appears light yellowish green at the hot spot (Fig. 4-10).

In addition to shadow hiding, other factors may contribute to the hot spot (Lynch and Livingston, 1995). Small rounded grains (sand and pebbles) may act as minute reflectors that collect light and send it back toward the sun. Likewise, mineral crystals within rocks may function as internal corner reflectors, and crystal faces may act as tiny mirrors. Light may be back reflected from tiny liquid droplets, such as dew or tree sap. Finally, coherent backscatter may contribute to the opposition effect. The combination of these factors with shadow hiding creates marked hot spots in many situations for small-format aerial photography.

The hot spot is commonly observed in oblique views taken opposite the sun; it is evident less often in vertical views. As with sun glint, the presence of the hot spot in vertical views increases with use of a wide-angle lens, for late spring or early summer imagery, or from low latitudes. In the authors' experience, hot spots are most noticeable for terrain in which the ground cover is relatively homogeneous, such as forest or prairie canopy, agricultural fields, and fallow or bare ground. In these cases, the color and brightness of the subject are more-or-less uniform, so the hot spot is conspicuous. For terrain with more complex land cover, the hot spot may not be so obvious. This applies often to urban scenes that contain large variations in the intrinsic colors and brightness of objects.

The foregoing discussion suggests that the appearance of the landscape changes dramatically with different viewing directions relative to sun position. This leads to a general assessment of the visual quality of oblique or wide-angle vertical images acquired with small-format aerial photography (Aber et al., 2002). In general, better oblique images are acquired in the azimuth range 50°–160° relative to the sun position (Fig. 4-11). This viewing range represents a balance of shadows and highlights with more-or-less uniform brightness levels. Views toward the sun (<50° azimuth) suffer from excessive shadowing, high contrast, and poor depiction of color; furthermore, sun glint may be present. On the other hand, views directly opposite

FIGURE 4-10 Hot spot at scene center in canopy of pine-spruce forest. The hot spot has a pale yellowish-green color compared to the normal dark-green color of the surrounding forest. Kite aerial photograph by SWA and JSA; coastal Vormsi, Estonia, August 2000.

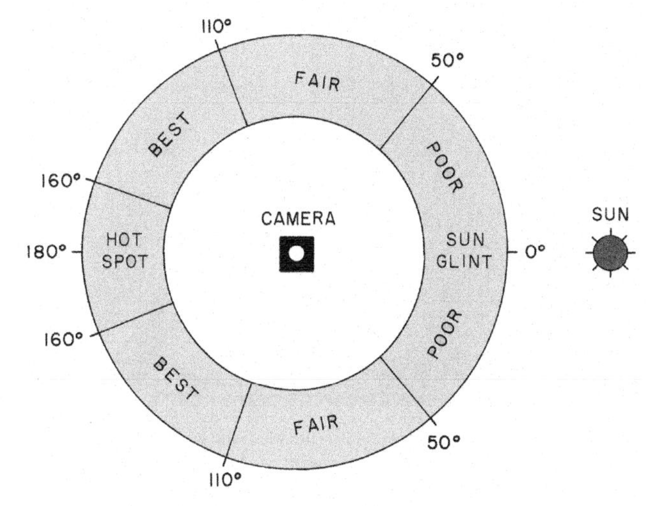

FIGURE 4-11 Schematic azimuthal (plan) diagram of lighting conditions for oblique small-format aerial photography relative to the sun position. The indication of image quality according to direction is a subjective visual interpretation. Adapted from Aber et al. (2002, fig. 5).

the sun may include the hot spot, wherein details are hardly visible due to lack of shadows. Figure 4-12 demonstrates visible changes in image quality with different viewing directions over a coastal setting in western Denmark.

FIGURE 4-12 Effect of viewing direction on oblique image quality relative to azimuth of the sun. (A) View toward ~160°; note shadow of lighthouse. (B) View toward ~120°; note shadow of tractor to left. (C) View toward ~40°; notice sun glint along right edge. A and B depict good color with moderate brightness contrast. C is heavily shadowed and displays excessive contrast. Kite aerial photographs by IM, SWA, and JSA; Bovbjerg, North Sea coast of Denmark, September 2005.

4.3. BIDIRECTIONAL REFLECTANCE DISTRIBUTION FUNCTION

Qualitative variations in scene brightness and contrast depending on viewing angle and sun position are commonplace in small-format aerial photography, as noted above. The goal of the bidirectional reflectance distribution function (BRDF) is to model these variations quantitatively (Lucht, 2004). The BRDF is based on viewing and illumination angles as well as complex geometrical and optical properties of objects in the target scene. Considerable effort has been made to understand BRDF better for various types of land cover, particularly different vegetation canopies. One approach is to model mathematically the reflective plants and canopy geometry. Some models are based on radiative properties of plants (Nilson and Kuusk, 1989), whereas others depend on geometric-optical considerations (Schaaf and Strahler, 1994).

In either approach, the models can be tested against actual sensor measurements, in other words ground truth. Such field experiments begin with ground-based goniometers (Fig 4-13). A radiometer moves around the target object in order to collect reflectivity values from all azimuthal and zenithal directions at close range (2 m). From slightly higher vantages, NASA's Parabola radiometer can be operated from a vehicle-mounted boom up to 5 m high or suspended from cables attached to towers up to 30 m high (Bruegge et al., 2004). Parabola is capable of panning and tilting to acquire measurements in all look directions over a target area. Airborne BRDF instruments include POLDER and ASAS, which are typically flown at heights of 3000 m or higher (Schaaf and Strahler, 1994; Leroy and Bréon 1996).

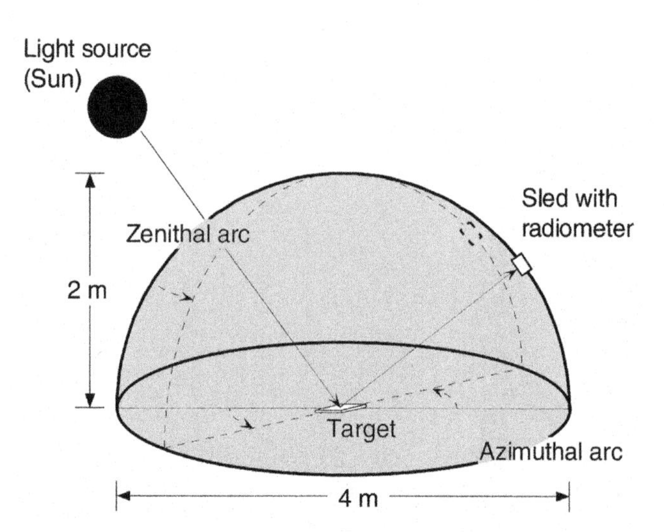

FIGURE 4-13 Schematic diagram of the field goniometer system (FIGOS) for multiview-angle, closeup measurement of reflectivity for targets under natural conditions. Adapted from Bruegge et al. (2004, fig. 5-10).

Representative ASAS results of such field measurements are presented for a spruce forest in Maine (Fig. 4-14). Three general observations have come from field observation of multiview-angle reflectivity of vegetation canopies (Murtha et al., 1997).

- Canopy reflectance is greatest when the sun is behind the sensor, in other words the camera records light that is backscattered from the canopy toward the sun. Maximum reflectance is located at the antisolar point, which produces a hot spot in the image due to shadow hiding.
- Canopy reflectance increases overall with higher solar zenith angles.
- Canopy reflectance is relatively uniform for vertical or near-vertical images, particularly those acquired with narrow fields of view.

One goal of BRDF research is to relate ground and aerial multiview-angle effects to observations collected from space-based remote sensing systems. Between ground-based and relatively high airborne BRDF instruments, however, the height range from 30 m to 1000 m is a little-explored interval for multiview-angle reflectivity. Small-format aerial photography is particularly well suited within this range to acquire all viewing angles from vertical to high oblique. Thus, SFAP represents an effective and inexpensive means for ultralow-height field experiments and testing of BRDF models. This is an application for which small-format aerial photography may play an increasingly important role in the future.

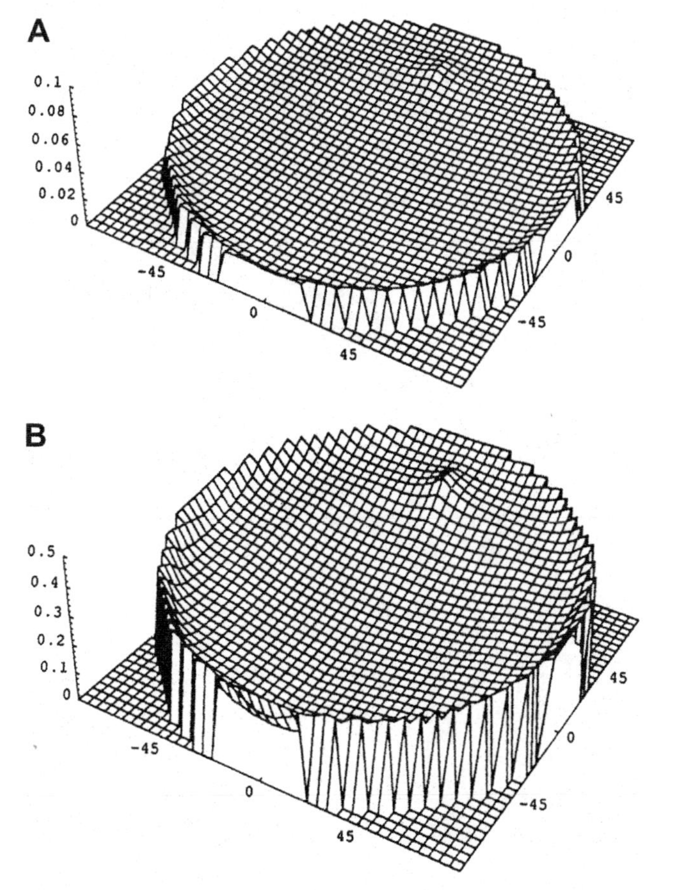

FIGURE 4-14 Multi-directional reflectivity for spruce forest in red (A) and near-infrared (B). In each case, the peak in reflectivity represents the hot spot at the antisolar point. The bowl of low reflectivity values indicates increased shadowing looking toward the sun. Adapted from Schaaf and Strahler (1994, fig. 3).

4.4. MULTISPECTRAL EFFECTS

As humans, we view the world through a narrow range of electromagnetic radiation, the visible spectrum (0.4–0.7 μm wavelength; see Fig. 2-1). Film photography extends this range from near-ultraviolet to near-infrared (0.3–0.9 μm). Black-and-white infrared film was developed in the 1920s and was utilized for aerial photography already in the 1930s (Colwell, 1997). World War II spurred a great need for aerial camouflage detection, and color-infrared (CIR) film was perfected. Nowadays both CIR film and digital cameras are available for small-format aerial photography.

The "color" of objects depends upon what parts of the spectrum are observed. This is obvious, for example, when comparing panchromatic (gray tone) images with equivalent color images. However, objects that appear similar in visible light often have quite different appearances in other portions of the spectrum. Diffuse reflectivity or albedo for most common surficial materials ranges from less than 10% for clear water to nearly 100% for fresh snow. Photosynthetically active vegetation typically has an albedo of 50–70% in the near-infrared portion of the spectrum (Fig. 4-15).

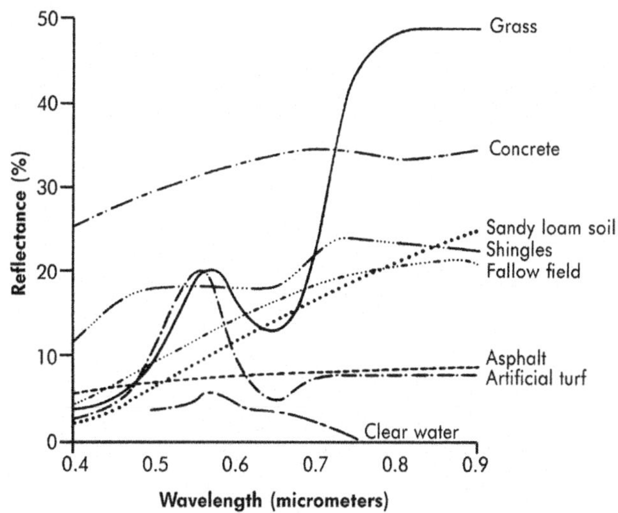

FIGURE 4-15 Comparison of reflectivity for common surficial materials in the visible and near-infrared portions of the spectrum. Note the distinctive spectral response curves for grass vs. artificial turf. Taken from Haack et al. (1997, fig. 15-5).

Different parts of the spectrum may be photographed by using various combinations of films, electronic detectors, and filters. Photographs are routinely taken in b/w panchromatic, b/w minus blue, b/w infrared, color-visible, color-infrared (minus blue), and multiband types. Near-ultraviolet photography is also possible for special applications. As an example, color-infrared film is exposed to green, red, and near-infrared wavelengths, which are depicted, respectively, as blue, green, and red in the photograph. This shifting of bands to visible colors is called false-color imagery (see Fig. 2-5).

The case of vegetation is most instructive. Compare, for example, grass with artificial turf (see Fig. 4-15). Both appear nearly identical green in visible light; however, grass is highly reflective in near-infrared, whereas artificial turf is not. Distinguishing between real and artificial turf requires photographs in the near-infrared portion of the spectrum. Photosynthetically active "green" vegetation has a unique spectral signature (Fig. 4-16). Leaves selectively absorb blue and red light, weakly reflect green, and strongly reflect near-infrared radiation. No other materials at the Earth's surface have this spectral signature. On this basis, CIR imagery plays a key role for analysis of all types of vegetation—crops, prairie grass, emergent aquatic plants, and forests.

Water, likewise, has a distinctive appearance in CIR images. Although clean and turbid water differ in their visible reflectivity (Fig. 4-17), both strongly absorb near-infrared radiation. Thus, water bodies typically appear dark blue or black in CIR photographs, unless photosynthetically active vegetation is floating on the water surface. CIR images are typically taken using a yellow or orange filter in order to remove blue light from the image. This renders shadows much darker, as scattered blue light illuminates

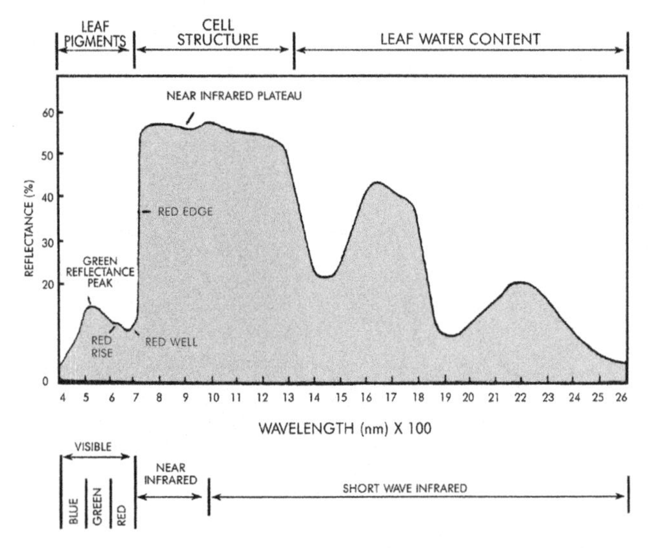

FIGURE 4-16 General spectral reflectance curve for a green leaf. Note blue and red absorption, weak green reflection, and strong near-infrared reflection. Adapted from Murtha et al. (1997, fig. 5-11).

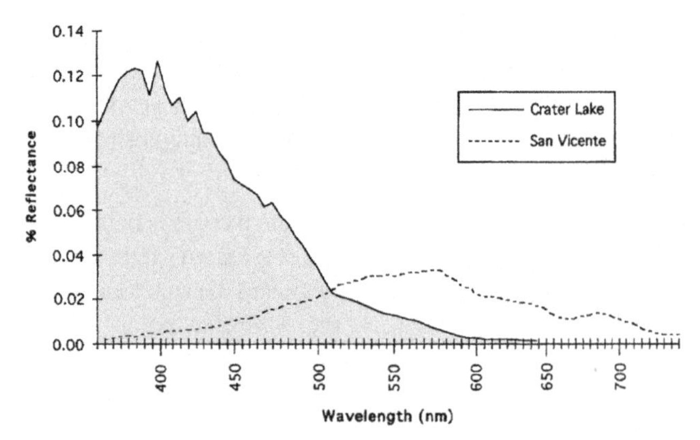

FIGURE 4-17 Visible spectral response curves for two lakes. Crater Lake (gray) is clear, deep water; San Vicente (dashed) is turbid, phytoplankton-rich water. Note that both trend toward zero reflectivity in the near-infrared (>700 nm). Adapted from Wiesnet et al. (1997, fig. 6-3).

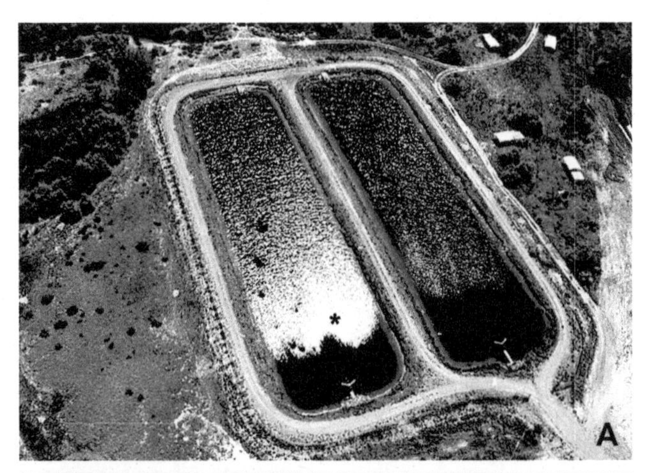

FIGURE 4-18 Special lighting effects are enhanced in color-infrared imagery. (A) Sun glint (*) and glitter from fish hatchery ponds; water is dark blue. Pueblo, Colorado, May 2003. (B) Hot spot at scene center on canopy of deciduous trees; shadows are black. Elkhorn Slough, California, November 2002. Kite aerial photographs by SWA and JSA, United States.

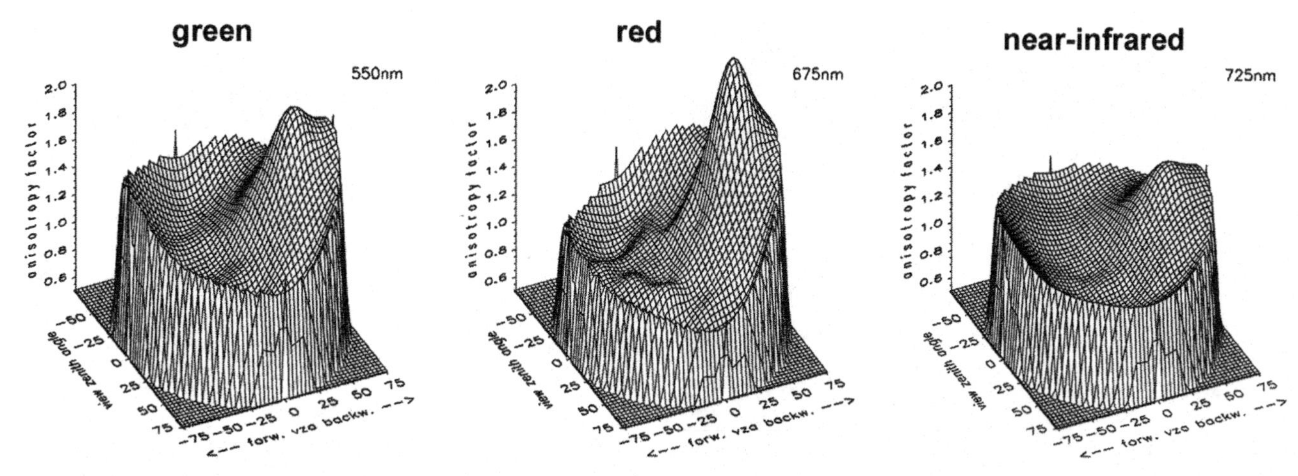

FIGURE 4-19 Anisotropy factors for three wavelengths measured in perennial ryegrass with a field goniometer. Red displays the strongest BRDF, green is intermediate, and near-infrared is weakest. Adapted from Sandmeier (2004, fig. 3-3).

shadowed zones. Thus, CIR images that contain active vegetation, water bodies, and shadowing are often high contrast—quite bright and dark portions with little mid-range of brightness. Furthermore, darkening of water and shadows tends to exaggerate the appearance of sun glint and the hot spot (Fig. 4-18).

Numerous laboratory and field studies have demonstrated that the BRDF effect displays strong spectral variation for vegetated surfaces (Sandmeier, 2004). This is a consequence of multiple scattering effects within vegetation canopy and selective absorption of certain wavelengths. In general, high-absorbing red light shows the strongest response, whereas high-reflecting near-infrared is the weakest, and green is intermediate (Fig. 4-19).

4.5. LATITUDE AND SEASONAL CONDITIONS

The position of the sun is a critical factor for controlling the amount and quality of light available to illuminate the Earth's surface. Latitude, day of year, and time of day determine where the sun would be located for any site. The normal expectation of toplighting for small-format aerial photography means the sun should be relatively high in the sky to avoid excessive shadowing of the landscape, and a further criterion is usually cloud-free sky for best illumination of the ground. This combination of clear sky and high sun position is the ideal for most SFAP applications. In addition, it is usually desirable to avoid sun glint and hot spot views.

For some parts of the world, however, these conditions are rarely or never realized. At Tromsø, Norway (~69.7°N), for example, even at noon on summer solistice, the sun is not high in the sky (Fig. 4-20). At the other extreme, the sun is nearly overhead at mid-day in the tropics, which greatly increases the chances for sun glint or the hot spot in vertical photographs. Even in middle latitudes during summer, it is possible at noon to capture the hot spot and sun glint in the same vertical view (Fig. 4-21). To avoid this predicament, SFAP should be conducted in the mid-morning or mid-afternoon, before or after the sun has reached its full height. Similar seasonal and time of day considerations apply in other circumstances in order to achieve optimum sun position for a given site or application.

The character of clouds is the second main factor for successful small-format aerial photography. Although cloud-free conditions are often assumed best, uniformly indirect lighting from high clouds or overcast sky is preferable to direct sunlight for some applications. Many parts of the world have persistent, even perennial cloud cover. This includes much of the tropical region. For example, the deltas of the Amazon River and Zaire (Congo) River are rarely photographed without cloud cover by astronauts (Amsbury et al., 1994). Tropical mountains are especially prone to cloud

FIGURE 4-20 Mid-day view of Kvaløysletta near summer solistice showing long shadows. Note poles, sign post and trees in foreground. Kite aerial photograph by JSA; near Tromsø, Norway at ~69.7° N latitude, June 1998.

FIGURE 4-21 Vertical blimp aerial photograph that includes both the shadow point (left) on aquatic vegetation and sun glint (right) on still water in a wetland channel. Image acquired near noon, a few days before summer solistice at ~40°N latitude. Photograph by S. Acosta and JSA; Squaw Creek National Wildlife Refuge, northwestern Missouri, United States, June 2003.

cover. Many tropical regions experience monsoon seasons with continual clouds and rain for months. Another setting known for extended cloud cover is mid- and high-latitude maritime environments, for example coastal British Columbia in Canada, southern Alaska in the United States, the British Isles, and Norway. On some islands, the sun is seen only a few days a year—Faeroe Islands (northwest of Scotland) and Kergulian Islands (southern Indian Ocean). On the other hand, deserts and semi-arid regions have abundant sunshine most of the year. However, dry climate combined with strong wind gives rise to frequent dust and sand storms (Amsbury et al., 1994).

FIGURE 4-22 Winter, leaf-off vertical image taken for property survey purposes. Picture depicts houses, garages, docks, and other lake-shore structures. Note the small cabin at scene center. During the growing season, this building is completely hidden beneath the tree canopy. Kite aerial photograph by SWA and JSA; Lake Kahola, Kansas, United States, December 2002.

Many regions of the world experience a regular progression of seasonal conditions that influence ground cover and human land use. Most noticeable are seasonal changes in water bodies and vegetation, both natural and agricultural. For many SFAP applications, particular criteria of ground cover may be necessary. For example, land surveys for property appraisal or population census may specify leaf-off conditions, in order that human dwellings and structures are seen clearly beneath deciduous trees (Fig. 4-22). On the other hand, agricultural monitoring must be conducted periodically during the growing season. Thus, time of year may be dictated by requirements of the SFAP mission.

As this discussion suggests, achieving optimal lighting for small-format aerial photography is often difficult or may be practically impossible in some situations. Logistical considerations provide further limitations, as people and equipment must be in place to take advantage of favorable lighting and atmospheric conditions (see Chapter 9). Every platform has its own requirements regarding weather conditions, especially wind, and the time of day for a photographic survey may be dictated by the presence or absence of wind (see Chapter 8). Given the reality of SFAP, it is sometimes necessary to proceed with less than ideal conditions in order to complete a mission. Some of the effects of cloud cover and other atmospheric disturbances are elaborated below.

4.6. CLOUDS

Clouds play a critical role for effective small-format aerial photography. The nature and optical properties of clouds vary enormously from high, thin cirrus clouds of ice crystals to dense ground fog. In addition to ice and/or water droplets, clouds may consist of dust, smoke, and other minute debris in the atmosphere, which are derived from both natural and human sources. The particles of clouds range in size from <1 μm diameter to a maximum of ~ 100 μm. Clouds are intrinsically white, as they absorb practically no visible light and all wavelengths are scattered equally (Lynch and Livingston, 1995). The impact of clouds for SFAP depends on their altitude relative to the camera. Clouds positioned above the camera mainly affect the amount and quality of incoming radiation and how it illuminates the ground; whereas, clouds below the camera influence both incoming radiation and light reflected upward from ground.

Beginning with clouds above the camera level, cloud cover has two main impacts on the visual appearance of aerial photographs, namely reduction of shadows and poor color definition, which result overall in low-contrast images (Fig. 4-23). As clouds become thicker and less transparent, the ground beneath becomes increasingly dark and uniformly illuminated, such that shadows of individual objects disappear. Shadows are visual clues to

FIGURE 4-23 Effect of cloud cover on image quality for a Scandinavian coastal scene. (A) High-oblique view under mostly sunny conditions. Colors are well defined and the image has good brightness contrast. (B) Similar view taken a few minutes later under cloud cover. Colors are dull and image has low contrast. Kite aerial photographs by SWA and JSA; Fegge, island of Mors, Limfjord district of northwestern Denmark, September 2005.

FIGURE 4-24 Dark cloud shadow obscures the center of this high-oblique view of wind turbines standing 90 m tall. Kite aerial photograph by SWA and JSA; Gray County, southwestern Kansas, United States, April 2006.

FIGURE 4-25 The foreground is well illuminated, and the background is shadowed by clouds. This visual contrast directs the viewer's attention to the main subject—the recreational development adjacent to the lake. Helium blimp aerial photograph by A. Uttinger and JSA; Lake Wabaunsee, Kansas, United States, June 2006.

relief and texture of features in the photograph, and without any shadows these objects may be difficult to identify. This affects vertical aerial photographs in particular, which often have a "flat" appearance without shadows.

The color in shadowed zones below clouds is rich in blue light and poor in green and red, which gives rise to the dull blue-gray color palette of cloudy weather. Under heavy cloud cover, color largely disappears from the landscape. Nonetheless, even in the case of moderate cloud cover good SFAP can be conducted in many situations, if the ground is uniformly illuminated. In the case of wispy cirrus clouds, however, the visual impact for most SFAP is minimal. High, thin clouds slightly reduce solar brilliance without significantly degrading shadows or color. This could be advantageous, in fact, for scenes that have high intrinsic brightness contrasts for ground objects, for example dark green conifer forest next to white beach gravel. Thus, altitude and

FIGURE 4-26 High-oblique view over the Gulf of Finland with patchy cloud cover. Note dark cloud shadows on water surface as well as bright zones from cloud reflections. This effect is noticeable over homogeneous surfaces, such as water, but is not recognized so often on land. Kite aerial photograph by SWA and JSA; island of Vormsi, Estonia, August 2000.

FIGURE 4-27 Poor visibility is typical of the eastern United States during the summer. (A) Nearby tow plane is clearly defined, but the ground below is depicted poorly through the haze, and the background is hardly visible. Hand-held photograph taken by JSA from a manned glider; Elmira, New York, July 2005. (B) Afternoon ground fog moves in from the ocean in this hazy view over Martha's Vineyard, Massachusetts. Kite aerial photograph by SWA, and JSA, July 2005. (C) Fog drifts in from the Atlantic Ocean over the Laudholm beach and the mouth of Little River; Wells, Maine. Helium blimp aerial photograph by JSA, SWA and V. Valentine, August 2009.

thickness of clouds render considerable variation in colors displayed at the Earth's surface.

Discontinuous or patchy cloud cover results in a mosaic of illuminated and shadowed zones on the ground. This is usually the worst possible lighting situation for effective SFAP, as the bright areas may be overexposed and the dark spots are underexposed with little of the image in the mid-range of brightness (Fig. 4-24). Still, patchy cloud cover may be utilized successfully in some instances to draw attention to illuminated foreground and central portions of the scene, which are of particular interest, while leaving the periphery and background in cloud shadows (Fig. 4-25). In addition to casting shadows, clouds also may reflect white light to brighten the ground in places. This effect is visible on homogeneous surfaces, such as lakes and seas (Fig. 4-26).

The examples of clouds discussed thus far are positioned above the camera. Clouds located below the camera—between the camera and the ground—may have all the same effects as higher clouds by limiting incoming solar radiation. In addition, low clouds scatter light reflected upward from the ground, thus further degrading image quality. Although such conditions should be avoided normally for SFAP, it may be necessary to cope with low clouds in some situations. For example, summer in the eastern United States and southeastern Canada is a time of hazy sky due to a combination of high humidity and industrial pollution. Thus, poor visibility is frequently the case, even on "clear" days (Fig. 4-27).

Smoke and smog are widespread around the world. Natural fires are started by lightning strikes, spontaneous combustion, and volcanic eruptions. However, most fires

FIGURE 4-28 Regional smoke from widespread burning of agricultural waste and crop stubble creates a hazy appearance in this high-oblique view over a bog landscape. Kite aerial photograph by SWA and JSA; Mannikjärve Bog, Estonia, October 2000.

FIGURE 4-29 Controlled prairie burning is a spring ritual to maintain tallgrass habitat in the Flint Hills of eastern Kansas. (A) Smoke billows up from a wall of fire. Note shadow from smoke cloud. (B) Smoke surrounds the camera rig. In spite of the smoke, a diffuse hot spot is visible in upper center of picture (*). Kite aerial photographs by M. Lewicki, SWA, and JSA; Lyon County, Kansas, United States, April 2000.

and smoke nowadays are related to human activities—clearing forests, prairie fires, burning agricultural waste, oil and gas fires, burning fossil fuels for transportation and industry, etc. Far from being an isolated problem, smoke is a common seasonal or perennial condition that may affect small-format aerial photography. In some cases, smoke spreads regionally from many small fires, for example from burning agricultural waste and crop stubble at the end of the growing season (Fig. 4-28) or spring burning of the tallgrass prairie to maintain grassland habitat (Fig. 4-29). In other situations, smoke has a significant point source, such as a coal-fired electric power plant or cement factory.

4.7. SHADOWS

The issue of shadowing in an image is obviously strongly connected to the latitude, time of day, and cloud aspects discussed above. With direct illumination by the sun, all three-dimensional objects or surface features cast a shadow if they project into the paths of the light rays between sun and ground. Shadowing in aerial images is most prominent at early morning and later afternoon and increases with higher latitudes. The darkening effect of shadowing increases with longer wavelengths as these are less subject to atmospheric scattering. For the same reason, shadows on a clear and dry day are much more pronounced than with slightly overcast conditions and a dusty or humid atmosphere.

Shadowing may occur at various scale levels in an image: a steep slope may cast a shadow onto a valley floor and opposite valley side; the shadow of a tree may obscure the ground beneath; shadowing between clods of earth may reveal the texture of a freshly plowed field. Shadows might

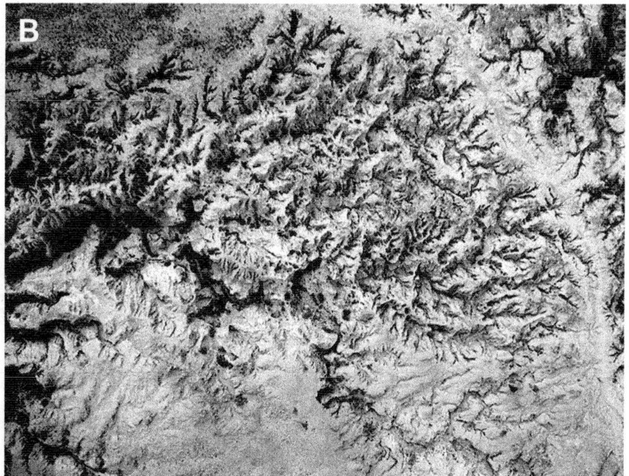

FIGURE 4-30 Vertical images of gully erosion on an abandoned Moroccan field. (A) Shadows emphasize the various depths and degrees of ruggedness of the geomorphologic forms carved by surface and subsurface erosion processes. (B) Image taken a few minutes later under cloud cover. The indirect lighting improves the visibility of details in all parts of the gully, but takes away most depth impression from the image, making it impossible to distinguish different levels of incision. Kite aerial photographs by IM and JBR; Talaa, Taroudant, Morocco, March 2006.

FIGURE 4-31 Every tree appears twice in this image of a burned pine forest near Castejón de Valdejasa, Province of Zaragoza, Spain. The long shadows cast by the afternoon sun are much larger from the vertical vantage point than the same-colored, soot-blackened skeletons of the pines, obscuring great parts of the image. The forest fire, sparked by a car accident in August 2008, destroyed 2200 ha of woodland. Model airplane photography by C. Claussen, M. Niesen, JBR, and IM, February 2009.

FIGURE 4-32 Low-height, vertical view of "Big Brutus" at the Mining Heritage Museum near West Mineral, Kansas, United States. One of the world's two largest power shovels, it stands 50 m tall with a working weight of 5000 metric tons. Shadow creates a silhouette side view of the machine. A normal-sized power shovel is visible at lower left; note people for scale standing in upper left corner. Picture taken with a compact digital camera from an unmanned, tethered, helium blimp (Aber and Aber 2009, fig. 72).

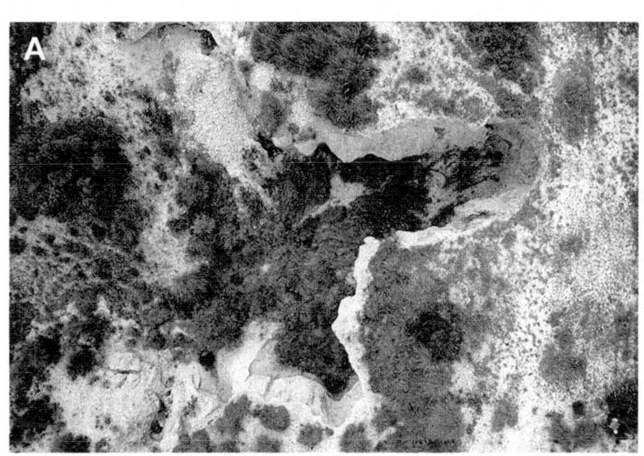

FIGURE 4-33 Bank gully in southern Spain. (A) Indirect lighting from an overcast sky uniformly illuminates all areas in this image, which completely lacks shadows. Details of the gully bottom are clearly visible. (B) Strong sun light casts a dark shadow from the steep gully wall onto its bottom. Image classification or stereoscopic photogrammetric analysis in this area is not possible and comparison between the two monitoring dates is difficult. Hot-air blimp photographs by IM and JBR; Rambla Salada, Murcia Province, Spain, April 2002 and 2004.

both be desired or unwanted depending on the purpose of the image: they may emphasize as well as obscure objects and surfaces in a scene (Fig. 4-30).

For visual interpretation of aerial images, shadowing can be extremely helpful as it offers clues to the third dimension, which the two-dimensional image lacks. The shadows can help to identify form and function of buildings or types of trees in vertical images (Fig. 4-31), disclose the course of a power transmission line, or reveal archaeological and historical landscape features. Shadows may be quite dramatic in some cases (Fig. 4-32).

Geomorphologic and geologic features and landforms at various scale levels—rock crevices, erosion rills, fault lines, dells or dolines, dunes or coastal cliffs—could be rendered nearly invisible or at least unfathomable when melting into the landscape background on an image without shadowing. In contrast, late afternoon sun might model even subtle variations of surface height, if the angle of incidence is more-or-less orthogonal to the relief structures.

In digital image analysis, shadows are usually undesirable as they change the spectral response of objects with otherwise homogeneous or identical reflectivity. For spectral classification algorithms, this results in misclassified or unclassified areas (see Chapter 11.5). To avoid this effect, masking techniques or topographic normalization may be applied (see Fig. 11-17; Colby, 1991; Zhan et al., 2005). Photogrammetric analysis both by visual stereoscopy as well as automated image matching is strongly hampered in dark shadow areas (Fig. 4-33) and may lead to considerable measurement errors (Giménez et al., 2009). The resulting spatially varying data quality of image classifications and surface models subsequently also afflicts time-series analyses and monitoring by remote sensing.

On the other hand, shadows might actually enable automatic detection of objects or surface types by adding

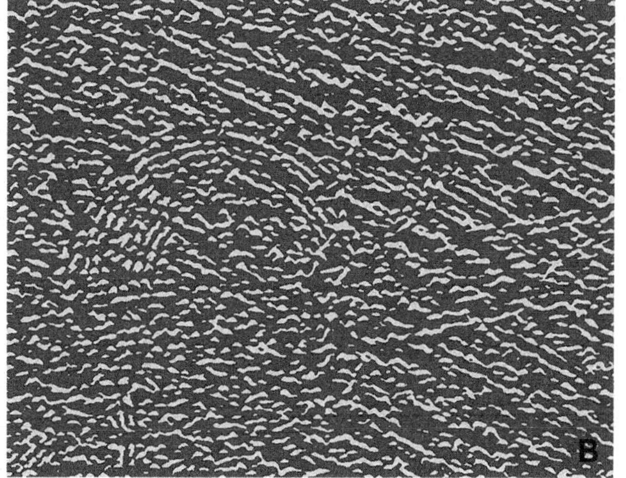

FIGURE 4-34 Fallow field in northern Spain. (A) After several drought years, remains of the pattern created by plowing five years previously are still visible in the early morning sunlight. (B) Digital texture and Fourier analysis were used to create this map of different micro-topographic surface positions (ridge and furrow), corresponding to different types of soil crusts. Field of view ~24 m across; taken from Marzolff (1999, maps 1A and 4A).

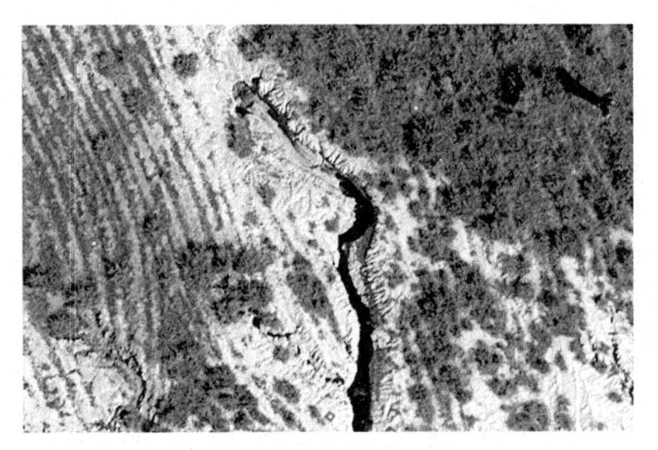

FIGURE 4-35 Changing its direction by 135° in the center of this vertical image, an erosion rill on a fallow field turns from an inconspicous feature melting into the bare-soil background into a starkly prominent incision. Note sun azimuth indicated by the kite flyer's shadow in the upper right. Kite aerial photograph taken in the Bardenas Reales, Province of Navarra, Spain, by IM and JBR, February 2009.

typical patterns or structures to the image, which can be enhanced and extracted by filtering, image segmentation, and texture analysis (Fig. 4-34; Marzolff, 1999; Shackelford, 2004). A difficult problem both for intentionally achieving and avoiding shadowing is that the degree of shadowing is not only dependent on sun elevation and object height, but also on the relative position of the objects or surface structures to the incident rays of light. Similar to multiview-angle effects, the same structures may have different appearance throughout the image depending on their orientation (Fig. 4-35; see also Giménez et al., 2009).

4.8. SUMMARY

Small-format aerial photographs typically are taken under clear sky when the sun is relatively high to provide for good toplighting of the Earth's surface, although uniformly indirect lightning from high clouds or overcast sky is preferable in some situations. The position of the camera relative to the ground and sun plays a key role for determining the nature of reflected solar radiation reaching the camera. Anisotropic variations in reflectivity, depending on sun position and angle of viewing, give rise to the bidirectional reflectance distribution function (BRDF). Within the solar plane, two positions yield special lighting effects. Sun glint occurs when the angle of solar incidence is equal to the angle of reflection directly toward the camera. This is common in oblique or wide-angle vertical views looking toward the sun. The hot spot is located at the antisolar point, which is the spot on the ground in direct alignment with the camera and sun. The hot spot is seen often in oblique or wide-angle vertical views in the direction opposite the sun.

Small-format aerial photography is done routinely in the visible and near-infrared portions of the spectrum. Many types of spectral combinations may be photographed, ranging from conventional panchromatic to color-infrared. All objects display spectral signatures, which are the bases for recognizing their compositions. Vegetation is particularly instructive. Active "green" vegetation absorbs blue and red light, weakly reflects green, and strongly reflects near-infrared radiation; this spectral signature is unique among materials at the Earth's surface. In color-infrared photographs, active vegetation appears in pink and red colors.

Achieving the optimum lighting conditions for SFAP may prove difficult or even impossible depending on many factors—latitude, time of year, local weather conditions, special mission requirements, logistical limitations, platform capability, etc. The presence of shadows in an image may be desired for interpretation or automatic analysis purposes, but it also may be inconvenient and obscure valuable information. In practice, SFAP must be done often when sun position, cloud cover, and other factors are less than ideal. Even under unfavorable lighting conditions, it is still possible in many situations to acquire effective small-format aerial photographs by carefully selecting viewing directions, time of day, variable cloud cover, or other factors. One of the great advantages of small-format aerial photography is its flexibility for adapting to a wide range of lighting and weather conditions in the field.

Chapter 5

Photographic Composition

The natural world reveals patterns, form and structure on all scales, especially when familiar objects are seen from an unusual perspective.

(Eastaway, 2007)

5.1. INTRODUCTION

Photographic composition is much more than simply aiming the camera at the subject and taking a picture that is in focus and properly exposed for the ambient lighting conditions. Photographic composition has to do with the subjective reaction of people who view and interpret the image, in other words the aesthetic characteristics of the photograph. Many books have been written about landscape photography from an artistic point of view, for example Caulfield (1987), Shaw (1994), and Zuckerman (1996). In general, *good photographic technique is identical regardless of the camera used* (Shaw 1994, p. 5). However, digital cameras offer some advantages as well as limitations compared with traditional film cameras (Wildi, 2006). Modern professional digital cameras meet or exceed film quality (Eastaway, 2007). In any case, nearly all literature on this subject is based on the *frog's-eye view* from the ground. The *bird's-eye view* of SFAP opens new vistas and considerations for effective composition.

The most important variable for successful outdoor photography is correctly exposing the film or electronic detector to the available light. The roles of toplighting, sun/camera position, shadows, multispectral effects, and atmospheric conditions were discussed in the previous chapter. Technical aspects of SFAP camera operation are reviewed in the next chapter. This chapter focuses on the end results, that is, the basic elements that comprise attractive pictures, with an emphasis on images acquired with digital cameras.

5.2. BASIC ELEMENTS OF PHOTOGRAPHIC COMPOSITION

Photographic composition begins with color or gray tones; it is through color variations that we see and interpret what is present in the picture. The visible objects in the scene are further defined based on their sizes, shapes, patterns, textures, and contrast with each other. Additional visual elements include the relative placement and balance between objects in the picture. Photographic composition has much in common with landscape painting, and many general guidelines have been elaborated over the centuries for creating effective visual imagery in both media. Like all such rules of thumb, however, exceptions often lead to dramatic pictures. Further considerations apply to small-format aerial photography because of its unusual vantage point from above.

5.2.1. Oblique and Vertical Views

The high-oblique view is similar to ground-based landscape photography, because the horizon appears. The position of the horizon is generally the most important visual element of such pictures. To begin with, it is extremely important that the horizon appears in nearly level position rather than tilted at an odd angle (Fig. 5-1). This is quite simply the way people are accustomed to seeing the world, and it is what they expect when the horizon is visible (Wildi, 2006). For low-oblique views, the horizon is not visible, and it is therefore generally not so important that the scene appear in horizontal position.

The general *rule of thirds* is often applied to the placement of the horizon in landscape photographs, in other words the horizon should appear approximately one-third of the vertical distance from the top (or bottom) of the picture (Caulfield, 1987). Seldom, if ever, should the horizon divide the picture in half (Wildi, 2006). For SFAP, however, the sky itself is rarely of interest; the emphasis is on ground features, so the horizon is often better placed less than one-quarter the distance from the top of the scene in order to minimize the empty sky portion of the picture (Fig. 5-2).

For vertical SFAP, shadows are often of paramount importance for recognizing objects and terrain relief. The human eye makes an unconscious assumption that the light source comes from the upper left corner of the scene, so that shadows fall toward the lower-right side of raised objects.

FIGURE 5-1 High-oblique views showing the horizon in (A) tilted and (B) level positions. The latter is preferred in nearly all situations. Red Hills, Kansas, United States. Kite aerial photos by SWA, June 2006.

FIGURE 5-2 Two high-oblique views of the Danish landscape. A prehistoric wall, Ramme Dige, crosses the left side of scene, and wind turbines are visible in the background. (A) Ideal rule of thirds for horizon position. (B) Horizon positioned only one-tenth of the distance from the top. The latter is preferred for most SFAP applications, as the sky adds little to appreciation of the scene. Kite aerial photos by IM, SWA, and JSA, September 2005.

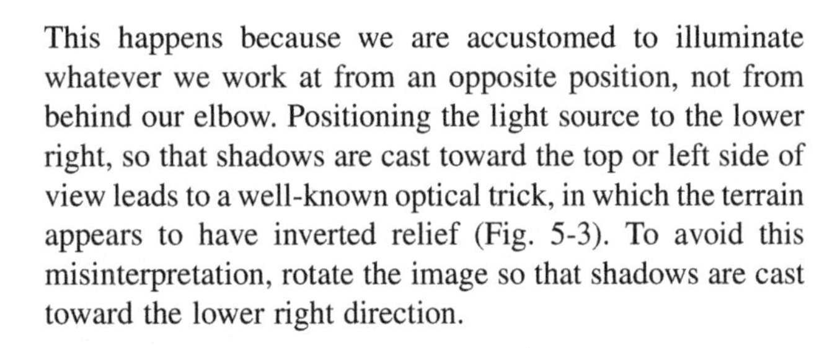

This happens because we are accustomed to illuminate whatever we work at from an opposite position, not from behind our elbow. Positioning the light source to the lower right, so that shadows are cast toward the top or left side of view leads to a well-known optical trick, in which the terrain appears to have inverted relief (Fig. 5-3). To avoid this misinterpretation, rotate the image so that shadows are cast toward the lower right direction.

5.2.2. Linear Features

Linear features are commonplace on the Earth's surface, both natural and man-made in origin. Such features may be straight or curved, continuous or discontinuous, and are depicted by differences in topography, land cover, or land use. The treatment of linear elements may have a strong impact on the photo composition. Generally straight linear features should not run vertically or horizontally across the picture, especially near the center of view; diagonal arrangement is much more pleasing (Fig. 5-4). In some situations, linear features may be the dominant visual elements present in a scene, and their diagonal placement may create dramatic views (Fig. 5-5).

Most straight linear objects on the Earth's surface are man-made structures, ranging from fencelines to airport runways (Fig. 5-6). Straight, linear features of natural origin are generally not so common. Fractures of various types in soil, rock, or ice often appear as straight features. When viewed from above, distinct linear patterns may be seen (Fig. 5-7). Curved linear features are still more pleasing to view (Wildi, 2006), such as meandering streams or wandering roads (Fig. 5-8). The S-shaped curve is a classic form, although finding an ideal S-shaped curve from the aerial vantage is easier said than done (Caulfield, 1987).

5.2.3. Image Depth

An important element for creating a 3D impression in an image is depth of field—focusing sharply only on portions of a scene, while the foreground or background remains

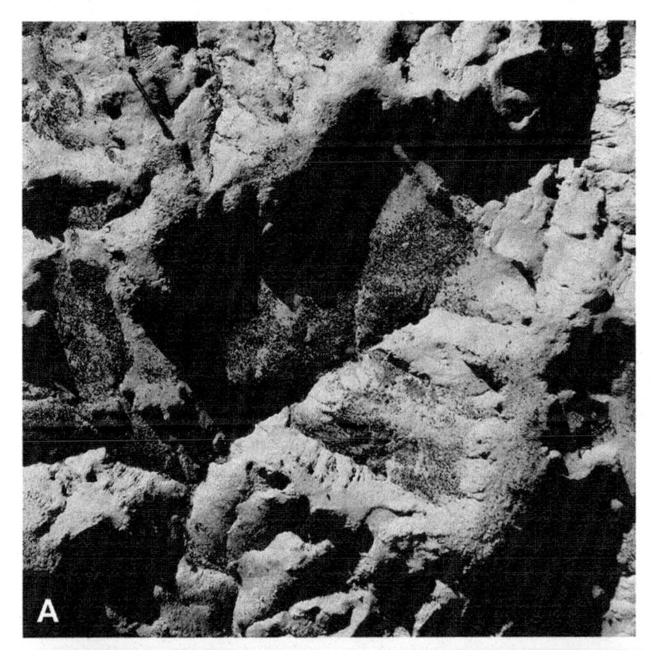

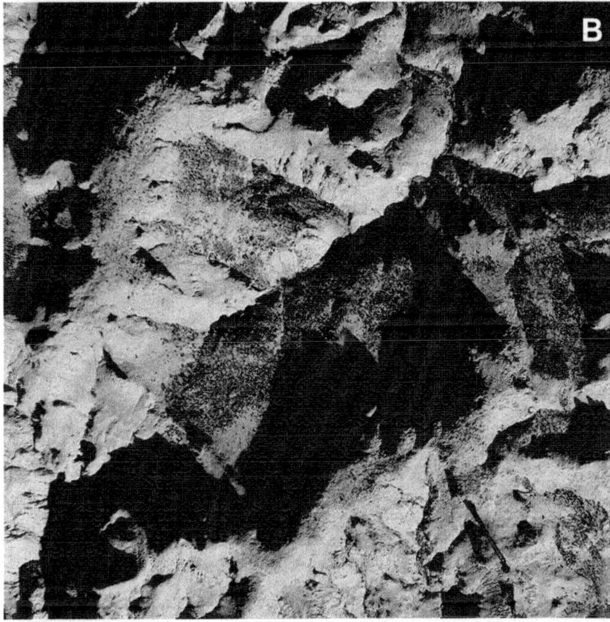

FIGURE 5-3 Vertical view of deeply eroded badlands in a sedimentary river terrace near Foum el Hassane, South Morocco. (A) Shadows cast from upper left to lower right emphasize the mesa-like structures remaining of the original surface and the V-shaped erosion channels between; note shadow of kite flyer standing on badland surface, upper left. (B) Same image rotated so that shadows fall in opposite direction. Many viewers may see inverted relief in this version, with sharp ridges and flat, sunken depressions. Kite aerial photograph by IM and JBR, March 2006.

softly blurred. However, this is not applicable for SFAP, where the subject distance is always large and the lens focus set to infinity. Thus, image depth has to be created by the composition itself. Linear features are an excellent element for adding dimensions to an oblique aerial photograph, as the figures in the preceding section have shown.

FIGURE 5-4 Two high-oblique views over the small town of Liebenthal in western Kansas, United States. (A) Road running vertically across the image is visually dominant and detracts from the rest of the picture. (B) Slight shift in camera vantage and viewing direction with diagonal roads. The latter draw the eye across the scene, so the viewer looks at all parts of the picture. Kite aerial photos by SWA and JSA, May 2006.

Another possibility is to structure the image into foreground, middleground, and background. This, too, is not easy to achieve in SFAP. The opportunities for using distinct objects as foreground elements are rare owing to the high vantage point (Fig. 5-9). Instead, the three depth zones can be composed from horizontally aligned changes in landscape color or patterns and the horizon as background element (Fig. 5-10). In vertical aerial photography, depth is not an essential compositional issue, although interesting effects can be achieved with shadows (see Chapter 4).

5.2.4. Pattern and Texture

Pattern refers to the arrangement of discrete objects, which individually are visible distinctly and form some regular arrangement with each other, for example vehicles in a parking lot, fruit trees in an orchard, crop rows, or

FIGURE 5-5 Strong diagonal elements may be utilized for visual emphasis. Fall River Lake, southeastern Kansas, United States. (A) Overview of dam and roads. (B) Closeup view of dam and spillway. Kite aerial photographs by JSA, June 2006.

FIGURE 5-6 The same effect that reveals the famous Nazca lines in Peru tells of men and animals crossing their ways in cars and on foot in northern Burkina Faso. The stones covering the ground surface—pisolith from laterite crusts—are shifted off the beaten track, exposing the light sandy soil beneath. Kite aerial photograph by IM, JBR, and K.-D. Albert, November 2001.

FIGURE 5-7 Mulitple sets of cross-cutting linear fractures are exposed in this granite outcrop at Point Pinos on the shore of Monterey Bay, Pacific Grove, California, United States. Vertical kite aerial photograph by SWA and JSA, November 2006.

FIGURE 5-8 Curved linear features. (A) The road and small stream channel marked by trees form parallel diagonal elements that lead the eye from the lower left corner toward the village with mountain backdrop in the upper right corner. Foreslope of the Tatra Mountains, Stráne pod Tatrami, northern Slovakia. Kite airphoto by JSA and SWA, August 2007. (B) Meanders of the dry channel of the Cimarron River, southwestern Kansas, United States. Such almost perfectly symmetrical S-shaped curves are rare in nature. Kite airphoto by SWA, November 2006.

FIGURE 5-9 The edge of a table mountain topped with remains of a thick laterite crust acts as a foreground element in this oblique airphoto taken in Burkina Faso's Sahel. The change of colors between the inselberg and the glacis in the background additionally increase the depth effect. Kite aerial photograph by IM, JBR, and K.-D. Albert, November 2001.

FIGURE 5-11 Near-vertical view of headstones arranged in rows of a cemetery. Note distinctive shadows cast by stones and grave markers. Liebenthal, Kansas, United States. Kite airphoto by SWA and JSA, May 2006.

gravestones in a cemetery (Fig. 5-11). Texture, on the other hand, involves elements that are too small to appear clearly as individual objects but still impart a distinct grain or fabric to the picture, such as waves on water, trees in a dense forest canopy, roof shingles, or prairie vegetation. The difference between pattern and texture is largely a matter of spatial resolution. Features that create texture at small scale may appear as distinct pattern elements at larger scale.

Patterns in an image can be pleasing and enjoyable to the observer, but also may be boring if their alignment and orientation are too regular and their content too monotonous. Patterns often have a most satisfying effect if combined with other elements breaking their uniformity or with other patterns of different scale (Fig. 5-12; see also Section 5.3 below).

FIGURE 5-10 Roughly following the rule of thirds, this view of the Bardenas Reales Natural Park in the Province of Navarra, Spain, is divided into a foreground zone of eroded rangeland, a middleground of green fields and hills, and a background zone where the Tertiary structural platforms of the Ebro Basin create a silhouette before the sky. Kite aerial photograph by IM and JBR, February 2009.

5.2.5. Color

Stereoscopic color vision is the most important human sense (Drury, 1987). The nature of color is, thus, a key element in photography. A huge amount of subjective discourse and quantitative research has been devoted to the subject of visible color, human perception of color, and color in nature (Lynch and Livingston, 1995). A full discussion is well beyond the scope of this book, so only a few basic aspects of color are reviewed here. For a more in-depth discussion of color in photography, see for example Langford and Bilissi (2007).

The primary colors—blue, green, and red—are also called additive colors, because they may be combined in various amounts to create all visible colors of the spectrum. Subtractive colors, also known as complementary colors, are created by removing (subtracting) an additive color from white light (Table 5-1). The human eye, video display, and digital photography depend on additive colors; whereas, printing and film photography are based on subtractive colors. In printing a color image from a computer display, for example, the printer must translate additive colors of the monitor into corresponding combinations of subtractive colors for the printer.

Several quantitative means exist to define color. One approach is based on the proportions of primary colors and their intensities. Another well-established means for defining color is the *Munsell Color* system, which is based on three attributes of color that are often illustrated in

FIGURE 5-12 A variety of patterns and textures discriminates different land uses in this vertical image of a landscape near Baza, Province of Granada, Spain. Flat, narrow valleys are dotted with olive trees of different age and size, north-exposed slopes and abandoned fields at hilltops show the mottled texture of steppe vegetation, and south-exposed slopes and fallow fields are lined with erosion rills and tillage patterns. On a flat surface in the lower right, a newly planted olive grove has the most regular pattern in the image. Cars pinprick the lower right corner with tiny yellow color splashes. Model airplane photograph by C. Claussen, M. Niesen, and JBR, September 2008.

a wheel diagram (Fig. 5-13). It is widely used in the geosciences for describing soil and rock colors, and is similar to the HSL or IHS color space implemented in many graphics programs.

TABLE 5-1 Relationship of primary and subtractive colors.

Primary colors	Wavelength	Subtraction from white
Blue	0.4-0.5 μm	Yellow (green + red)
Green	0.5-0.6 μm	Magenta (blue + red)
Red	0.6-0.7 μm	Cyan (blue + green)

- *Hue*: actual spectral color such as red, yellow, green, blue, etc. Hue is designated by a number and letter: 5R (red), 10YR (yellow-red), or 5GY (green-yellow), etc.
- *Value*: lightness or brightness of the color. Value ranges from zero for pure black to 10 for pure white.
- *Chroma*: intensity or saturation of the color. Chroma begins with zero for neutral (gray) and increases with no set upper limit.

Common rock and soil colors, for example, are moderate yellowish brown (10YR 5/4), light olive gray (5Y 5/2), and pale red purple (5RP 6/2). Moderate yellowish green (10GY 6/4) and brilliant green (5G 6/6) are typical vegetation colors (GSA Rock-Color Chart, 1991).

Colors may be described in general as hot (warm) or cold (cool). The former include red, orange and yellow; the latter are blue, cyan, and green. Hot colors represent longer wavelengths; cold colors are shorter wavelengths (see Fig. 2-1). Most of the Earth's surface is covered by objects of cool or neutral colors: green vegetation, blue-green water, tan-brown soil, etc. Hot colors are much less common and, so, tend to stand out in aerial photographs, especially when the hot colors are bright compared with the background (Fig. 5-14). It is because hot colors are generally lacking in

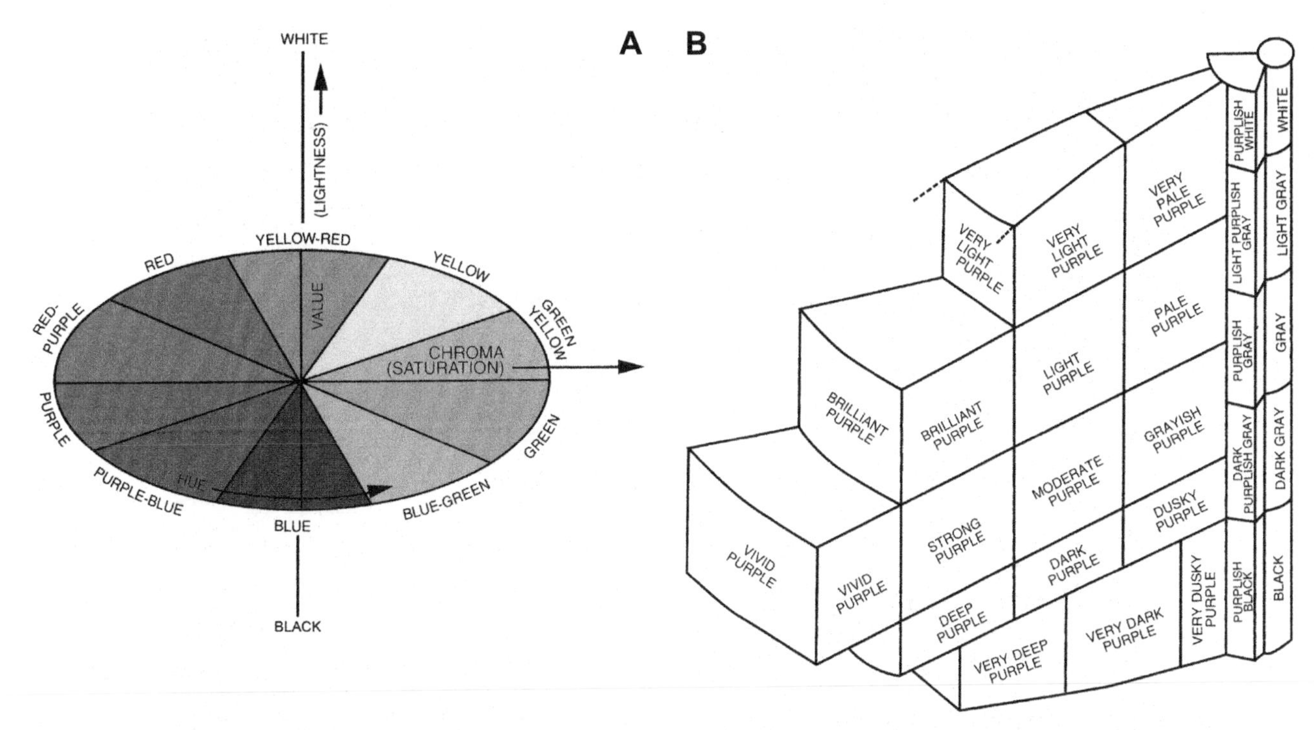

FIGURE 5-13 Schematic illustration of the *Munsell Color* system. (A) Wheel arrangement of hue, value, and chroma. (B) Wedge representing the purple hue section of the wheel. Adapted from GSA Rock-Color Chart (1991).

nature, that people are attracted to autumn foliage, painted-desert scenes, flower gardens, and similar bright, warm colors (Fig. 5-15). In fact, hot and cold colors induce a pseudo depth perception—hot colors appear closer and cold colors seem farther away from the viewer (Eyton, 1990; Stroebel, 2007).

FIGURE 5-14 Vertical view of summer cabins and boat docks on a recreational lake. The bright red vehicle stands out in this otherwise dull winter scene dominated by neutral gray-brown and cool colors. Without the red splash of color, this picture would be quite unattractive. Lake Kahola, Kansas, United States. Kite airphoto by SWA and JSA, February 2002.

5.3. COMBINING COMPOSITIONAL ELEMENTS

The combination of multiple visual elements creates the most dramatic photographs from an aesthetic point of view. The concept of compositional balance refers to the placement and relative visual impact of objects in the picture (Wildi, 2006). Most pictures consist of a main subject and secondary subjects arranged within a less conspicuous background. In general, the main subject should not be located at the geometric center of the photograph. The main subject should be offset toward the top, bottom, or one

FIGURE 5-15 Oblique view of autumn foliage with red and golden colors. Oak forest and small lake in the Chautauqua Hills, Kansas, United States. Helium blimp photo by SWA and JSA, November 2007.

FIGURE 5-16 Vertical view of formal rose garden in early summer. Loose Park, Kansas City, Missouri, United States. Helium blimp aerial photo by JSA, June 2006.

side—following the rule of thirds—to create a more dynamic image. In some cases, no main or secondary subjects exist, for which repetition of similar elements leads to visual balance. The following examples demonstrate the potential of multiple visual elements to create striking images.

The vertical photograph of a formal rose garden depicts strong diagonal and circular elements (Fig. 5-16). The straight diagonal walkways intersect at the circular fountain, which forms the visual main subject, and which is offset below the geometric center of the image. Most of the scene is medium green lawn grass with strong contrast between the bright paved walkways and dark tree shadows. Red and yellow rose flowers form tiny dots of color in the otherwise green, tan, and gray scene. Shadows fall toward the lower right corner, but obscure little of the ground area. Smaller pattern and texture elements include the rectangular lattice of arbors, arrangement of individual rose plants, and faint mowing lines in the lawn. A few people stand in random positions on the walkways or in the lawn. Finally, sun glitter from waves in the fountain adds a sparkling touch.

The portrait format of bicycle riders emphasizes the curving path of the trail and stone wall leading the eye from the near right corner to top left edge of this scene (Fig. 5-17). Most of the picture consists of neutral gray, green, and tan colors. Brightly colored wind-surfing kites stand out prominently, and shadows are perfectly

FIGURE 5-17 Low-oblique view of recreational park and bicycle path. Foster City, San Francisco Bay, California, United States. Kite aerial photo by SWA and JSA, October 2006.

proportioned to depict silhouettes of the bicycle riders. Sand, rocks, vegetation, and water waves display a rich variety of textures and small pattern elements.

Sun glint from waves on the water surface forms distinctive, intersecting curved patterns in this view of a small lagoon (Fig. 5-18). Most of the picture consists of nearly monochromatic gray and green-gray colors with high brightness contrast. The linear dock with right angles extends from the upper left corner toward the scene center, but without reaching the center. The dock also has neutral gray colors, except for a tiny splash of red clothing on a person, and shadows fall toward the bottom of the view. The juxtaposition of artificial linear and natural curved elements creates a visual conflict that draws the eye back and forth between the two patterns.

Linear elements of different qualities are overlaid with an irregular punctual pattern in this vertical photograph of a wide gully cutting into a near-flat *glacis d'accumulation* in Burkina Faso's Sahel (Fig. 5-19). Small rills forming dendritic networks unite in progressively broader fluvial channels; the course of the main drainage line roughly mirrors the slight curve of the dirt track running alongside the gully. The comparatively homogeneous glacis area on the left side is delimited by the cauliflower-like outline of the gully edge. From left to right, the size and density of bushes and trees dotting the scene increase. The position of the gully, whose shape is suited by the 3:2 format of the image, respects the rule of thirds with both the main drainage line and its left edge.

The sinuous boundary between lake and land is the dominant geometric element that leads the eye from the near edge toward the background, and the distant horizon enhances the visual expanse of this open prairie landscape (Fig. 5-20). The curved lake shore is reinforced by a single, parallel line of small trees that cast shadows toward the right. Most of the scene consists of neutral gray, green, and tan colors. The lake displays considerable variation in water color with subtle wave patterns. Most of the land has mottled vegetation textures, but agricultural fields can be seen in the distance. In the left foreground, the overturned boat with red bottom, although small, is a critical human element in this composition.

Raised bogs may display spectacular autumn colors (Fig. 5-21). The bright green, gold, and red colors represent various species of *Sphagnum* moss, which contrast with dark water pools and dull gray-green dwarf pine trees in this Estonian bog. This scene lacks a main subject; rather it is composed of color, pattern and textural elements that are repeated throughout the image more or less uniformly in distribution. Shadows fall in the lower right direction. Nothing of human origin is visible, and no objects of known size or shape are present; thus, the photograph has no scale reference. This image demonstrates the remarkable spatial complexity of a completely organic, natural environment.

FIGURE 5-18 Low-oblique picture of lagoon and fishing dock. Grand Isle State Park, Louisiana, United States. Kite aerial photo by JSA and SWA, March 2004.

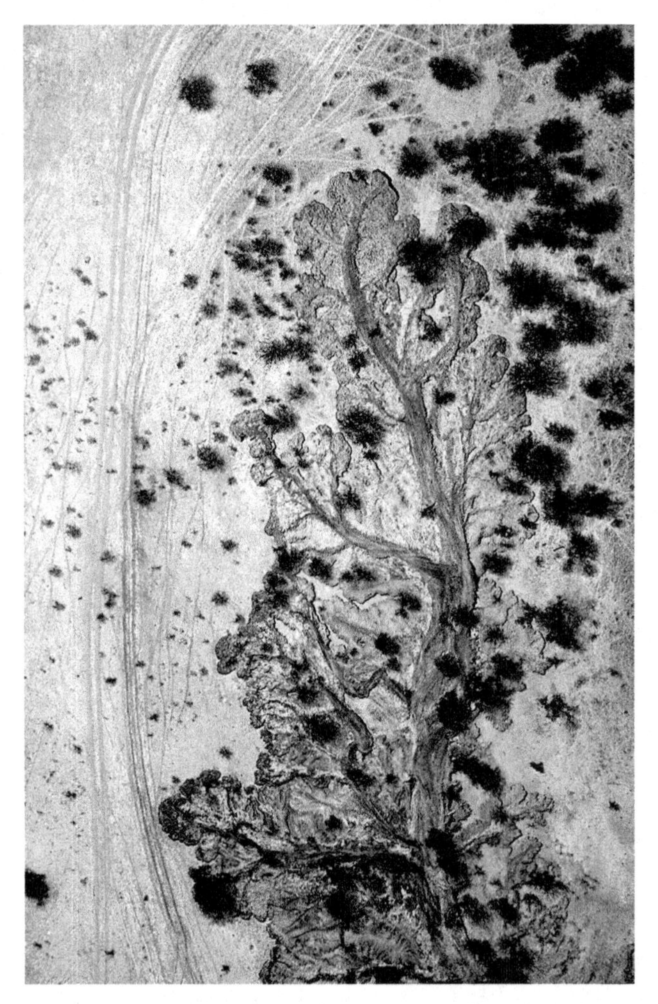

FIGURE 5-19 Vertical view of gully erosion near Gorom-Gorom, Province of Oudalan, Burkina Faso. Kite aerial photograph by IM, JBR, and K.-D. Albert, November 2001.

5.4. PHOTOGRAPHS VS. HUMAN VISION

Photographs are often considered to be accurate portrayals of visual scenes. However, photographs do not record images in the same way the human eye would respond to the identical scenes. For the following discussion, we ignore special techniques, such as filters, infrared, etc. The discussion focuses on natural-light photography under sunlit conditions. There are several fundamental ways in which photographic images differ from what is seen by the human eye.

- *Field of view*: Camera lens focal length determines the field of view which is focused onto the film or electronic detector array. A focal length of ~50 mm (for 35 mm format film) approximates the central zone of the human visual field. A wide-angle lens compresses more field of view onto the image plane; whereas a telephoto lens severely limits (crops) the field of view. The human eye provides for an extremely wide angle of peripheral vision, up to ~160° in most people. This peripheral vision is good for recognition of gray-tone patterns and is highly sensitive for detection of movement, but has almost no color capability. Photographs do not provide a similar peripheral field of view—the softly edged elliptical viewing field of the photographer is sharply trimmed to the rectangular extent of the image frame.
- *Latitude*: This refers to the range of highlights and dark features that are properly exposed in a photograph. Color-slide film has a total latitude of about five *f*-stops (Shaw, 1994). Assuming the mid-tone of the scene is correctly exposed, this means that brighter or darker features also appear correct within ±2.5 stops. Features more than 2.5 stops brighter would be "burned out" (pure white) and objects more than 2.5 stops darker are all black. Digital cameras generally have more restricted latitude and a greater tendency to burn out for bright objects compared with film. The human eye, in contrast, has much greater latitude equivalent to 12–14 *f*-stops.
- *Color saturation*: Our visual system adapts itself to changes in light intensity and color temperature, but the color rendition of color film may be different from the human eye. People tend to prefer vibrant, rich, saturated colors in photographs. Filmmakers have responded to this preference in different ways. Some manufacturers favor color-saturated films, whereas, other films render color as close as possible to reality (Zuckerman, 1996). Similar differences are apparent in color recorded by digital cameras (see Chapter 6).

Given these and other factors, it should be clear that a photographic image of a scene is different in several ways from the human visual impression of that same scene. The photograph represents a selection of certain elements—field of view, exposure range, and color saturation, which are in general more limited than a human observer would sense. On the other hand, a photograph is a permanent record of the scene, while the human visual impression is stored as memory that cannot be reproduced fully or accurately for analysis or sharing with others.

5.5. SUMMARY

Aerial photography shares many common attributes with ground-based photography. However, the bird's-eye view afforded by small-format aerial photography opens new vistas for image composition. Most important is a consideration of toplighting for illuminating the ground scene as viewed from above. For high-oblique views, generally the horizon should be nearly level and little sky should be visible. For vertical views, shadows are often critical; for best effect, shadows should fall from the upper-left toward the lower-right corner of the image.

FIGURE 5-20 High-oblique view across an ephemeral lake and prairie landscape. Dry Lake, western Kansas, United States. Kite aerial photo by JSA and SWA, May 2007.

FIGURE 5-21 Vertical shot of raised bog displaying autumn color. This picture won the photography award in the Science and Engineering Visualization Challenge from the U.S. National Science Foundation (AAAS/NSF 2005). Field of view ~100 m across. Mannikjärve Bog, Estonia. Kite aerial photo by SWA and JSA, September 2001 (see Aber et al., 2002).

Linear features, both straight and curved, draw the viewer's eye into the picture, and the placement of linear objects has a strong influence on the overall visual impact of the image. Patterns, especially at different scales or combined with other elements, may be an important aspect for making an image interesting and enjoyable to the observer. Color is the fundamental basis of image recognition, and much quantitative research and qualitative evaluation have been done on this subject. In general, people react more favorably to warm (red, orange, yellow) colors than to cool (blue, green, violet) colors; warm colors stand out and may create a pseudo depth perception. The combination of multiple visual elements creates the most dramatic photographs from an aesthetic point of view.

Photographs differ in several significant ways from human vision. In general, humans have greater field of view, latitude (light to dark), and color range. Furthermore, humans normally perceive the world in stereoscopic vision from ground level. Small-format aerial photographs represent a permanent image record, whereas, human vision is stored in memory that cannot be reproduced fully or accurately for analysis or sharing with others.

Chapter 6

Cameras for Small-Format Aerial Photogrammetry

I hate cameras. They are so much more sure than I am about everything.

John Steinbeck in a letter to his editor, 1932 (Steinbeck, 1975)

6.1. INTRODUCTION

A plethora of cameras is available for small-format aerial photography (SFAP), based on film and electronic recording techniques. In general, these cameras are relatively compact, lightweight, and capable of operating in largely automated modes. In fact, any camera designed primarily for hand-held use on the ground may be adapted for manned or unmanned SFAP from a variety of platforms (see Chapter 8). SFAP cameras range from inexpensive, disposable film cameras to high-end, professional-grade digital cameras.

In spite of what appear to be great differences between various cameras, all have certain basic components—lens, diaphragm, shutter, and image plane within a light-proof box. Most also have either an optical viewfinder or a monitor screen to depict the image. Geometry of the lens and image format determine the scene area focused onto the image plane. The diaphragm and shutter control the amount of light to expose each photograph.

Conventional film cameras are designed to accept film of certain format or width—for example, 35 mm, 70 mm, 5 inches, or 9 inches. Most popular cameras are for 35-mm film, whereas professional photographers tend to employ 70-mm format cameras. The still larger formats are utilized mainly for scientific and engineering applications. For example, the *Linhof* camera used by space shuttle astronauts is a 5-inch (125 mm) format, which produces remarkable images (Fig. 6-1). Modern metric cameras for aerial surveys have 9-inch (23 cm) format with film magazines holding rolls up to 150 m in length. These and other similar formats exist for digital cameras as well. The following discussion begins with film cameras and proceeds to digital cameras with an emphasis on those aspects that are most useful for SFAP. Additional camera features important for SFAP, such as image downlink to a ground receiver and remotely controlling camera functions via a USB interface, are discussed elsewhere (see Chapter 7).

6.2. FILM CAMERA BASICS

The basis for traditional photography are light-sensitive chemicals in the film emulsion. Such photochemical imagery depends upon the reaction to light of silver halide crystals, which undergo a chemical change when exposed to ultraviolet (UV), visible, or near-infrared (NIR) radiation. This photochemical change can be developed into a visible picture. The sensitivity of chemical photography ranges from about 0.3 μm to 0.9 μm wavelength. The lower limit is based on available UV energy and strong atmospheric scattering; film spectral sensitivity determines the upper limit.

Different parts of the spectrum may be photographed by using various films and filters. In fact, many film-and-filter combinations have been developed for routine and special purposes in aerial photography (Table 6-1). Photographs are routinely taken in b/w panchromatic, b/w extended red, b/w infrared, color-visible, color-infrared (Fig. 6-2), and multiband types.

Multiband photography is taking simultaneous photos in different portions of the spectrum. Early attempts at color photography in the 1890s were, in fact, based on separate b/w photographs in blue, green, and red bands that when viewed together simulated natural color (Osterman, 2007; Romer, 2007). Another variation is four-band photography with separate b/w photographs in blue, green, red, and NIR bands. As an example, the *Apollo* 9 mission in 1969 included a photographic experiment in which four coaxially mounted *Hasselblad* 70-mm film cameras were used to simulate multiband imagery. These results provided a proof of concept for the technology employed in the early Landsat multispectral scanner (Lowman, 1999).

6.3. DIGITAL CAMERA BASICS

During the 1990s, digital still cameras were developed for general and scientific use, and the capability of digital cameras advanced rapidly in the early years of the

Small-Format Aerial Photography

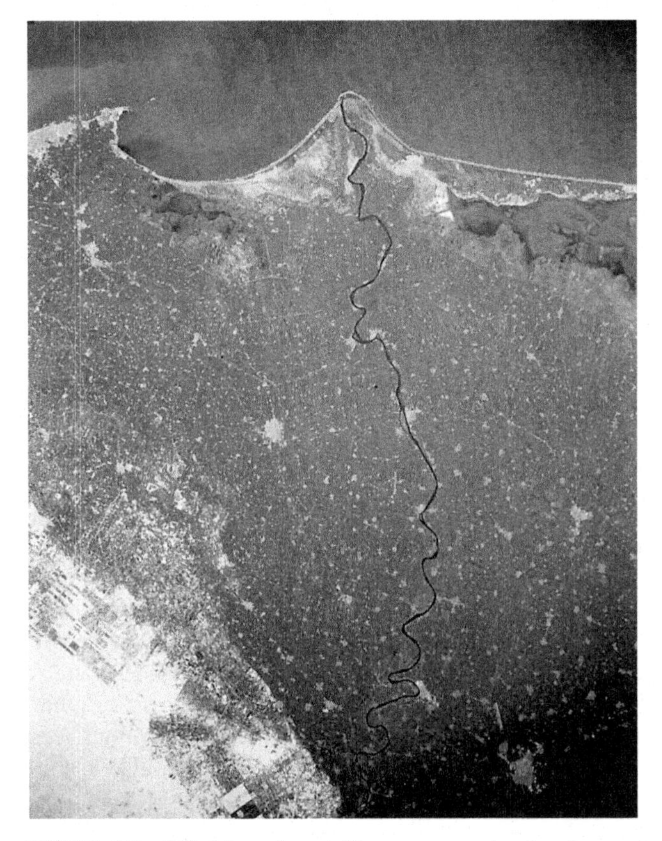

FIGURE 6-1 *Linhof* large-format film camera used onboard space-shuttle and space-station missions. Near-vertical shot of the Nile River and Mediterranean coast in the vicinity of Alexandria, Egypt. Natural color, March 1990, STS36-151-101; image obtained from NASA Johnson Space Center.

FIGURE 6-2 Color-visible (A) and color-infrared (B) 35-mm film photographs of the Arkansas River valley below Lake Pueblo, Colorado, United States. In (B), active vegetation appears in pink, red, and maroon colors on the floodplain. Kite aerial photos by JSA and SWA, May 2003.

twenty-first century. These cameras employ electronic devices to sense light. The sensor is essentially a microscopic array of semiconductors that measures light intensity in a raster grid. The picture is made up of many small squares—also called cells or pixels (picture elements).

TABLE 6-1 Common film types and uses in conventional aerial photography.

Film type	Uses of film
b/w visible	panchromatic, normal visible
b/w extended red	panchromatic, haze penetration
b/w infrared	ultraviolet, visible or infrared
color visible	normal color: blue, green and red
color infrared	false color: green, red, near-infrared

b/w, black-and-white (gray tone) photos.

Early digital cameras of the 1990s could not deliver the high spatial resolution possible with conventional film. However, within the past few years, digital cameras with megapixel sensors have come on the market at moderate cost. These cameras produce images that rival the sharp detail of 35-mm and 70-mm films. Digital cameras are quickly becoming more popular, and their capabilities are increasing dramatically. In fact, digital technology has replaced film for most photography, except for the low-end disposable-camera market and special high-end applications.

6.3.1. Types of Digital Cameras

The largest choice exists for digital compact cameras, which are small, light, and easy to use. Technically and in terms of image quality, they are less advanced than the second-largest group, the digital single-lens reflex (DSLR) cameras. These cameras are based on the design of analog single-lens reflex (SLR) cameras and feature parallax-free optical viewfinders, interchangeable lenses, and larger sensors than compact cameras. With the mirror-and-pentaprism system and the large body and lenses, they are far heavier than compact cameras and therefore often a difficult choice for SFAP. Between these main camera types, several other designs exist. The so-called bridge cameras with their large zoom ranges have an appearance and manual exposure control possibilities similar to DSLRs, but non-interchangeable lenses and the smaller sensors of compact cameras. The last few years have seen a new type of mirror-free system camera, the so-called Micro Four Thirds standard, with similarly sized sensors as DSLRs but smaller interchangeable lenses and lower weight, which makes them an interesting camera type for SFAP applications. At the high-price and high-quality end, there are also a few digital cameras of the classic rangefinder type (compact design, manual focusing mechanism, fixed single-focus lens) that was best known for analog cameras in the mid-twentieth century.

Still a drawback for many digital cameras, especially compact models, is the comparatively long shutter delay and slow frame rate. Although all cameras feature a continuous drive mode, the image repeat speed may be as slow as 1 fps, and maximum shots in a row are usually limited to 3–10 images. In normal mode, the interval required between repeated exposures is often as long as 3–4 seconds for compact cameras with standard storage cards, which is problematic for fast-moving platforms like model airplanes (see Chapter 8.5.2). An interesting alternative for SFAP are industrial cameras that are small, light, and designed for high frame rates (e.g., Deneault, 2007; Prosilica, 2009). However, only a few models have a comparably high spatial resolution (>8 MP) now customary for good compact and DSLR cameras, and the prices exceed 10-fold that of a DSLR. Industrial cameras need to be attached to a separate computing unit, which increases costs further, but enables to take large numbers of images during an SFAP survey without having to change the storage card. Custom-built onboard computers such as employed in the new *MAVinci* autopiloted airplane (see Chapter 8.5.2) can now store up to 256 GB and are as small as 6 × 6 × 4 cm.

TABLE 6-2 **Comparison of lens focal lengths for 35-mm film, compact digital, and digital single-lens reflex cameras for various fields of view.**

Field of view	Focal length			Light reaching film or sensor
	35-mm film	Digital SLR*	Compact digital**	
Fish eye	15 mm	9 mm	4 mm	Very high
Wide angle	28 mm	18 mm	7 mm	High
Normal	50 mm	31 mm	13 mm	Moderate
Telephoto	200 mm	125 mm	50 mm	Low

Relative amount of light reaching the film or detector is indicated to the right.
**Crop factor of 1.6 (sensor size: 23.7 mm × 15.7 mm), typical of DSLR cameras.*
***Crop factor of 4, typical for compact digital cameras.*

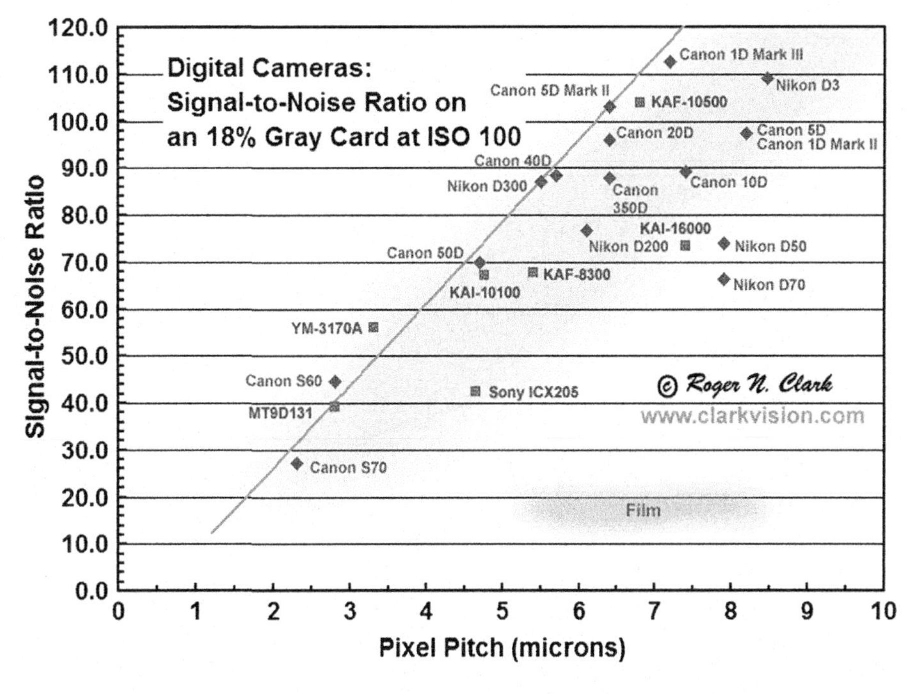

FIGURE 6-3 The signal-to-noise ratio on an 18% gray card for various camera models (brown diamonds) and sensor types (blue squares) with different pixel sizes. Note the clear trend of increasing signal-to-noise ratio with increasing pixel size. Reproduced with permission from Clark (2008a, fig. 2); see there for further explanation and details about the sensor performance model shown with orange line.

6.3.2. Image Sensors

Picture quality for film cameras is determined mainly by the type of lens, because the same film can be used in many cameras. For digital cameras, however, the lens as well as sensor capability determine picture quality, because both vary greatly in different camera models (Meehan, 2003). Digital cameras typically employ lenses of shorter focal lengths because nearly all electronic detectors have smaller dimensions than 35-mm film (Table 6-2).

Point-and-shoot digital cameras mostly display a picture format length-to-width ratio of 4:3, same as standard computer monitors, and the sensor is typically only 1/30 the size of 35-mm film. DSLR cameras usually have the 3:2 format ratio, same as 35-mm film, with sensor sizes typically 40–100% the size of a 35-mm negative. The largest sensors used in DSLR may have pixel densities below 3 MP/cm^2, while the smallest sensors currently built into compact cameras are only around 6 × 4 mm in size and have pixel densities up to 43 MP/cm^2. The much smaller pixel size of the point-and-shoot camera sensors is the main reason for the difference in image quality (noise and dynamic range) between digital compact cameras and DSLRs (Langford and Bilissi, 2007; Clark, 2008a, b; see also Fig. 6-3).

Two main types of electronic detectors are employed in digital cameras, nowadays, known as the charge-coupled device (CCD) and complementary metal oxide semiconductor (CMOS). Both employ an array of semiconductors (usually silicon) to detect light intensity; their differences are in terms of architecture of semiconductor structure (Kriss, 2007; Langford and Bilissi, 2007). According to a filter placed on top of each sensor, each semiconductor in the array detects light energy for only one primary color—red, green, or blue. The result is an image mosaic of three colors; the missing color values for each pixel are later interpolated from neighboring pixels by demosaicking algorithms. The most popular arrangement for filters is the Bayer pattern, which has twice as many green filters as red and blue filters (Fig. 6-4). This takes into account the green peak in solar energy and enhanced green perception in the human eye (Fig. 6-5; Padeste, 2007).

The main disadvantages of these sensors are associated with the difficulties of demosaicking. The resulting image may contain artifacts, especially in areas of fine color patterns, and is strongly influenced by the algorithms used to reconstruct missing pixel color values and deblur the image. Numerous solutions have been published on dealing with these problems (Trémeau et al., 2008). The image-processing software is proprietary for each camera model, and individual digital cameras may produce noticeably different versions of the same scene. The influence of different Bayer demosaicking algorithms on the geometric quality of the resulting images—crucial for photogrammetric measurement—has been discussed by Perko et al. (2005) and Shortis et al. (2005).

As an alternative to mosaic-patterned CCD or CMOS sensors, the *Foveon* X3 CMOS sensor is designed in the

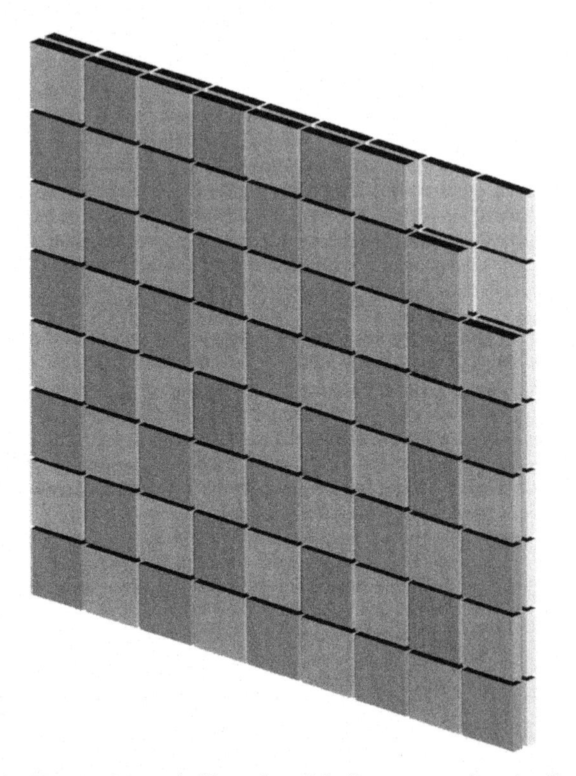

FIGURE 6-4 Schematic illustration of the Bayer pattern for color filters employed in CCD and CMOS digital image arrays. Taken from Padeste (2007, fig. 3).

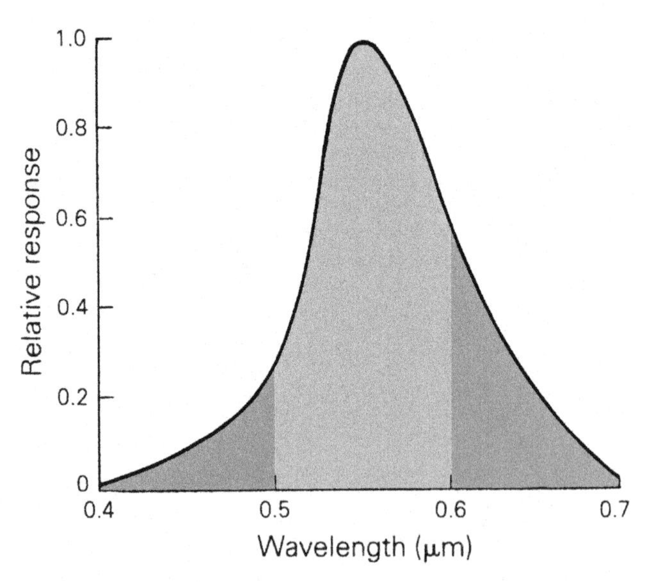

FIGURE 6-5 Overall sensitivity of the human eye to photopic (daylight) vision. Blue and red sensitivities are much less than green. Adapted from Drury (1987, fig. 2.5).

same way as photographic color film, with an array of vertically layered sensor cells (Fig. 6-6). Because this full-color sensor records all three primary colors for each cell, it creates sharper images, which are largely devoid of interpolation and sharpening artifacts. However, with the increasing image quality of Bayer pattern images resulting from the recent advances in algorithm development, the *Foveon* X3 sensor has so far not been a serious challenge to traditional digital sensors and is used in only a few cameras.

Single-array CCD and CMOS sensors have several advantages compared with other types of electronic imaging systems. They capture still and moving images, detect many types of light, operate in milliseconds with ambient light, have no moving parts, and are quite compact and robust (Padeste, 2007). Many digital cameras allow the user to select different ISO settings for detector sensitivity to light (see below). For SFAP, a high ISO (800) or sports-mode setting helps to minimize the effects of camera motion (Meehan, 2003).

6.3.3. Image File Formats

For most digital cameras, the default setting for image capture is an automated within-camera processing, which allows saving the images in 8-bit JPEG format. JPEG is a standard image format with excellent compression capacities, fast access times, and immediate usability for viewing, printing, and web posting. However, such in-camera processing and image compression reduce the original information present in a scene in several ways. The original measurements of each sensor cell, typically taken in 12-bit radiometric resolution, are referred to as RAW image values.

FIGURE 6-6 Schematic illustration of the *Foveon* X3 direct image sensor with three layers of sensor cells.

The RAW format represents the complete and lossless image information as recorded by the sensor and can be saved as such by some compact cameras and by all DSLR cameras. This option allows the photographer to control more finely the processing of the data when converting the raw measurement values to an image file, for example by custom histogram adjustment for images with both bright and dark areas (see below and Chapter 11). The drawbacks of the RAW format are much larger storage size, slower access, and the necessity for post-processing using the camera manufacturer's proprietary decoding software.

6.4. CAMERA GEOMETRY AND LIGHT

A geometric relationship exists between the lens focal length, image format, area (angle) of view, and the amount of light that reaches the film or electronic sensor. Some common lens focal lengths are given in Table 6-2. Many cameras are now equipped with zoom lenses that allow the user to vary the focal length. The lens aperture and shutter speed are fundamental controls for how much light reaches the film or image sensor. Film speed or ISO rating determines how much light is required for correct exposure of the photographic emulsion or detector elements. Each of these factors is examined in turn.

6.4.1. Focal Length

All but the simplest camera lenses are compound lenses made of multiple glass elements (Kessler, 2007). Rays of light entering the lens are refracted within the lens in such a way that they seem to converge in a single point on the optical axis (the front nodal point) and leave from a single point on the optical axis (the rear nodal point). The distance between the rear nodal point when the lens is focused at infinity and the focal plane or image plane is called focal length (Wolf and Dewitt, 2000). Shorter focal lengths result in wider angles of view and vice versa (Table 6-2; see Chapter 3). A so-called normal lens captures a view in a similar angle as our eyes would in the primary field of vision (~50°).

As the angle of view depends both on the focal length and on the image area (see Fig. 3-11), a normal angle for a digital camera with a small sensor chip, e.g., 6.2 mm × 4.6 mm, is achieved with a shorter focal length than a normal angle for an analog small-format film negative (36 × 24 mm). Nevertheless, lenses for digital cameras are often characterized by their 35-mm film-equivalent focal lengths, because many photographers are familiar with traditional small-format cameras, and digital sensor chips are not standardized in size as is 35-mm film. With few exceptions, compact cameras today have zoom lenses with variable focal lengths, whereas lenses with single focal lengths as well as zoom lenses are available for SLR cameras.

6.4.2. Lens Aperture

The aperture of the lens opening, which is controlled by the diaphragm, is given by a value called *f*-stop, which is defined as lens opening diameter divided by lens focal length. *F*-stop is thus a fraction. On most cameras, *f*-stop values are arranged in sequence, such that each interval represents a doubling (or halving) of light values. Smaller *f*-stop denominator values mean more light enters the camera; larger values indicate less light. The typical sequence of *f*-stops is given below (Shaw, 1994). Most lenses operate in the range *f*/2.8–*f*/22.

f/1, *f*/1.4, *f*/2, *f*/2.8, *f*/4, *f*/5.6, *f*/8, *f*/11, *f*/16, *f*/22, *f*/32

Although the *f*-stop values as fractions are useful as a comparable aperture definition for lenses of all sizes, the absolute amount of light entering the camera is still dependent on the lens diameter. Smaller lens diameters—as used in compact cameras in contrast to SLRs—allow fewer photons to reach the image plane during a given exposure time. For digital cameras, this in turn means higher image noise (Clark, 2008b).

6.4.3. Shutter Speed

The length of time the shutter remains open is called shutter speed. On most cameras, shutter speed values are arranged in sequence such that each interval is twice as long (or short) as the next. In other words, changing shutter speed by one interval either doubles or reduces by half the time interval. Longer intervals mean more light enters the camera; shorter intervals indicate less light. The typical sequence of shutter speeds is given below in seconds (Shaw, 1994). Intermediate values may exist depending on the camera characteristics, and some recent DSLRs reach even faster shutter speeds up to 1/8000.

1, 1/2, 1/4, 1/8, 1/15, 1/30, 1/60, 1/125, 1/250, 1/500, 1/1000, 1/2000, 1/4000

For SFAP, fast shutter speeds are indispensable as most platforms either move fast (model airplanes) or cause considerable vibration (kites, model helicopters).

6.4.4. Film Speed or ISO Rating

Different photographic emulsions on film vary greatly in their sensitivity to light—some require little, others need much light to properly record an image. This factor is known as film speed, which is indicated by a standard ISO rating. Low ISO values indicate "slow" films that need much light. High ISO values, in contrast, are typical of "fast" films that require little light. The latter have larger silver halide crystals in the emulsion, which tend to give a grainy structure to the image. Typical film speeds are displayed below. The main speeds are given in bold, and intermediate speeds are shown for one-third increments (Shaw, 1994).

12, 16, 20, **25,** 32, 40, **50**, 64, 80, **100**, 125, 160, **200**, 250, 320, **400**, 500, 640, **800**, 1000, 1280, **1600, 3200**

The same concept is used for indicating the sensitivity of digital camera sensors. The standard setting of ISO 100 corresponds to the light sensitivity of an ISO 100 film and may be increased to 200, 400, or up to 3200 for high-end DSLRs. A higher ISO setting amplifies the signal from the sensor, so less light is needed for the photograph. However, this procedure also amplifies the sensor noise, both for intensity and color, which in turn creates the impression of a grainier image, quite similar to high-speed photographic film. Many aerial photographs presented in this book were taken with ISO 100 film speed or sensor setting, although some were taken at faster speeds up to ISO 800.

6.4.5. Camera Exposure Settings

Settings for image exposure are best understood using the concept of stops. A stop is defined as doubling or halving of any factor that affects the exposure (Shaw, 1994). Notice that values for each variable—shutter speed, *f*-stop, ISO rating, or film speed—are arranged by doubling intervals. Among these variables there is a relationship called reciprocity. In other words, changing one variable by one stop can be matched exactly by changing another variable in the opposite direction by one stop.

For photographs under bright sun, which applies to most SFAP, the sunny *f*/16 rule can be employed (Caulfield, 1987). For a given film or sensor under full-sun conditions, the shutter speed should be the approximate inverse of the ISO rating at *f*/16. For example, ISO 200 film—or a digital sensor at ISO 200 setting—should be exposed at *f*/16 and 1/250 shutter speed under bright sun conditions.

The exposure factors have other influences on the resulting photograph. In general faster shutter speeds are desirable for SFAP in order to "freeze" the motion between the airborne camera and the ground. This necessitates using high ISO ratings (fast film) and/or lower *f*-stops. High ISO ratings (>400) tend to result in lower image quality, due to grainy appearance, in comparison to lower ISO ratings. Lower *f*-stops reduce the depth of field, which refers to the range of distance over which the image is in good focus. For ideal SFAP, a fast shutter speed should be combined with low ISO settings or high-quality (slow) film and a medium to high *f*-stop setting. However, this combination simply does not work in practice. SFAP, thus, represents a trade-off involving these factors.

Most modern cameras employed for SFAP utilize built-in light meters and automatic adjustment of shutter speed and *f*-stop. Such automated photography introduces certain artifacts in the picture-taking process. Inexpensive point-and-shoot cameras normally have a light meter that is separate from the lens or viewfinder; whereas the light meter in SLR cameras and recent mirror-free system cameras operates through the lens. The latter is clearly preferable, as the meter registers the light actually entering the camera through the lens.

Advanced point-and-shoot and most SLR cameras allow the user to set a priority for shutter speed or *f*-stop. For example, cameras with a high-speed (sports) mode select the fastest possible shutter speed in order to minimize blurring effects of camera or object motion. Faster shutter speed is highly desirable for effective SFAP, but represents a compromise with lower *f*-stops. Assuming the ground target is at infinite focal distance, depth of field and thus *f*-stop should be negligible factors. However, image sharpness usually is better at medium *f*-stops due to higher lens aberrations and diffraction at low and high *f*-stops (Langford and Bilissi, 2007).

Cameras normally determine exposure settings based on a medium gray tone—equivalent to the light reflected from an 18% gray card—to represent the average value sensed by the light meter. This works well for a scene in which most objects are uniformly lighted, for example blue sky and green trees or grass. However, the results may be unsatisfactory for a scene comprised of bright highlights and dark shadows, such as sunlit hill tops and valleys in shadows or for scenes that are generally brighter or darker than average. Some objects are intrinsically bright—snow, ice, concrete, chalk, deciduous trees, and grass; other objects are naturally dark–burned ground, basalt, asphalt, conifer trees, etc.

For scenes with mixed illumination, the bright features would be overexposed and washed out, and the dark objects remain nearly black and lacking in details (see Chapter 5.4). In the case of a generally bright scene, the image would be underexposed because the camera aims at a medium gray tone, while a generally dark scene could be overexposed for the same reason. These effects can be avoided with most cameras by adjusting the exposure correction factor accordingly.

6.4.6. Image Degradation

Photographs do not record the energy reflected off their motifs flawlessly. Chemical and physical characteristics of films and image sensors may introduce unwanted effects and artifacts into an image. Also, the lens is crucial for image quality, and some degree of distortion always is introduced into the optical paths by deviations from a perfect central perspective. The most frequent types of distortions and artifacts are briefly described in the following (see Chapter 11 for more details).

- Radial distortion slightly changes the image scale with increasing distance from the image center. Short focal lengths (wide image angles) typically show barrel distortion, which causes straight lines to bend outwards, while long focal lengths (telephotos) tend to exhibit pincushion distortion, which causes straight lines to bend inwards (Fig. 6-7). These effects tend to be stronger in zoom lenses than fixed-focus lenses. Radial distortion can be corrected with image processing.
- Chromatic aberration is caused by the inability of a lens to focus all wavelengths onto the same plane (longitudinal CA) and/or by varying magnification of different wavelengths (latitudinal CA). With increasing distance from the image center, the three primary colors become slightly offset, causing a color-fringe effect around contrasting edges (Fig. 6-8). Very wide-angle lenses and the extreme focal lengths of super-zoom lenses are most prone to chromatic aberration. The color offset in the image bands can be corrected with some image-processing packages.
- Similar-looking colored fringes also can be caused around overexposed image areas by overflowing electrical charge (blooming) or near high-contrast boundaries by aberration effects associated with microlens arrays placed onto some sensors (purple fringing; Langford and Bilissi, 2007). Blooming and purple fringing are not easy to correct, and the first is best prevented by avoiding overexposure.
- The term vignetting describes the gradual darkening of an image toward the edges and corners (Fig. 6-9) and may have different causes, all of which are related to the lens design (Ray, 2002; Kessler, 2007). Optical vignetting results from the reduction of the effective lens opening for oblique light rays, because the lens diaphragm is set back from the front rim of the lens tube. Thus, it can be reduced or cured by stopping down the lens to smaller apertures. Similarly, mechanical

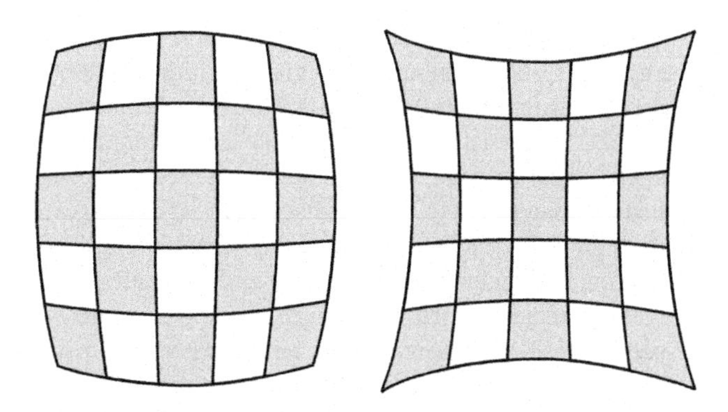

FIGURE 6-7 Barrel distortion (left) and pincushion distortion (right) are typical optical lens aberrations leading to the bending of straight lines.

FIGURE 6-8 Chromatic aberration causes colored fringes around the edges of long afternoon shadows cast by soil clods in a dry river bed in South Morocco. Subset of a kite aerial photograph taken by IM and JBR; aberration effect is strongly exaggerated for illustration purpose.

vignetting is caused by too long extensions of the lens (filters, lens hoods) that narrow the effective lens opening; in this case, the image corners will be blackened out. Finally, natural vignetting or natural light falloff, which is described by the $\cos^4$ law, occurs with all lenses but is more prominent for wide angles. The darkening effect by vignetting is usually quite small, but it can become more obvious when several images are stitched together in a mosaic. As wide-angle lenses are frequently used for SFAP and vignetting generally is at its worst when the lens is focused at infinity, some contrast and brightness adjustment may be necessary for correcting vignetting effects in aerial photographs.

- Other than the distortions listed above, image noise is independent of focal length and aperture. It is defined as the random variation of pixel values caused by fluctuations of the signal transmitted by the sensor. While there are several sources of noise (Langford and Bilissi, 2007), they all result in high-frequency brightness and color variations ("image speckles"), which are more pronounced at low signal-to-noise ratios. Thus, noise is most prevalent in dark image areas (e.g., hard shadows in SFAP images), for high ISO speeds (where the sensor signal is amplified to provide for poor light conditions), and for small sensor cells (compact cameras with high megapixel numbers; see Fig. 6-3). Image noise may be smoothed and made less conspicuous with image processing, but choosing a good-quality sensor with large pixels is probably the most important remedy against noise.

FIGURE 6-9 Vignetting of an aerial photograph showing experimental reforestation plots near Guadix, Province of Granada, Spain. Hot-air blimp photograph taken by IM and JBR, March 2002; vignetting effect (originally inconspicuous) introduced artificially for illustration purpose.

6.5. COLOR-INFRARED PHOTOGRAPHY

Color-infrared film is sensitive to visible and NIR portions of the spectrum. In normal practice, a yellow filter is employed to eliminate blue and UV wavelengths. In some cases, orange or red filters may be used to further restrict visible light from reaching the film. Color-infrared film carries no ISO number; nor do conventional light meters provide correct indications of NIR radiation. Without an ISO bar code on the film case, most cameras cannot make automatic settings. Therefore, taking photographs with color-infrared film requires manual settings for exposure based on estimates of available light. When using a film without an ISO rating, most cameras default to ISO 100 for setting adjustments (Table 6-3).

Given that most SFAP takes place under bright, sunny conditions, color-infrared film can be treated as ISO 200 according to Table 6-3. Following the sunny *f*/16 rule of thumb, the camera setting should be equivalent to shutter speed 1/250 and *f*/16. However, other camera–lens–filter combinations may produce different results. For example, best settings for a *Canon Rebel* SLR camera with a zoom

TABLE 6-3 **Manual compensation for SFAP color-infrared film for default value of ISO 100.**

Lighting conditions	Exposure correction
Bright sun, mid-day-clean, dry atmosphere	ISO 200 (+1 f-stop)
Light, but not bright sun-hazy, humid, dusty atmosphere	ISO 160 (+½ f-stop)
Slightly overcast, indirect light, thin clouds	ISO 100 (no correction)
Pale, diffuse light-early morning or late afternoon	ISO 80 (-½ f-stop)
Overcast, indirect light, rather dark-heavy clouds	ISO 50 (-1 f-stop)

Based on Pentax *SLR camera with 50-mm lens and orange filter. Adapted from Marzolff (1999, Table 4-2).*

lens and yellow filter are 1/250 shutter speed and *f*/11 for full sun and active vegetation (Aber, Aber, and Leffler, 2001). Apart from these empirical results, aerial photography with color-infrared film remains an uncertain proposition—considerable trial-and-error testing is necessary, and results cannot be predicted well. A final and nearly insurmountable problem is that since about 2005 most commercial photo laboratories no longer process color-infrared film.

For digital SFAP, the primary challenge is to identify a suitable color-infrared digital camera of relatively small size and weight at a cost that could be justified. One commercial camera that meets these requirements is the Agricultural Digital Camera (ADC) by *Tetracam*. This camera employs a 3.2 megapixel CMOS sensor, which operates in the spectral range of 0.52–0.92 μm wavelength (green, red, and near-infrared). A permanently mounted long-pass filter behind the lens blocks blue and UV light, and the camera has a robust machined-aluminium body (Fig. 6-10).

The primary applications for the *Tetracam* ADC camera are, as the name suggests, agriculture as well as forestry and other studies involving vegetation, soil, and water. The camera is designed to be operated on the ground or from manned or unmanned aircraft either by hand or remote control. Its size, shape, weight, and operating characteristics place this camera within the normal range for DSLR-type cameras. The camera produces results that are quite comparable with color-infrared film photography (Fig. 6-11), and functions well from remotely operated aerial platforms (Fig. 6-12).

FIGURE 6-10 *Tetracam* ADC digital, color-infrared camera in a remotely operated radio-controlled rig for kite or blimp aerial photography. R, radio receiver; P, pan servo and gears; B, NiMH battery pack; A, antenna mast; S, shutter miniservo; and T, tilt servo. Camera rig built by JSA.

FIGURE 6-11 Digital ground photographs of a late-summer garden scene in color-visible (A) and color-infrared (B) formats. Active vegetation appears in bright red-pink colors in the latter. Also some artificial fibers and dyes are highly reflective for near-infrared (Finney, 2007), as seen in the flags. Compare with Figure 6-2.

FIGURE 6-12 Color-visible (A) and color-infrared (B) digital images of marsh at the Nature Conservancy, Cheyenne Bottoms, central Kansas, United States. Active vegetation appears in bright red-pink colors in the latter. Kite aerial photographs from Aber et al. (2009, fig. 5).

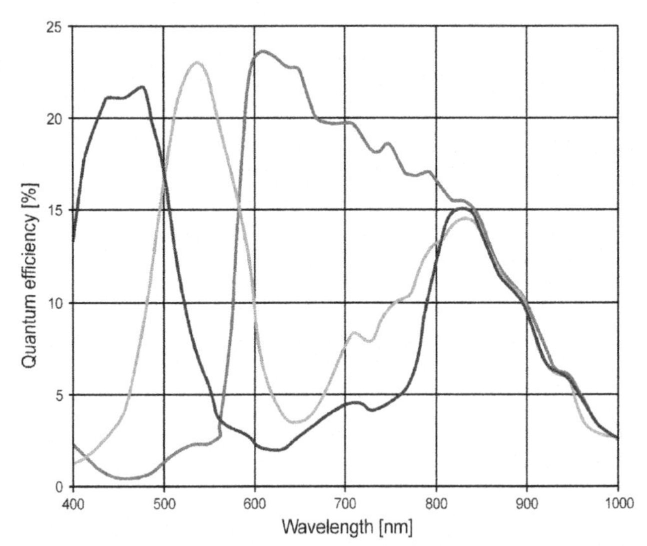

FIGURE 6-13 Typical response curve for CCD and CMOS image sensors without NIR blocking filter showing the transmission response after the light passes through the mosaic color filter over the image sensors. A "hot mirror" usually added to such a sensor blocks wavelengths above 700–750 nm. Adapted from sensor specifications given by Prosilica.com.

Another possibility for taking pictures in the NIR spectrum is the modification of a customary digital camera. All digital camera sensors are sensitive not only to visible, but also to NIR light (Fig. 6-13). In order to prevent NIR light from degrading the quality of normal color images, a blocking filter ("hot mirror") is placed in front of the sensor that allows only visible light to pass. By removing this hot mirror, the spectral sensitivity of the sensor cells to NIR light can be employed for photographs in two ways. The blocking filter is replaced either by an infrared (visible-light blocking) filter for pure NIR photography or by a clear filter for preserving the whole spectral sensitivity of the detector (UV to NIR). The latter option will merge NIR energy with each of the visible primary colors. While this may offer unusual artistic possibilities, it is of little interest for scientific use as it does not separate NIR reflectance in a single image channel like color-infrared film or the *Tetracam* ADC camera.

The modification for converting a camera for infrared photography is no trivial task and therefore increasingly offered as a commercially available infrared-camera conversion service. Successful SFAP with NIR-converted digital cameras is reported by Jensen et al. (2007) for crop yield studies and by Verhoeven (2008) for archaeological reconnaissance. Both studies used a double-camera system, combining the natural color image of the original camera with the NIR image of the modified camera for a four-band color-infrared image.

6.6. CAMERA CAPABILITIES FOR SMALL-FORMAT AERIAL PHOTOGRAMMETRY

Photogrammetric analysis of small-format aerial photographs presents additional challenges for the cameras employed. Because photogrammetry is based on precise measurements, reconstructing 3D objects from 2D photographs (see Chapter 3), the geometric stability of the camera becomes a key factor for accurate surface data collection from stereo images. Consumer-grade cameras lack many of the features included in metric cameras specifically designed for photogrammetry (e.g., fix-mounted, near-perfect lenses, calibrated focal lengths, and film-flattening image planes).

Contrary to the notion that recent technological development should increase the potential of consumer-grade cameras for sophisticated image analysis, the latest technical innovations in digital cameras actually may impede their value for photogrammetric applications. Image stabilizers (reducing the effect of camera shake during slow shutter speeds by sensor or lens counter movements) and dust-removal vibration, which are designed for improving image quality and sharpness, now introduce destabilizing parameters into the interior orientation of such cameras.

Nevertheless, off-the-shelf small-format cameras can be employed successfully for photogrammetric analysis (Warner et al., 1996; Fryer et al., 2007). Numerous studies have investigated the effects and accuracies associated with different types of cameras, lenses, and sensor chips (e.g., Chandler et al., 2005; Perko et al. 2005, Shortis et al., 2006; Rieke-Zapp et al., 2009). Most applications using consumer-grade cameras deal with terrestrial close-range photogrammetry rather than low-height aerial photogrammetry, but the principal problems associated with non-metric cameras are comparable. Some of the most important aspects to bear in mind when choosing a camera for aerial photogrammetric purposes are listed below.

6.6.1. Camera Lens

Lenses with a single focal length are clearly preferable to zoom lenses in terms of stability, accuracy, and precision (Shortis et al., 2006). Especially for calibrated cameras or when self-calibrating procedures are used, the focal length between images should be kept as invariable as possible. In order to prevent changes in focal length, it is preferable (if possible) to deactivate autofocus and manually focus the lens at infinity before mechanically fixing the focus ring with tape, screws, or adhesives. Regarding SLR cameras, detaching the lens from the camera body and thus changing the lens should be avoided as well as lenses with image stabilization systems.

6.6.2. Image Sensor

For digital cameras, larger image sensors with bigger sensor cells are usually preferable to smaller sensors. Larger pixels capture more photons during exposure, which means lower signal noise (Clark, 2008a). This not only allows larger ISO ranges, but also improves the image quality and the reliability of the light measurements between multiple images. Stereo-matching procedures for digital elevation-model extraction rely on measurements of correlation between pixel values of two images. Therefore, higher accuracies can be expected for images with low noise. However, to the best of the authors' knowledge there are no systematic investigations of consumer-grade digital camera sensors into this subject so far, although results from investigations into 3D measurement performances of stereo microscopes show the degrading influence of image noise (e.g. Marinello et al., 2008).

Most recent digital camera sensors feature some type of image stabilization system, and most recent DSLRs are equipped with sensor vibration systems for dust removal. Both features can change interior camera geometry and alter interior orientation values between consecutive images (Rieke-Zapp et al., 2009). As the effects on photogrammetric accuracy remain to be investigated, these stabilization and vibration systems should be avoided or at least permanently disabled for small-format aerial photogrammetry.

6.6.3. File Format

In order to avoid artifacts by image compression, which could degrade measurement accuracy and stereo-matching results, lossless image formats (RAW, lossless TIFF, lossless JPEG) are more favorable than lossy JPEG compression. Only RAW image storage allows preserving the full 12-bit image information and the customized post-processing that is desirable for sophisticated exposure corrections. The possibility of fully controlling the conversion process of RAW measurement values is considered an advantage over in-camera conversion by professionals, although Rieke-Zapp et al. (2009) could show that different RAW development software may have significantly different effects on the photogrammetric measurement accuracy yielded with the images.

6.6.4. Camera Type

As a consequence of the above items, SLR cameras and their recent mirror-free equivalents must be considered more suitable for small-format aerial photogrammetry than are compact cameras. At the time of writing, there were only two compact cameras on the market that could meet all of these requirements (*Leica* M8 and *Sigma* DP1/DP2), but they have other disadvantages for SFAP (high price and slow shutter speeds and frame rates, respectively). The disadvantage of SLR cameras, although they are becoming increasingly smaller and lighter, is their often considerably larger size and weight. This might—in addition to considerations of photogrammetric survey issues like stereo coverage, navigability, and stableness—also influence the choice of platforms used for small-format aerial photogrammetry (see Chapter 8).

6.6.5. Camera Calibration

Camera calibration (see Chapter 3) significantly improves the accuracy of photogrammetric analyses. Chandler et al. (2005) found that radial lens distortion errors effectively constrain the accuracies achievable, making accurate modelling of lens distortion an important issue for the use of consumer-grade digital cameras. Investigations into the temporal stability of a digital compact camera by Wackrow et al. (2007) confirmed the relative importance of inaccurate lens distortion parameters as compared to internal geometry variations, which were found to be remarkably low over a one-year period.

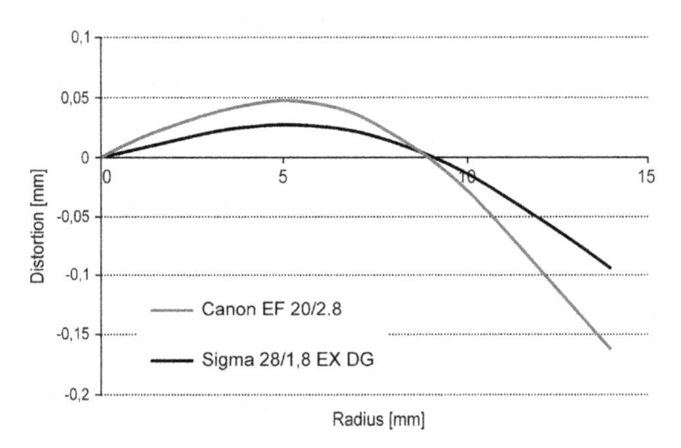

FIGURE 6-14 Typical radial distortions for two SLR lenses, resulting from the calibration described in Table 6-4. Both lenses show slight pincushion distortion near the image center (radius < 5 mm) and increasing barrel distortion toward the edges.

TABLE 6-4 Interior orientation parameters for two DSLRs focussed to infinity.

Interior orientation parameter	EOS 350D with Canon EF 20/2.8	EOS 300D with Sigma 28/1, 8 EX DG
Focal length	20.2207 mm	26.8299 mm
Principal point X	-0.100826 mm	0.035143 mm
Principal point Y	-0.009645 mm	0.171121 mm
Lens distortion parameter A1	-2.12E-04	-1.12E-04
Lens distortion parameter A2	4.0921E-07	1.89E-07

Calibrated by IM, M. Koch, and M. Kähler using the test field of Berlin's University of Applied Science and Pictran photogrammetry software.

For non-metric cameras, calibration reports are not provided by the manufacturer, but methods of camera calibration applicable to digital consumer-grade cameras have evolved rapidly over the last decades (Wackrow, 2008). Various calibration software, both commercial and non-commercial, exists as well as prefabricated 3D test fields with photogrammetric target points. However, these test fields are designed for close-range photogrammetry and usually are too small to be suitable for calibrating images focused at infinity (as in the SFAP case). For calibration of SFAP cameras, a larger test field is required, for example with high-precision targets mounted on the cornered walls and courtyard of a building. For small-format aerial photogrammetry, the commonly used alternative to test-field calibration is camera self-calibration during the actual project, where the elements of interior orientation are determined at the same time as the object points coordinates. The quality of the results, however, is highly dependent on the number, precision, and distribution of the ground control points involved (see Table 3-1).

Test-field calibration and field self-calibration procedures use the same concepts and methods as those outlined for object point reconstructions. The focal length, the position of the principle point and one to several lens distortion parameters are determined with iterative adjustment algorithms. Figure 6-14 and Table 6-4 show typical results for a consumer-grade DSLR camera.

6.7. SUMMARY

Any camera designed primarily for hand-held use on the ground may be adapted for small-format aerial photography. In spite of a tremendous range in cost and quality of such cameras, they all have certain basic components—lens, diaphragm, shutter, and an image sensor (film or electronic detector) within a light-proof box. Traditional cameras record photographs in the light-sensitive chemicals of the film emulsion. Spectral range is ~0.3–0.9 μm wavelength (near-ultraviolet, visible, near-infrared). Films of 35-mm and 70-mm formats are most commonly utilized for analog SFAP.

Digital cameras have come to dominate the market in the early twenty-first century, as cost has declined and quality has improved rapidly. Digital image sensors employ an array of tiny semiconductors to detect light intensity; two main types are the charge-coupled device (CCD) and the complementary metal oxide semiconductor (CMOS).

Three main controls of image exposure are the lens aperture (*f*-stop), shutter speed, and detector sensitivity (ISO rating). These factors are related by reciprocity, such that changing one variable by one stop can be matched exactly by changing another variable in the opposite direction by one stop. In most cameras for SFAP, automatic light settings are utilized. However, this may create problems for scenes with mixed illumination or highly contrasting bright and dark features. The geometrical and technical characteristics of the camera lens and sensor may have important influence on image distortions and artifacts.

Color-infrared photography normally utilizes the green, red, and NIR portions of the spectrum that are color coded, respectively, as blue, green, and red in the resulting false-color image. Both film and digital cameras may be utilized for color-infrared SFAP, although availability and processing of color-infrared film have become quite limited in recent years. The primary applications for color-infrared photography include vegetation types, soils, and water bodies.

Photogrammetric analysis of SFAP involves additional challenges. Among the important considerations are the camera lens, image sensor, file format, camera type, and camera calibration. In general DSLR cameras with single-focus lenses are most suitable for photogrammetric purposes and should be operated without image stabilization or dust removal functions.

Chapter 7

Camera Mounting Systems

Air photography is by no means simple. Much still remains to be done by way of adapting a camera to its peculiar demands.

(Lee, 1922)

7.1. INTRODUCTION

Camera mounts for small-format aerial photography (SFAP) vary tremendously depending upon the type of platform, camera, and functional requirements for camera operation. The mount could be as simple as a hand-held camera pointing out the window of a small airplane or helicopter, or as complex as a multi-camera array operating autonomously. The basic methods for mounting SFAP cameras apply to many types of manned and unmanned platforms and can be modified to fit a broad spectrum of flying machines (see Chapter 8). SFAP camera mounting is an undertaking with a great deal of experimentation and innovation during the past decade, particularly for unmanned platforms.

In addition to the primary imaging camera, many mounts include secondary cameras and ancillary equipment and functions, such as an altimeter, GPS unit, video downlink, and data logger. Each extra device adds weight, power, reliability, and integration issues for effective operation of the overall mounting system, and these complications should be taken into account when considering camera mounts to achieve the desired photographic results. SFAP mounting systems may be separate, attachable rigs, suspended from or quick-connected with the platform in various manners, or may be rigidly fixed on the outside surface or within the body of the flying machine. The mounts described in this chapter have been developed by the authors and their collaborators mainly for various types of unmanned platforms—blimp, kite, drone—in which the camera is controlled from the ground via a radio link or is programmed to function automatically.

7.2. CAMERA OPERATION

The main tasks of the mount are holding and operating the camera. The solutions to the latter may be as basic as a one-time mechanical trigger that has to be reset manually before each exposure (Fig. 7-1A, B), but in times of electricity and remote-control devices, repeated triggering via radio-controlled microservos (Fig. 7-1C) or electronic shutter release cables (see below) is the established standard. Many recent camera models (mostly digital single-lens reflex (DSLR) cameras) that support the picture transfer protocol (PTP) also offer remote capture via USB interface. This enables controlling virtually all camera functions (exposure intervals, shutter speed, aperture settings, ISO settings, white balance, etc.) using remote capture software via wireless connection to a field laptop or onboard computer.

Apart from image capturing functions, the mount may also control camera orientation, that is, the horizontal (pan) and vertical (tilt) position of the camera lens. Such variable camera orientation is usually only implemented in suspendable mounts (see below) and is less easy to realize for fixed mounts. Pan and tilt positions could be mechanically fixed for the duration of a photographic sortie, for example, in a vertical position, or may be variable via radio-controlled servos. The same remote-control device that is used for triggering the camera may be used to operate pan and tilt servos for camera orientation and, if applicable, for platform navigation functions in airplanes, helicopters, and hot-air blimps (see Chapter 8).

7.3. DETACHABLE MOUNTS

This category comprises camera mounts that hang below slow-moving or stationary (tethered or moored) platforms, such as blimps and kites. Detachable suspended camera rigs are necessary when the camera cannot be mounted into or onto the platform directly or must not move directly with a vibrating, swaying platform. Although it may seem easier to have a camera firmly fixed to the platform, removable mounts do have several advantages. Most importantly, they can be attached after the starting phase and detached before the landing phase of the platform, decreasing the crash risk for the camera. In addition, the camera position relative to the platform can be adjusted more freely with a separate rig than with fixed mounts. Finally, the camera rig may be used for several platforms and enable separate packing and maintenance of the sensor unit.

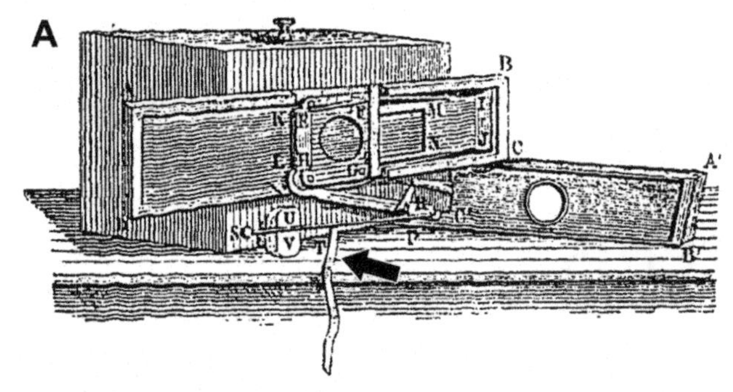

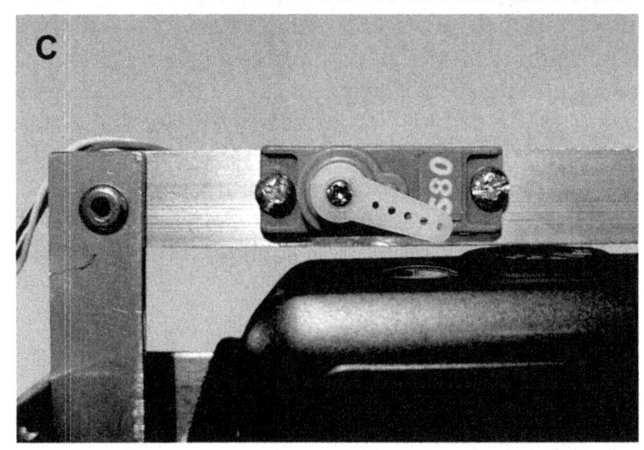

FIGURE 7-1 Various devices for triggering a camera. (A) Arthur Batut's wooden KAP camera with slow match wick (arrow) triggering mechanism (Batut, 1890, fig. 4). (B) Graupner Thermik timer (max. 6 min, here set to approx. 40 s) releasing a rubber-band trigger lever on a simple, electronic-free rig for disposable camera; built by IM after the design of "Brooxes Basic Brownie Box." (C) Radio-controlled microservo with lever for triggering a compact camera in the rig shown in Figure 7-3.

7.3.1. Single-Camera Suspended Rigs

Most single-camera mounting systems for kite or blimp aerial photography include the basic functions for camera position (pan and tilt) and shutter trigger, which are usually controlled by radio from the ground. In this relatively simple approach, no video downlink or other onboard equipment is involved in the mounting system. The camera position (pan and tilt) is estimated by visual observation of the camera from the ground, usually with binoculars, and by radio-control settings. Many pictures are taken to insure that the ground target is fully covered from appropriate viewing angles.

A typical, double U-shaped camera mount includes a cradle to hold the camera, a frame in which the cradle can be moved, and small servos to accomplish the movements (Fig. 7-2). Such a mount for a relatively small compact camera normally weighs around 0.5–1.0 kg, including the camera, batteries, and other components, with weight depending mainly on the specific camera model. Aluminum is the usual building material for the frame and camera cradle, although titanium, plastic, wood, fiberglass, or other materials may be used for lighter or stronger components (Fig. 7-3).

Another configuration for this lightweight approach involves a single vertical post as the main architectural element for the camera mount (Fig. 7-4). This rig has radio control of camera pan, tilt, and shutter trigger. The tilt servo and battery pack act as a counterbalance for the camera tilt mechanism.

The lighter-is-better theme reaches its ultimate development with a small electronic chip that is pre-programmed to change the camera position and trigger the shutter automatically. This may be done in a systematic manner or could be random in position and timing. In either case, this method completely removes any in-flight ground control of camera

FIGURE 7-2 Typical mounting system for a small digital camera. Aluminum frame and cradle held by a Picavet suspension. R, radio receiver and antenna mast (yellow); P, pan servo and gears; B, nickel-metal-hydride battery pack; T, tilt servo; and S, shutter-trigger microservo. Total weight of this rig with camera is just 0.6 kg. Originally built by B. Leffler (California, United States) with extensive modifications by JSA.

FIGURE 7-3 Mounting system for a mid-sized digital camera. Aluminum cradle for the camera is held in a titanium frame. The pan (P) and tilt (T) servos are relatively robust devices, and the shutter is triggered by a microservo (S). Total weight of this rig with camera is ~0.8 kg. Originally built by B. Leffler (California, United States) with extensive modifications by JSA.

FIGURE 7-5 Mounting system for a digital SLR camera. Aluminum frame and cradle with pan and tilt controls similar to previous examples. The primary difference here is an electronic interface to control the camera shutter (S), which eliminates one servo. In this configuration total weight of the rig and camera is 1.3 kg; with a larger lens, weight may exceed 1.5 kg. Rig built by B. Leffler (California, United States).

operation, which means that a radio receiver and antenna are not necessary on the camera rig.

Single-lens reflex (SLR) film or digital cameras are normally larger in size and heavier than point-and-shoot cameras and, thus, require larger mounting systems with stronger servos, bigger batteries, and other components. Total weight is typically >1.2 kg. Lens interchangeability means that longer lenses and filters may be employed. In fact, a large lens could be the heaviest single component of the camera rig, which might substantially affect the balance of the mounting system. In order to save moving parts that need precise adjustment, the shutter servo and lever may be replaced with an electronic shutter cable connected to a radio-controlled microswitch (Fig. 7-5). Alkaline camera batteries could be replaced with lithium-ion batteries, depending on camera models; lithium batteries weigh about 40% less than alkaline and provide longer-lasting power.

Primary design criteria for the camera mounts presented thus far are lightweight, rugged components, and reliable camera operation. One consequence of this approach is that all components are exposed with little or no protection to the elements. The mount, electronic parts, and camera are potentially vulnerable to dust, debris, and water as well as possible damage during impacts with the ground or obstacles (Fig. 7-6). In actual practice, the authors have experienced only a few mechanical failures or damage for these types of camera mounts.

In some situations, however, more robust mounting systems may be favored in order to protect the camera

FIGURE 7-4 Single-post mounting system for a small camera. P, pan servo and gears; A, antenna mast and radio receiver; S, shutter microservo; T, tilt servo and battery pack. Total weight of this rig with camera is less than 0.6 kg. Rig built by B. Leffler (California, United States).

FIGURE 7-6 Results of a hard crash of the rig pictured in Figure 7-3. Tilt servo gear broken, titanium frame bent, and antenna wire (yellow staff) pulled out of the radio receiver. The camera and kite were undamaged, and the broken rig could be repaired in the field. Photo by JSA, May 2009.

FIGURE 7-7 Two mounting systems with pan and tilt functions for digital SLR cameras designed for robustness, increased protection of all parts, and minimal requirements for disassembly. Frame and cradle from aluminum, camera triggered by electronic remote-control release. Rigs built by the technical workshop staff of Frankfurt University's Faculty of Geoscience and Geography.

apparatus better from blowing dust, salt, or sand as well as other harsh environmental elements or difficult flying conditions. Figure 7-7 shows two sturdy, warp-resistant SLR rigs with pan and tilt functions, electronic shutter release, and heavy-duty batteries; both weigh > 2 kg with camera. Battery packs, radio receiver, microswitches, and in rig B also the pan servo, are enclosed in boxes for protection against dust, sand, and dampness. The servo and radio-receiver batteries can be switched off when unused and charged without removing via the socket outlets, minimizing the need for disassembly and handling wear for delicate parts. These rigs have been used successfully for many years by IM and JBR in semi-arid and arid conditions, mastering several near-crash situations without damage.

7.3.2. Multiple-Camera Suspended Rigs

For some purposes, a single camera is not enough. Dual- or even multiple-camera mounts may be used for simultaneous images either in various spectral ranges, with different focal lengths/image scales, or from different vantage points for stereo imagery. Multiple-sensor mounts also may combine cameras with such non-imaging spectral measurement devices as spectroradiometers or thermal-infrared sensors in order to collect multispectral information about the viewing target (e.g. Vierling et al., 2006). Such rigs are by necessity larger in size and heavier than single-camera mounts, which makes flying them potentially more difficult. In order to minimize weight, some functions and extra devices may be omitted as described for selected mounts below.

For simultaneous stereo SFAP, two cameras need to be mounted some distance apart. The length of the boom determines the air base B that together with the flying height H_g control the base–height ratio and thus the amount of stereoscopic parallax (see Chapter 3)—the longer the boom, the better is 3D perception. In the example shown in Figure 7-8, the boom position is always mounted parallel to the kite line or blimp keel, in order to minimize wind resistance, and camera tilt angle is set on the ground prior to

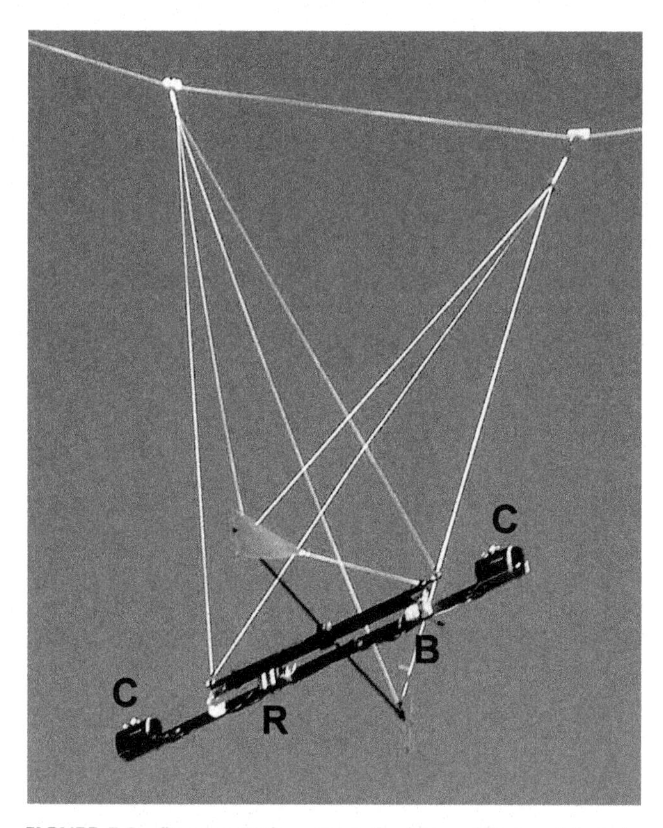

FIGURE 7-8 Stereo-mounting system for kite or blimp aerial photography. The boom is ~1 m long with cameras (C) at each end. The small dihedral wings (red) are designed to keep the rig parallel to the wind. R, radio receiver; B, battery pack. Total weight of this rig and cameras is ~0.9 kg. Rig built by B. Leffler (California, United States).

each flight. Once in flight, tilt and pan positions of the cameras cannot be changed. Microservos trigger each camera simultaneously to acquire pairs of overlapping images (see Fig. 2-9).

For multispectral systems, two cameras are commonly mounted side-by-side with minimal separation between the lenses. A typical configuration is for one camera to take color-visible or panchromatic photographs, while the other camera takes simultaneous color-infrared images (e.g. Jensen et al., 2007; Verhoeven, 2008). In the example shown in Figure 7-9, camera pan and tilt positions are set on the ground prior to each flight and cannot be changed once the rig is in the air. The camera shutters are triggered electronically to produce dual images of identical scenes (see Fig. 2-5). As the two images are usually to be combined into a single multiband image for analysis, careful mounting of the two cameras is important. Any further separation and any relative tilt between the cameras increase the parallax between the two images, which in this case should be avoided. Multiband imagery may be especially useful for certain applications involving vegetation, soils, and other environmental features.

FIGURE 7-9 Dual-camera rig for kite or blimp aerial photography. Two SLR film cameras are mounted bottom-to-bottom for simultaneous pictures of the same scene in color-visible and color-infrared formats. Total weight of this rig and cameras is 1.5 kg. Rig built by B. Leffler (California, United States).

7.3.3. Attaching Suspendable Mounts to a Platform

Suspendable camera mounts are connected to the platform via lines or poles that are free to swing, pivot, or flex, so that the camera is removed to some extent from erratic motions and vibrations of the lifting platform. Ideally the mounting system should hang in a stable, level position. Two different solutions are the pendulum and Picavet suspensions. Both may be combined with a cable-car system in kite aerial photography, or directly attached to a kite line or to a balloon or blimp envelope.

The Picavet suspension is a string-and-pulley cable system that attaches to a kite line or a blimp keel at two points (Fig. 7-10). It was devised by a Frenchman, Pierre Picavet, in 1912 as a self-levelling platform for a camera rig suspended from a kite (Beutnagel et al., 1995). Various methods may be used for threading the line through the pulleys or eye bolts of the Picavet cross (e.g., Hunt, 2002; Beutnagel, 2009) as well as various means of attaching the two end points to a kite line (Fig. 7-11). A typical Picavet suspension hangs 1–2 m below the kite line or blimp keel, and the two attachment points are spaced a similar distance apart, resulting in a triangular arrangement.

The other main suspension type is the pendulum, which may be an aluminum staff, carbon rod, or even flexible wire. Figure 7-12 shows the pendulum suspension used by IM and JBR together with a sledge-and-pulley system running on the kite line (see Chapter 8.4.3). Although a Picavet suspension may be attached to such a sledge as well, the resulting comparatively short distance between the suspension points to some degree decreases the stabilization

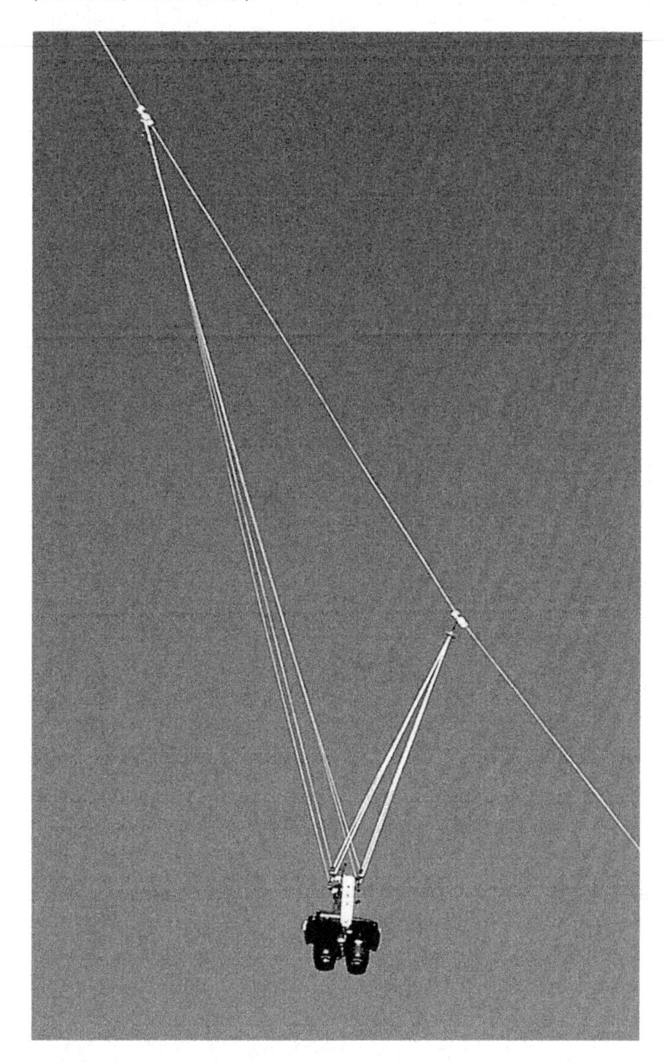

FIGURE 7-10 Picavet suspension system attached to kite line at two points. The rig holds dual SLR cameras shown in previous figure. Photo by JSA; see also Figure 7-8.

FIGURE 7-11 Simple methods for attaching a Picavet suspension to a kite line without the need to tie a knot. (A) Line wrapped around heavy bent wire. (B) Brooxes Hangup™, line wrapped around a plastic block, which is ~5 cm long. Photos by IM and JSA.

effect. There are two great advantages of the sledge system: (a) launching and landing of the kite can take place without the camera attached, and (b) the camera can be retrieved for changing exposure settings, film, storage card, or lens without having to bring the kite down.

By grievous experience, the authors have learned that aluminum is a good electrical conductor. Mysterious camera malfunctions, which were observed during several surveys in hot and dry environments, were finally found to be caused by electrostatic charging from the friction of the sledge wheels on the kite line. This problem was resolved by replacing the original aluminum wheels with ceramic wheels and inserting a plastic barrier into the pendulum staff.

FIGURE 7-12 A pendulum suspension system for kites with camera sledge made from aluminum. Sledge wheels made from ceramics, aluminum pendulum staff connected by hard plastic block to break electrical conductivity between kite line and camera rig.

7.3.4. Detachable Modular Unit Mounts

All rigs shown so far are suitable for suspension from kites, balloons, and blimps. Such free-swinging mounts are useful to counterbalance sudden movements or vibrations of a platform, but a firmer attachment of the camera may offer more stability for slow as well as fast platforms. An alternative combining the advantages of both detachable and fixed mounts are semi-fixed modular units where the complete camera mount is quick-connected with the platform.

The hot-air blimp presented in Chapter 8 is operated by a burner system suspended from a metal frame that offered itself as a receptacle for such a plug-in camera unit. Both single-camera (SLR or medium-format type) and double-camera cradles can be used (Figs. 7-13 and 7-14). The double-camera system may be used for multispectral imagery or different focal lengths. The cradle, which also supports a small video camera downlinked to a portable display, is suspended from a vibration-damped cardan joint fixed to a 360° radio-controlled turntable. This rotatable camera mounting is necessary because the blimp always aligns itself with the wind direction but the desired image orientation on the ground may be different. Polystyrene balls are fixed to the ends of two aluminum wires and are fastened to the camera cradle so they protrude from the protection box parallel to the long image axis (see also Fig. 8-17). These "direction signs" are quite visible from the ground even at high flying heights. This is particularly important if the same area needs to be photographed repeatedly in a multi-annual monitoring.

The cameras are protected by an open box or basket, which is inserted into the burner frame like an upside-down

FIGURE 7-13 Upside-down view of a box-type camera mount built from aluminum and polystyrene, designed as a plug-in module for the hot-air blimp presented in Chapter 8. V, video camera as navigation aid; T, pan-servo turntable; P, plug establishing connection to the batteries and radio receiver (see Fig. 7-15). Mounting system built by the technical workshop staff of Trier University's Faculty of Geography and Geoscience.

FIGURE 7-14 Camera mount similar to Figure 7-13, but for a double-camera system with simultaneous image capture (small video camera may be attached to the left side of the rig). Note the two aluminum booms indicating the orientation of the image format to the photographer on the ground. Basket box made from plywood, wicker, and aluminum. Mounting system originally built by *GEFA-Flug GmbH*, with extensive modifications by the technical workshop staff of Frankfurt University's Faculty of Geoscience and Geography.

FIGURE 7-15 The camera mount unit, secured by a locking pin, slides into rails fixed to the blimp burner frame, plugging into the battery and radio receiver connection.

FIGURE 7-16 Simple gimbal mount for a free-flying paraglider with combustion engine. R, pivoting axis in flight direction (roll); N, brackets swinging in flight direction (nick); A, oil-pressure shock absorbers; S, shutter-trigger microservo; M, motor. Drone frame and mount built by *ABS Aerolight Industries*, photo by V. Butzen and G. Rock.

cupboard drawer (Fig. 7-15). When pushed home, a plug establishes connection to the batteries and radio receiver. Owing to the recessed and protected camera position, oblique images are not possible with this mount, which is specially designed for vertical imagery. The verticality of the weight-balanced cradle is guaranteed by gravity. Here, also, the great advantages of the detachable mount are the lower risk for the camera system during the launching and landing phases (the camera mount unit is attached and removed when the blimp is floating just 1–2 m above the ground) and the possibility of quick removal for changing exposure settings, film, memory card, or lenses.

7.4. FIXED MOUNTS

Suspended or semi-fixed mounts are not suitable for free-flying, aerodynamic SFAP platforms if they jeopardize the stability of the platform, swing with the platform movements, or take too much room. The autopiloted model airplane and the paraglider (see Chapter 8) feature fixed mounts, because they fly too fast for suspended rigs. Also, a modular plug-in mount would not make sense for free-flying platforms, as it could not be attached after launching.

Two problems need to be addressed for achieving sharp images with a fixed mount on motorized platforms: ensuring verticality or other intended angles of the images and avoiding vibration transmission from the motor. The powered paraglider (Chapter 8.5.3) is an aerodynamic platform with considerable accelerating force and rather variable flight attitude; thus, it calls for a well-damped, self-balancing camera mount in order to avoid heavy vacillations. Figure 7-16 shows the gimbal-mounted system constructed from a suspension staff pivoting on the flight-direction axis (roll) and swinging between double brackets in flight direction (nick). Two oil-pressure shock absorbers (model-making supply) ensure limited and slow swinging, which is especially useful regarding the fitful and jerky flight behavior of the drone. The camera is fixed to the swinging pipe by the tripod screw together with an aluminum bracket supporting a shutter-trigger microservo. Despite the gimbal mount, many instances happen during a flight when the camera is not hanging vertically, and many images turn out as low-oblique shots. This is best compensated for by repeated overflights of the area.

The paraglider is powered by a gasoline engine whose vibrations are inevitably transmitted to the mount and camera, making it difficult to achieve sharp photographs. Vibration absorbers are needed between flying machine, camera mount, and camera. Here, all joints between motor and motor bracket and between frame, shock absorbers, mount, and camera are furnished with small rubber grommets. This multiple damping has proven to allow focused images in spite of the motor vibrations, although the number of blurred images remains considerable. As the motor revolution speed and flight attitude change during the flight, so do vibrations, and complete absorption would require much more sophisticated cushioning.

FIGURE 7-17 Foam-padded DSLR camera (*Canon* EOS 300D with 28 mm *Sigma* lens) fitted into the body of the autopiloted model airplane shown in Figure 8-42. Port (left) side of plane removed; camera lens pointing through hole in the bottom. Photo by C. Claussen.

For a model airplane, room in the hull is usually too confined for a gimbal mount with pan and tilt functions, so verticality of the images depends on the ability to keep the flight attitude steady. Figure 7-17 is a view into the body of *der Bulle* model airplane shown in Figure 8-42. The existing hollow in the Elapor material was enlarged to accommodate the DSLR camera damped with foam rubber. The camera is triggered via an electronic shutter release cable by the onboard computer in a pre-defined time interval, which may be adjourned if a chosen tilt-angle limit is exceeded (see Chapter 8.5.2). The camera battery is replaced by a connection to a larger 11.1 V battery that also supplies the electric motors and GPS/INS; this has the advantage of better power management, fewer charging tasks and accessibility of the battery from the airplane cockpit.

Electric motors as used in model airplanes do not cause extreme vibrations if they are well balanced. Because of the appreciable weight of the camera, the foam padding is sufficient to avoid propagation of motor vibrations; however, the smaller airplane with the lightweight compact camera (see Fig. 8-42) is somewhat more prone to vibration-blurred images.

7.5. SUMMARY

Camera mounts serve for attaching one or more cameras or other sensors to a platform and operating them. The most basic function is triggering the camera, usually with remote-controlled microservos or electronic shutter release cables, and additional functions may include pan and tilt orientation, remote image capture, and control of ancillary devices such as GPS, altimeter, or video eye.

Small-format aerial photography (SFAP) mounting systems may be rigidly fixed or separate, detachable mounts, where the camera is not directly fixed to the platform but in some way suspended or quick-connected. Suspended rigs are used for slow-moving or stationary platforms such as kites, balloons, or blimps; they are usually attached after launching and thus decrease the crash risk for the camera. Detachable modular mounts are a convenient solution for slow-moving platforms carrying a rigid framework (e.g., hot-air blimps or balloons) to which they may be quickly connected before takeoff. Fixed mounts, on the other hand, are necessary for fast-moving aerodynamic platforms (all types of drones) that must not be burdened with swinging, suspended payloads.

Most mounts feature some sort of self-balancing mechanism, keeping the camera orientation vertical or at another intended angle. This can be accomplished with self-balancing pendulums, Picavet suspensions, gimbal mounts, or cardan joints. Dual- or even multiple-camera mounts allow the simultaneous capture of two or more images either in various spectral ranges, with different focal lengths/image scales, or from different vantage points for stereo imagery.

Chapter 8

Platforms for Small-Format Aerial Photography

One sky, one world: the wind knows no borders, and the molecule that hits my kite today was probably flying over Chile yesterday and will be in Mongolia next week.

N. Chorier (Chorier and Mehta, 2007)

8.1. INTRODUCTION

Mankind has devised diverse kinds of flying machines that reach into the atmosphere and low-space environment. Since the Chinese invention of kites centuries ago, an irresistible urge has led people to fly higher and faster above the surface of the Earth. Ranging in size and complexity from small paper kites to the International Space Station, virtually all these flying platforms may be adapted for small-format aerial photography (SFAP). The emphasis here is on relatively low-flying platforms using unconventional aircraft, both manned and unmanned.

Platforms fall into several primary categories—manned or unmanned, powered or unpowered, and tethered to the ground or free flying. The number of possible combinations is quite large. For example, balloons may be manned or unmanned and tethered or free flying; helicopters vary from large manned vehicles to small unmanned aircraft. The term drone is generally applied to unmanned, powered, free-flying platforms. Other terms referring to small, remotely controlled aircraft include UAV (unmanned aerial vehicle), MAV (micro air vehicle) and RPV (remotely piloted vehicle).

The most basic distinction is whether the platform is manned or unmanned. The former is necessarily large enough to lift a person safely along with photographic equipment. The latter may be relatively small. Manned lifting platforms, such as airplanes, helicopters, hot-air balloons, and ultralights, are fairly expensive to operate and normally require a trained pilot, ground-support crew, and some kind of launching and landing facilities. Unmanned platforms are much more variable in their technical specifications. The requirements for lifting capability and platform safety are much less stringent with unmanned platforms; the cost and technical expertise needed to operate such systems also vary greatly. Unmanned platforms may operate in an automated fashion once airborne, or they may be controlled remotely by a person on the ground.

A further distinction can be made between powered and unpowered platforms. The former become airborne through some kind of artificial thrust provided by a motor or engine. Airplanes, autogyros, helicopters, and rockets utilize a variety of wings, blades, rotors, and fins to create lift and/or stabilize the vehicle in flight. These aircraft move with moderate to high velocity; even the helicopter that appears to hover is in fact rotating its blades rapidly against the air. All powered platforms vibrate and move relative to the ground.

Unpowered platforms achieve their lift either through neutral buoyancy (balloons and blimps) or via resistance to the wind—kites and sailplanes. Regardless of the kind of platform, any vibration or movement of the camera relative to the ground creates the potential for blurred imagery. Balloons, blimps, and gliders that drift with the wind move slowly relative to the ground and have minimal mechanical vibration. Tethered platforms—balloons, blimps, and kites—tend to vibrate and swing with the wind. The following sections, which present selected examples of SFAP platforms, also include discussions on their characteristics affecting image acquisition, quality and other properties, and exemplary uses taken from the literature.

8.2. MANNED LIGHT-SPORT AIRCRAFT

The arena of gliders and ultralight aircraft, rich with experimentation since the 1960s, has evolved into light-sport aircraft (LSA). For innovative pilots, mechanics, and photographers, LSA has become the successor of earlier *Cessna, Piper,* and other small airplanes of the mid-twentieth century. LSA is defined by the U.S. Federal Aviation Agency (FAA) according to these criteria, which apply to both powered and unpowered vehicles.

- Maximum takeoff weight of not more than: (a) 660 pounds (300 kg) for lighter-than-air aircraft, (b) 1320 pounds (600 kg) for aircraft not intended for operation on water, or (c) 1430 pounds (650 kg) for an aircraft intended for operation on water.
- Maximum airspeed in level flight with maximum continuous power (VH) of not more than 120 knots CAS under standard atmospheric conditions at sea level.
- Maximum never-exceed speed (VNE) of not more than 120 knots CAS for a glider.
- Maximum stalling speed or minimum steady flight speed without the use of lift-enhancing devices (VS1) of not more than 45 knots CAS at the aircraft's maximum certificated takeoff weight and most critical center of gravity.
- Maximum seating capacity of no more than two persons, including the pilot.
- Single, reciprocating engine, if powered.
- Fixed or ground-adjustable propeller if a powered aircraft other than a powered glider.
- Fixed or autofeathering propeller system if a powered glider.
- Fixed-pitch, semi-rigid, teetering, two-blade rotor system, if a gyroplane.
- Non-pressurized cabin, if equipped with a cabin.
- Fixed landing gear, except for an aircraft intended for operation on water or a glider.
- Fixed or repositionable landing gear, or a hull, for an aircraft intended for operation on water.
- Fixed or retractable landing gear for a glider.

FIGURE 8-1 *Challenger* II light-sport aircraft (LSA) for potential small-format aerial photography. (A) Aircraft is light enough for one person to move it on the ground. (B) Overhead wings and large open windows allow good views for the photographer. Photographs ©W.S. Lowe; used here with permission.

8.2.1. Powered Light-Sport Aircraft

LSA are supplied nowadays mainly by small companies. The relatively low costs of purchasing, building, and operating manned LSA have proven popular. One suitable model, for example, is the *Challenger* II produced by Quad City Challenger of Moline, Illinois, United States. It is a two-seat LSA and ultralight trainer with an overhead wing and pusher motor (Fig. 8-1). The long-wing version increases the glide ratio and is the optimum configuration for stable, slow-speed SFAP. Wheeled or float landing gear allow dry-land or amphibious operation, respectively.

The SFAP capability of the *Challenger* II was field tested over the Ocala National Forest near Deland in central Florida, United States. An experienced pilot flew the LSA, and W.S. Lowe was the co-pilot/photographer. He employed a hand-held *Canon* A590 IS, 8-megapixel camera, with 4× zoom lens and optical image stabilization. The flightpath was designed to place the sun behind the plane and photographer for oblique views toward ground features of interest (Fig. 8-2). This test flight demonstrated the advantage of manned SFAP for selecting targets of opportunity and positioning the aircraft to best advantage for acquiring suitable photographs.

FIGURE 8-2 Oblique view of a glider airpark and sports complex near Pierson in central Florida, United States. Taken from the *Challenger* II light-sport aircraft at a height of ~340 m (~1100 feet). Image ©W.S. Lowe; used here with permission.

Among the most daring SFAP conducted from manned LSA is the motorized paraglider operated by G. Steinmetz of the United States. His flying machine consists of a large parafoil kite from which he hangs with a small gasoline-powered propeller strapped to his back (Tucker, 2009). The whole contraption weighs less than 45 kg and is highly portable. Steinmetz has flown extensively in Africa and China, although he has suffered some serious crashes that resulted in wrecked aircraft and personal injury (Anonymous, 2008). In spite of his undoubted photographic success, it seems unlikely that many others would follow Steinmetz's lead for this high-risk means of manned SFAP.

8.2.2. Unpowered Light-Sport Aircraft

Unpowered LSA include manned gliders, also known as sailplanes, which may be launched via ground or aerial towing to achieve sustainable height for continued flight. Closely related manned aircraft include hang gliders, paragliders, motorgliders, and autogyros. Gliders are essentially large, manned, untethered kites. Modern gliders are highly sophisticated aircraft that have evolved during the past century, just as powered airplanes have developed. Soaring is a popular sport, and many associations and museums exist around the world to support this type of flight.

As platforms for SFAP, gliders have considerable potential compared with other types of manned aircraft. The most important advantages are maneuverability combined with quiet operation. For two-seat gliders, the co-pilot can concentrate on photography while the pilot rides the air currents (Fig. 8-3). Pictures with hand-held cameras are mostly limited to oblique views (Fig. 8-4), unless the glider is put into a steep roll for a look straight down. A remotely operated camera rig could be placed in a special compartment in the nose or body of the glider, which would provide more control for aiming the camera. The main disadvantage of using gliders, aside from the high cost associated with most manned aircraft, is their need for towing to become airborne. Aerial range is variable, depending on weather conditions, but gliders can be transported effectively on the ground in specially built trailers (Fig. 8-5).

FIGURE 8-3 Two-seat gliders. (A) Conventional sailplane on the ground. Overhead wing allows good side and downward views from the second seat. (B) High-performance glider during pre-flight preparations. Wing to rear of seats provides great lateral views. Note small windows in canopies; windows can be opened during flight for taking photographs. Photos by JSA and SWA, July 2005, Harris Hill National Soaring Museum near Elmira, New York, United States.

Various uses of manned LSA are documented in the literature spanning a wide range of applications. A microlight aircraft was used by Mills et al. (1996) as a slow-flying alternative to faster light aircraft, allowing them to achieve the necessary stereoscopic overlap for photogrammetric aerial surveys in Great Britain. Henke de Oliveira (2001) employed SFAP taken from a light aircraft for analyzing urban land use, vegetation, and settlement patterns in Luis Antônio City, Brazil. For precision farming and urban surveys in Germany, Grenzdörffer (2004) developed a digital airborne imaging system capable of direct georeferencing, which is based on a global positioning system (GPS) linked with a digital single-lens reflex (DSLR) camera mounted in a *Cessna*.

A relatively conventional approach for manned SFAP was employed by Li, Li et al. (2005) for high-resolution imagery and analysis of informal settlements or shantytowns in South Africa. They used a *Piper Arrow* 200 light aircraft, in which a custom-built camera mount was fitted into the porthole below the passenger seat. Imagery was collected with a *Kodak* DSC460c color digital camera at a flying height of 520 m, which resulted in ground sample distance (*GSD*) of

FIGURE 8-4 Low-oblique view of drive-in movie theater with the Chemung River in the background. Few active outdoor movie theaters are still operating in the United States. Photo taken from glider shown in Figure 8-3B by JSA, July 2005, near Elmira, New York, United States.

FIGURE 8-5 Trailer for transporting a glider cross country. Photo by SWA, July 2005, Harris Hill National Soaring Museum near Elmira, New York, United States.

0.18 m. The U.S. Department of Agriculture, Rangeland Resources Research Unit in Cheyenne, Wyoming, has developed a manned ultralight airplane to acquire very large-scale aerial imagery (Hunt et al., 2003). The aircraft is a *Quicksilver* GT 500 single-engine, single-pilot plane that flies a mere 6 m above the ground. The ultralight is equipped with a laser altimeter for precision height measurement. Various high-speed film and digital cameras are triggered by a computer interfaced with a GPS unit according to a pre-programmed flight plan and photo coordinates.

8.3. LIGHTER-THAN-AIR PLATFORMS

Lighter-than-air platforms comprise various balloons and blimps that may be manned or unmanned, tethered or free flying, and powered or unpowered. Contrasting with the spherical shape of a balloon, a blimp has an elongated, aerodynamic shape with fins. Strictly speaking, the term blimp refers to a free-flying airship without internal frame structure, which is kept in shape by the overpressure of the lifting gas. However, the term blimp is often, as in this book, also used for tethered or moored airships, including those where the lifting medium is hot air.

Balloons and blimps are widely employed nowadays for meteorological sounding, commercial advertising, sport flying, and other purposes. Manned, hot-air balloons are, in fact, quite popular (Fig. 8-6), but lack positional control unless tethered. Near-calm conditions are necessary for launching and landing, which often restrict flights to dawn and dusk times of day with consequences for lighting conditions (Fig. 8-7).

Owing to their comparatively low level of high-tech components, unmanned, tethered lighter-than-air platforms have been employed for SFAP for many decades. Their suitability for aerial surveys using various instruments from compact cameras to multi-sensor systems is documented by a wide range of publications. Among the most basic versions, a simple plastic bag containing the lifting gas was employed by Ullmann (1971) for SFAP of bogs in Austria. Examples for the use of helium balloons include the coastal and periglacial geomorphology studies by Preu et al. (1987) and Scheritz et al. (2008), the photogrammetric documentation of archaeological sites by Altan et al. (2004) and

FIGURE 8-6 Hot-air balloon, *Dragon Egg*, of Jane English lifting off from a field near Mt. Shasta, northern California, United States. This is a relatively small balloon that can carry four people. Photo by JSA, May 1999.

FIGURE 8-7 Early morning picture looking south toward Mt. Shasta taken from the hot-air balloon (in previous figure) about 150 m high above Shasta Valley, northern California, United States. Note morning fog in the distance and heavy shadowing of ground. Photo by JSA, May 1999.

Bitelli et al. (2004), and vegetation studies in wetlands and semi-arid shrubland conducted by Baker et al. (2004) and Lesschen et al. (2008).

For an investigation on salt-marsh eco-geomorphological patterns in the Venice lagoon spanning different spatial scales, Marani et al. (2006) suspended a double-camera system (VIS and NIR) from a helium balloon operating at low heights of around 20 m, and a similar system is used by Jensen et al. (2007) for wheat crop monitoring. In order to sidestep costly satellite imagery and conventional aerial photography with a low-cost remote-sensing method applicable in developing countries, Seang and Mund (2006) have used SFAP taken from a hydrogen balloon for various applications in regional and urban planning, monitoring of degraded forest, and basemap compilation for infrastructure projects in Cambodia.

Tethered spherical balloons are, unfortunately, highly susceptible to rotational and spinning movements (Preu et al., 1987). For SFAP, more aerodynamically shaped airships, namely blimps, are preferable because they align themselves with the wind direction. As an alternative to spherical balloons, various researchers have employed helium blimps: Pitt and Glover (1993) for the assessment of vegetation-management research plots, Inoue et al. (2000) for monitoring various vegetation parameters on agricultural fields, Jia et al. (2004) for evaluating nitrate concentrations in agricultural crops, and Gómez Lahoz and González Aguilera (2009) for 3D virtual modelling of archaeological sites. The suitability of blimps even for heavy high-tech sensors at greater flying heights (up to 2000 m) was proven by Vierling et al. (2006), who used a 12-m helium blimp as a multi-sensor platform for hyperspectral and thermal remote sensing of ecosystem level trace fluxes.

Hot-air blimps as camera platforms have been employed for many years by two of the authors for investigations on soil erosion and vegetation cover (Marzolff and Ries, 1997, 2007; Marzolff, 1999, 2003; Ries and Marzolff, 2003). The same hot-air blimp and its prototype predecessor were used for archaeological applications documented by Hornschuch and Lechtenbörger (2004) and Busemeyer (1994).

Because of their lightweight and comparatively large surface area, lighter-than-air platforms, especially balloons, are relatively difficult or impossible to fly in windy conditions. This wind susceptibility is even increased and thus positively exploited by the so-called *Helikite*, a hybrid of helium balloon and kite combining the advantages of both platforms. Examples for investigations conducted with this rather unusual platform are given by Verhoeven et al. (2009) for aerial archaeology and Vericat et al. (2009) for monitoring river systems.

8.3.1. Lifting Gases

Several lighter-than-air gases could be employed as the lifting medium—hydrogen (H_2), helium (He), methane (CH_4), and hot air (Federal Aviation Administration, 2007). Hydrogen is the lightest of all gases and is used occasionally for SFAP (Keränen, 1980; Gérard et al., 1997; Seang et al., 2008). However, hydrogen and methane are both explosive and highly flammable; they are not considered further for obvious safety reasons, which leaves helium and hot air as the gases of choice for most balloon and blimp applications.

Helium has a lifting capacity of about 1 g/L; whereas, hot air lifts only about 0.2 g/L. Thus, for a given volume, helium lifts approximately five times more weight than does hot air. In addition to a larger and heavier envelope, a hot-air platform also needs to carry a gas tank and burner in order to stay aloft for more than a few minutes. This added weight reduces the potential payload a hot-air system could carry.

Helium is created as a byproduct of radioactive decay within the solid Earth. Continental crust, which is enriched in uranium and other radioactive elements, is a constant source for helium. Because it is inert, helium does not combine with minerals in the crust, but it does readily dissolve into fluids such as ground water and natural gas; the latter typically contains 0.2–1.5% He by volume. Eventually the helium reaches the surface, for example in hot spring water (Persoz et al., 1972), and is released into the atmosphere. Earth's gravity is too weak to retain the helium molecule (single He atom), so it ultimately escapes into space.

Helium was little known prior to the twentieth century. This changed with the discovery that helium is a significant component of natural gas in some situations. Throughout most of the twentieth century, helium was regarded as a strategic resource for military and industrial purposes in

FIGURE 8-8 Sign at entrance to helium plant. Crude helium is piped from natural gas fields in Texas, Oklahoma, and Kansas for further purification and liquefaction at processing plants in central Kansas, United States. Photo by JSA, May 2007.

the United States. In the 1990s, however, all helium production was privatized (Natural Academy of Sciences, 2000). Nowadays helium extracted from natural gas is the only commercial source, and the United States is the major supplier (Fig. 8-8). Helium is also produced in Algeria, Poland, Qatar, and Russia. As an industrial commodity, compressed helium is widely available at modest cost in the USA in steel cylinders that can be purchased or rented from gas distributors (Fig. 8-9).

FIGURE 8-9 Helium tank and balloon filler valve. (A) Helium cylinder mounted on a hand truck for easy transport. This tank weighs ~65 kg when full and contains ~7 m^3 of helium. It is shown here with the safety cap in place (top of tank). The cap is required whenever the tank is transported or stored. (B) Close-up view of valve and nozzle for inflating balloons or blimps. The valve is opened by bending the black tip. This valve must be removed for transportation and storage of the tank. Photos by JSA.

Helium has significant lift and definite safety advantages compared with hot air, but helium is either quite expensive or simply not available in many countries around the world. Conversely propane, cooking gas, or other types of natural gas for firing a hot-air balloon or blimp can be found just about everywhere. Because commercial gas tanks are too large and heavy for model airships or do not allow extracting liquid gas, they need to be decanted into special gas bottles. It has proven extremely important to use a fine-meshed gas filter (Fig. 8-10) in this process to prevent the transfer of small dirt and rust particles from commercial gas tanks into the flight bottles, where the debris can turn into a severe hazard when blocking valves and burners. The difficulties the authors initially had with such impurities were numerous and endangered several field surveys. Filtering is also advisable for the gasoline used as fuel for small combustion engines. Even if the degree of fuel purity is standardized nowadays, old tanks and barrels are often rusted inside, and a spluttering motor in the air may terminate not only the flight but also the existence of the aircraft and even the camera.

8.3.2. Helium Blimp

The small helium blimp used by one of us is 4 m long and has a gas capacity of ~7 m^3 (Aber, 2004). The blimp has a classic aerodynamic shape with four rigid fins for stability in flight. It has a payload lift of ~3 kg, which is 2–3 times the weight of camera rigs and has proven to be an adequate margin of safety for blimp operation. The camera rig, which is the same system utilized for kite aerial photography (KAP) (see Chapter 7), is attached to a keel along the bottom of the

FIGURE 8-10 Small gas filter used for preventing dirt and rust particles from entering the special hot-air blimp gas bottle during filling from commercial gas tanks. Photo by IM.

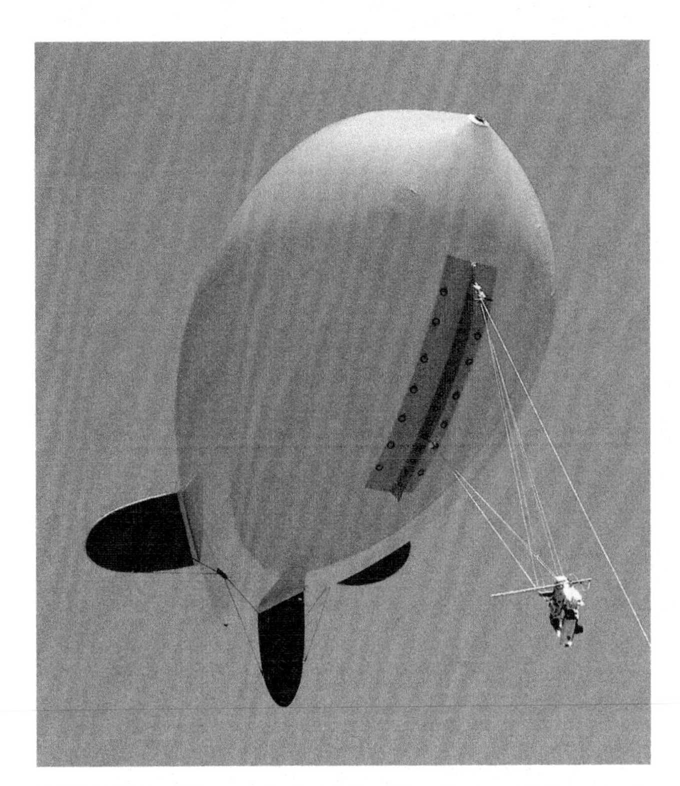

FIGURE 8-11 Blimp ready for flight with radio-controlled camera rig and Picavet suspension attached to keel. Blimp is 4 m (13 feet) long and contains $\sim$7 m^3 of helium; tether line extends to lower right. Photograph courtesy of N. Hubbard.

blimp (Fig. 8-11). The blimp is secured and maneuvered using a single tether line of braided dacron with a breaking strength of 90 kg (for knots and bends, see Section 8.4.2).

All equipment can be transported in the back of a small truck or trailer including one large helium tank that holds $\sim$7 m^3, which is just enough to inflate the blimp one time. Field operation is relatively simple; first, a large ground tarp is laid out, and the blimp is inflated on this tarp (Fig. 8-12). Once inflated, a camera rig is attached and tested. The whole preparation and inflation procedure takes about half an hour and requires only two people. As with other tethered platforms, the blimp may be flown up to 500 feet (150 m) above the ground in the United States without filing a flight plan with the nearest airport (see Chapter 9.8.2). The tether line is marked at 500 feet, and a laser altimeter is used to confirm blimp height.

The blimp may be sent aloft and brought down repeatedly to change camera rigs or to move to new locations around the study site. In principle, the blimp could remain aloft until the helium gradually leaks out, a period of several days. In practice, blimp aerial photography is conducted normally at a single study site in one day, from mid-morning until mid-afternoon, when the sun is high in the sky. At the end of each session, the helium is released, as there is no practical means to recover it in the field.

The blimp has been utilized under conditions ranging from completely calm to moderate wind speeds in rural and urban settings (Figs. 8-13 and 8-14). The blimp has proven quite stable in calm to light wind and remains relatively stable in wind up to 10–15 km/h. Under light wind, the blimp can be maneuvered quite precisely relative to ground targets. Stronger wind, however, tends to push the tethered blimp both downwind and down in height, so that surrounding obstacles may become troublesome.

This particular blimp is relatively small and easy to handle, which results in excellent portability for reaching any site accessible to a four-wheel-drive vehicle. Furthermore, small blimps of this type are widely available for advertising purposes and are low in cost. Many larger helium blimps have been employed by other people for SFAP. In some cases, a permanently inflated blimp is stored and transported in a big trailer, for example a horse trailer. In this case, none or

FIGURE 8-12 Inflating the blimp on a canvas trap. The helium tank, deflated blimp, camera rigs, and all other equipment are transported in the back of the small four-wheel-drive truck, which can reach remote locations. Photo of SWA by JSA, April 2007.

FIGURE 8-13 Overview of Shoal Creek with agricultural fields in valley bottom to right and forested uplands to left. Launching and operating the helium blimp under near-calm conditions was feasible from the narrow opening between the stream and forest in lower right corner of view. Schermerhorn Park, Cherokee County, Kansas, United States. Taken from Aber and Aber (2009, fig. 74).

only some of the helium has to be released. This reduces the costs significantly and enables several ascents on successive days. In other cases, a blimp may remain in the air over a study site for long periods of time—days or weeks. Special fabrics and wind-resistant designs are necessary for success with such extended usage. Larger and more robust blimps mean higher costs for equipment and operation and are generally less portable in the field.

8.3.3. Hot-Air Blimp

Spherical hot-air balloons experience the same stability problems as helium balloons. If they are free flying, they are simply uncontrollable, and in tethered mode they start turning erratically even at quite low wind velocities. In addition, the internal pressure inside the hot-air balloon needs to be relatively high in order to prevent deformation of the envelope in light wind. Therefore, hot-air balloons are not well suited as camera platforms. For these reasons, it is no surprise that this kind of camera platform is almost forgotten nowadays, although it was the first platform used for aerial photography by Gaspard Tournachon (see Chapter 1). In Jahnke (1993) valuable hints on the construction and operation of model hot-air balloons can be found.

Tethered hot-air systems as camera platforms have emerged since the late 1970s particularly in archaeology. In this field of research, a high need exists for quickly producible, detailed images for documenting excavations in remote areas (Fig. 8-15; Heckes, 1987). The hot-air blimp

FIGURE 8-14 Urban, industrial scene, Kansas City, Kansas, United States. Missouri River on right side. Blimp was launched from a small park next to the river. Taken from Aber and Aber (2009, fig. 58).

FIGURE 8-15 In research cooperation of the German Archaeological Institute (DAI) with the German Mining Museum Bochum (DBM) and Frankfurt University, the hot-air blimp is used as a medium-format camera platform for documenting archaeological excavations at the Sabaean city of Sirwāh, Yemen. Photo by U. Kapp.

FIGURE 8-16 Small 100 m^3 hot-air blimp built by *GEFA-Flug*. (A) Carrying capacity and wind susceptibility of this model reach their limits in the alpine conditions of the Spanish Pyrenees. (B) Small packing size makes it easy to transport to remote regions such as high mountain ranges; no single part is heavier than 15 kg. Photos by IM, 1996.

introduced by Busemeyer (1987, 1994), which was also used by Wanzke (1984) for the documentation of excavations in Mohenjo-Daro in Pakistan, is the prototype on which all such blimps constructed by *GEFA-Flug* (Aachen, Germany) are based. These hot-air blimps combine the principle of an open hot-air system with an elongated egg-shaped blimp form equipped with tail empennages.

In comparison to the spherical shape of balloons, the blimp is considerably more stable in the air. Owing to the streamlined shape and tail empennages, the blimp is much more aerodynamic and aligns itself with the wind. Thus, the air vehicle becomes much easier to control from the ground even with a light breeze. In case of increasing wind, the envelope may get pushed in at the nose, but a safe landing is still possible as this affects the uplift properties only slightly. The danger for an ignition of the envelope owing to a wind gust blowing the fabric into the burner flame is rather low for the blimp in contrast to a balloon. The concept and

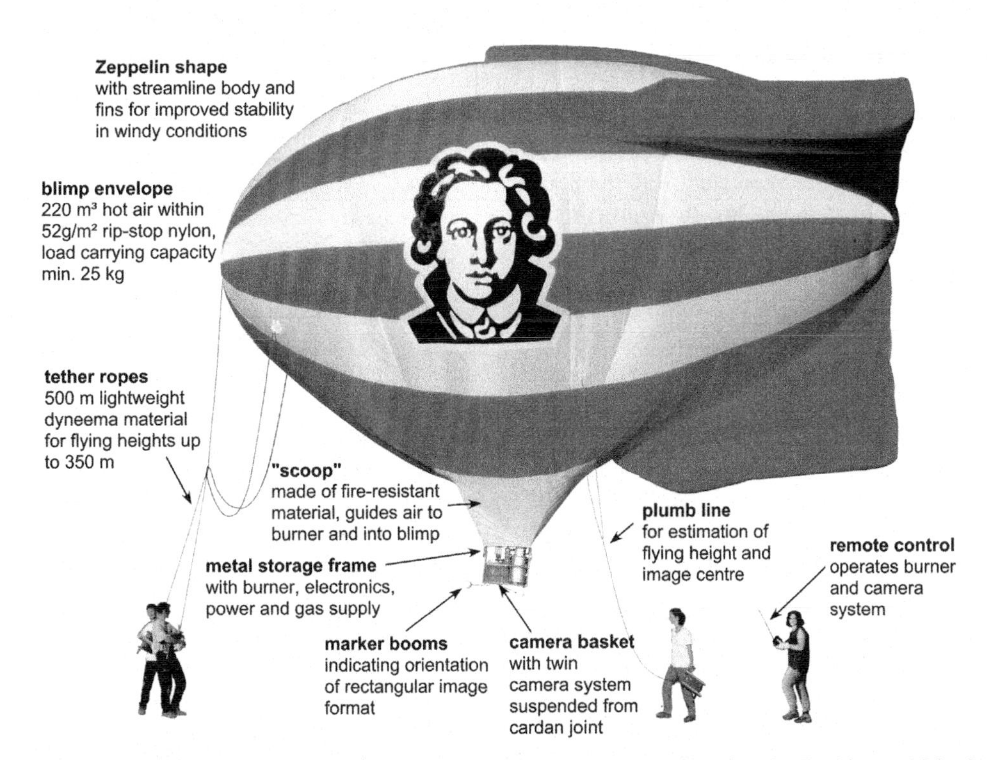

FIGURE 8-17 The hot-air "Goethe monitoring blimp" carries the logo of Johann Wolfgang Goethe University, Frankfurt am Main, Germany. Blimp and frame constructed by *GEFA-Flug* and by the technical workshop staff of the Faculty of Geoscience and Geography. Adapted from Ries and Marzolff (2003, fig. 2).

functionality of this airship, its construction, and its use for aerial photographic surveys are presented and discussed in detail by Marzolff (1999).

Two of the authors, IM and JBR, have experimented during the last 15 years with two different sizes of hot-air blimps, a smaller one with about 100 m^3 and a load capacity of 5 kg and a larger one with 220 m^3 providing a carrying capacity of approximately 25–40 kg. The smaller version ranges at the lower border for model airships, but has the crucial advantage that the individual parts can be transported by one person each (Fig. 8-16). Only four people are necessary to carry the equipment even to remote study areas. Unfortunately, the smaller model turns out to be comparatively susceptible to wind influence because the net-lifting capacity of only 5 kg does not leave a wide scope for "heating against the wind."

In the following sections, the most important components and their functioning are explained using the larger hot-air blimp—the "Goethe monitoring blimp" belonging to the Department of Physical Geography of Frankfurt's Johann Wolfgang Goethe University (Fig. 8-17). The monitoring blimp consists of four main components—the envelope, burner frame with camera mounting, remote-control device, and tether ropes. Additional equipment for the inflation phase includes a tarpaulin, a large inflation fan and miscellaneous tools.

Blimp envelope: The envelope consists of tearproof, polyurethane-coated, 52 g/m^2 heavy, rip-stop nylon and has a length of 12 m, a height of 6 m, and a width of 5 m when inflated. The airship has a volume of 220 m^3 and does not need any additional inner stabilization. For the three tail empennages a specific airflow is necessary. On top of the blimp the hot air is drawn off into a tube that is externally sewn onto the envelope. Through this tube, the air reaches the distributing chamber which is separated from the main body in the rear end. From there the hot air is distributed into the three empennages. These fins are characterized by an unfavorable ratio of large cooling surface to low volume; therefore, they can be kept in shape only with a constant influx of hot air.

At the bottom of the combustible envelope, the so-called scoop is attached, which is made of fire-resistant Nomex material and connects the blimp body to the burner frame. It is open to the front in order to allow an inflow of outside air to compensate for the permanent air loss through the envelope seams. At the scoop, two steel cables encompassing the envelope end in metal chains to which the burner frame is fixed with carabiners.

Burner frame with camera mount: The burner frame consists of a steel-tube frame with a collapsible burner panel, an exchangeable gas bottle, and a plug-in camera unit (Fig. 8-18). Four 100,000 kcal/h Carat liquid phase burners produce flames of about 90 cm height. They are capable of heating up the air inside the envelope to a maximum of 140 °C. The burners are screwed on the burner panel beneath which the electronic control with two magnetic valves is situated. The gas is kept in an aluminum bottle with a net weight of 5.6 kg and a filling capacity of 11 kg of gas. The gas bottle is strapped underneath the burner but easy to retrieve. The gas is extracted in the liquid phase with an immersion pipe and led to the valves through a steel-reinforced tube equipped with a gas filter. Additionally, the burners are supplied with a mixture of propane and butane from a commercially available 0.7 L multigas cylinder. This gas is used to produce an approximately 20-cm-high pilot flame, which is permanently ablaze in all burners and can be ignited by an electronically controlled piezo igniter.

FIGURE 8-18 JBR testing the blimp burner before the survey. All burner functions such as ignition, valves, and pilot flame, as well as leak-tightness of the gas tubes need to be thoroughly checked. G, main gas bottle; P, pilot gas bottle; C, control box with receiver; B, battery packs. The empty space next to the gas bottle is for the camera unit (see Fig. 7-15). Photo by A. Kalisch.

As for conventional manned balloon-flight (Federal Aviation Administration, 2007), uplift is achieved by intermittent, not permanent heating. When the magnetic valve of the burner is opened by remote control, liquid gas

FIGURE 8-19 (A) Goethe blimp during a survey in the Hoya de Baza, Province of Granada, Spain. (B) Blimp is navigated with two tether lines usually kept parallel. In difficult terrain and during landing, the two pilots (here JBR and M. Seeger) may take turns in holding the airship. In order to keep the lines taut, constant reeling and unreeling is required while the blimp changes altitude. (C) IM with the remote-control device operating ignition, heating, and camera functions. Photos by JBR, IM, and A. Kalisch.

flows through the burner-coils, streams out the orifice, and is ignited by the pilot flame. The burner valves and the piezo igniter are powered by 12-V rechargeable batteries, whereas the remote-control receiver and the pilot-flame valve are supplied by 6-V rechargeable batteries. The burner frame is equipped with a manual control that enables opening of the piezo igniter and the main burner valve during the inflation process. The remote control is not used until the blimp floats and is ready for departure. The gas supply is sufficient for approximately 45 minutes of flight.

The net carrying capacity of the hot-air blimp ranges between 25 and 40 kg depending on ambient temperature and the filling level inside the gas bottle. The system is particularly suitable for different flight altitudes, especially regarding the fact that it becomes increasingly lighter during the flight owing to gas consumption. Accordingly, in the beginning of the flight the lower, and later on the higher flying heights should be scheduled. The camera mount is inserted into the burner frame as a separate plug-in unit (see Chapter 7.3.4); it sits underneath the burner and next to the gas bottle. Power supply for the mount is established automatically with a plug connection. The cameras are protected from damage in rough landings by a robust wicker basket or polystyrene-walled aluminum frame, open at the bottom to ensure free sight for the cameras (see Figs. 7-13 and 7-14).

Remote control: Piezo igniter, pilot flame, and main burner valve as well as camera servos and camera trigger are all operated with a commercial remote-control device (e.g. *Graupner* mc 16/20). The remote control is equipped with a pulse code modulation (PCM) fail-safe control that is programmed to close all valves automatically in case of problems with radio reception, external signals, or a failure of the transmitter. This is quite important in order to make sure that uncontrolled burning can be avoided—permanent burner blasts would lead to overheating and possible inflaming of the envelope. The radio signals are transmitted on the frequency band locally used for model aircraft (e.g. 40 MHz in Spain).

Tether lines: Being a captive airship, the blimp position can be controlled from the ground only by means of two captive tethers (Fig. 8-19). For this purpose extremely light ropes made from PE-coated *Dyneema* are used, an extremely strong polyethylene fiber with low elongation and twist and high breaking strength. The ropes are fixed to the envelope via a trapezoid of lines attached to the bow and front flanks of the blimp body. A rubber expander in the

FIGURE 8-20 Launching the blimp. (A) Envelope is filled with cold air. At least four people are needed to hold the nose, tail, and back to the ground. A light breeze is enough to make the blimp writhe like a stranded whale. (B) Once enough space has filled inside the body, the burner is positioned to face the blimp spine and the air is heated with intermittent blasts. (C) As the envelope is released and rises, the most critical phase of the launch begins. The uplift force is already strong, but the blimp still needs to be held tightly and heating discontinued until the burner frame is securely fixed to the scoop. (D) Fully inflated airship in launch position. After testing all valves again, the camera unit is attached and tested. The blimp is quickly heated then and lifts up to a safe height of >30 m. Photos by JBR and IM.

FIGURE 8-21 The blimp is turned upside down after the survey for deflation, and the delicate envelope is swiftly recovered and packed to avoid further exposure to the sun and wind. Photo by V. Butzen.

trapezoid minimizes the jerky pulls that are passed onto the airship during fast maneuvers by the ground crew; thus, constant nicking of the blimp nose can be prevented. The rope of 500 m length is wound on a wooden reel and permits a flying altitude of approximately 350 m. When in flight, the blimp's bow always points toward the wind; this enables stable maneuvering and free influx of fresh air into the blimp. Additionally a third rope, the so-called plumb line, hangs down vertically from the envelope. It is marked every 5 m in order to allow an estimation of the flying height and indicates the current position of the airship (approximate image center) to the crew on the ground.

Launching and landing: For cold inflation, the blimp envelope is spread out on two fabric-reinforced, 50 m^2 plastic tarpaulins and inflated with fresh air by means of a 5.5 PS fan of the kind frequently used for manned balloons (Fig. 8-20A). Afterwards the air is heated up with the burner, and as soon as the blimp floats upright in the air the camera mount unit is inserted into the burner frame (Fig. 8-20B–D). A minimum of five people is advisable for helping with these tasks—one each holding the envelope at bow, top, and rear for ensuring that no "free-floating" fabric is blown into the burner flames, and two at the mouth of the blimp for holding it open and operating the fan and later the burner system. Particularly in the hot inflation phase the absence of wind is crucial in order to avoid damage by fire.

After the survey, the blimp is landed on the tarpaulin, first removing the camera mount, then the burner frame. Consequently, the blimp is relieved of its carrying burden and needs to be secured well by the tether lines again, especially in light wind. The hot air keeps the weight of the envelope aloft long enough for a change of gas bottle if the survey is to be continued. If not, the blimp is quickly turned upside down to let the hot air escape (Fig. 8-21). A velcro-fastened opening in the rear allows releasing the air from the empennages, and the air in the main body is carefully squeezed out through the mouth before the envelope is packed into its canvas bag.

8.4. KITE AERIAL PHOTOGRAPHY

Kites were among the earliest platforms used for aerial photography, beginning in the 1880s (Batut, 1890; see also Chapter 1). More recently, kites have experienced a renaissance for SFAP as a fascinating hobby for photographers around the world, who share their techniques, experiences, and images via the Internet (e.g., Benton, 2009; Beutnagel, 2009). In relation to these activities, scientific applications of KAP are comparatively rare, but of increasing technical sophistication. Several researchers, including the authors, have used kites for geomorphological investigations (Bigras, 1997; Boike and Yoshikawa, 2003; Marzolff et al., 2003; Giménez et al., 2009; Marzolff and Poesen, 2009; Smith et al., 2009), vegetation studies (Gérard et al., 1997; Aber et al., 2002, 2006), archaeological documentation (Bitelli et al., 2001), and other applications (Aber et al., 1999; Tielkes, 2003).

KAP is highly portable and flexible for field logistics, adaptable for many types of cameras and sensors, relatively safe and easy to learn, and has among the lowest costs overall compared with other types of SFAP. These characteristics explain the growing interest in KAP from the popular and sport level to high-end scientific applications.

The basic deployment of KAP is depicted in Figure 8-22, which illustrates the typical setup for the kite, camera rig, and ground operation. Central to KAP are the types of kites, necessary kite-flying equipment, and ground operations, which are described in the following sections based primarily on the authors' experience. KAP camera rigs are detailed in Chapter 7.

8.4.1. Kites for SFAP

Many types of kites may be employed, but no single kite is optimum for KAP under all circumstances. Various kites are utilized depending on wind conditions and weight of the camera rig. The goal is to provide enough lift to support the payload—normally ranging from 1 to 3 kg. For a given camera rig, large kites are flown for lighter wind and smaller kites for stronger wind. Kite designs fall in two general categories.

- *Soft kites* have no rigid structure or support to maintain their shape. The kite inflates with wind pressure and forms an airfoil profile, like the wing of an airplane, which provides substantial lift. Soft kites have several advantages for KAP. They have quite low weight-to-surface-area ratios, they are exceptionally easy to prepare and launch, and they are a breeze to put away—just stuff the kite into a small bag. For light-weight travel or backpacking, soft kites are the type of choice. Soft kites do have a tendency to collapse when the wind diminishes, so a watchful eye is necessary while in flight.
- *Rigid kites* employ some type of hard framework to give the kite its form and shape. Traditional supports of wood and bamboo are replaced in most modern kites by graphite rods and fiberglass poles. Their weight-to-surface-area ratios are intrinsically greater than soft kites, but rigid kites do have some important advantages for KAP. The primary advantage is the ability to fly well in light breezes without the danger of deflating and crashing. The frame maintains the kite's proper aerodynamic shape regardless of wind pressure. Although frame members may be disassembled, rigid kites can be troublesome for packing and travelling.

Among the most popular SFAP soft kites is the *Sutton Flowform*, invented and patented in 1974 by Sutton (1999). The flowform design employs venting to reduce drag and control air pressure within the kite body; these kites are known as smooth and stable flyers under moderate to strong wind (Fig. 8-23). Many other types of soft airfoils are utilized for SFAP. Various types of parafoils, vented or unvented, range in size up to >10 m^2 and can loft substantial payloads (Fig. 8-24).

Many rigid kite designs are suitable for KAP, including delta and delta-conyne types. Delta kites have a basic triangular shape, and the wing dihedral provides stability in flight (Fig. 8-25). The delta-conyne is essentially a triangular box kite with wings for added lift (Fig. 8-26). Both these styles are easy to fly and not likely to crash unexpectedly. They are good choices for both beginning and advanced kite flyers. However, they suffer from relatively high weights compared with their surface areas, particularly the delta-conyne, which may limit the payload capacity.

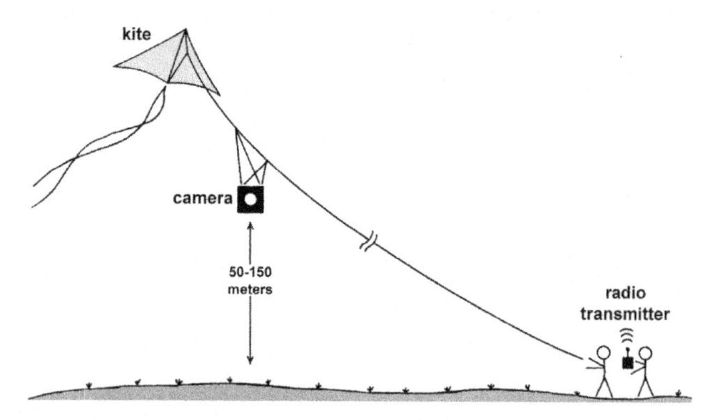

FIGURE 8-22 Cartoon showing the typical arrangement for kite aerial photography. The camera rig is attached to the kite line, and a radio transmitter on the ground controls operation of the camera rig. The kite line normally is anchored to a secure point on the ground. Not to scale; adapted from Aber et al. (2003, fig. 1).

FIGURE 8-23 The *Sutton flowform* is a wind-inflated airfoil that flies well in moderate to strong winds. This side view shows the kite with two 4.5-m streamer tails, which greatly improve kite stability. Flowforms come in several sizes; this one has about 1.5 m^2 of surface area and weighs about 0.3 kg (without tails). Photo by JSA.

The rokkaku is the favorite rigid kite of the authors (Fig. 8-27). This traditional Japanese kite has a low weight-to-surface-area ratio, and through centuries of design improvements it has achieved an elegant status among kite flyers. It provides the greatest intrinsic lifting power compared with other types of rigid kites in our experience. Rokkakus may be somewhat unstable in near-surface ground turbulence, but once aloft they are remarkably smooth and powerful fliers. We utilize rokkakus for the majority of our KAP, choosing for each survey from a selection of different sizes and tether lines depending on the current wind conditions.

FIGURE 8-24 This *SkyFoil* measures 3 m wide by 2.6 m long giving it nearly 8 m^2 surface area. It is a powerful lifter; in a moderate wind it can pull a person off the ground. It must be flown on a 500-pound (225 kg) line. Photo by JSA.

8.4.2. Kite-Flying Equipment

Beginning with items attached to the kite itself, the tail is an important option (see figures above). Tails generally increase the stability of kites, but at a price of increased weight and drag, which could make the kite fly at a lower angle. Under light, steady wind, the tail may not be needed, particularly for the rokkaku. However, a tail may be essential for strong, gusty, or turbulent wind conditions. When to use a tail is based on experience; test flying the kite without a camera rig is a good idea to judge the wind conditions and possible need for a tail.

Kite line is the next critical component. Having broken a 200-pound line early in his KAP career, JSA now typically uses 300-pound (135 kg) braided dacron line for most kites and 500-pound (225 kg) line for some larger kites. The line is wound on a simple hoop or reel, which is firmly anchored (Figs. 8-28 and 8-29). For the large rokkaku kites used by IM and JBR, where the kite line also serves as the rail track for pulling up the camera sledge (see below), even stronger and more rigid lines made from PE-coated *Dyneema* are used. *Dyneema* is a light, extremely strong polyethylene fiber with low elongation and twist and high breaking strength. Some kite flyers prefer *Kevlar* line, which is strong, thin, and light; however, we consider it rather

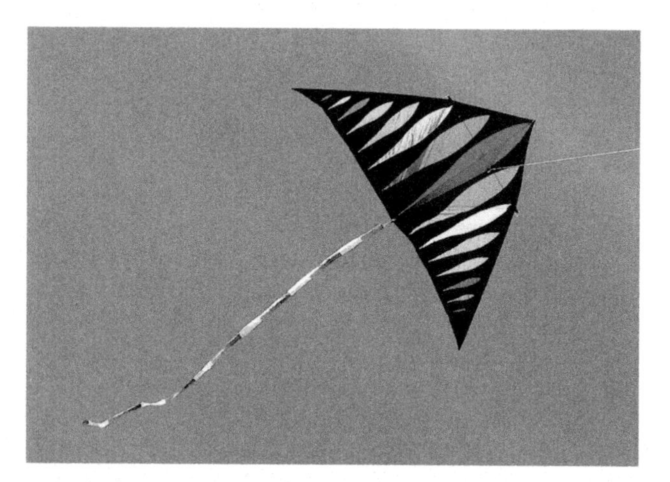

FIGURE 8-25 Giant delta has a wingspan of 5.8 m and a total surface area of 8.2 m^2. It flies beautifully on 300-pound (135 kg) line in a gentle breeze and has excellent lifting power. Seen here with a 6-m-long tube tail; it folds into a 1.2-m-long case and weighs ~2.25 kg. Photo by JSA.

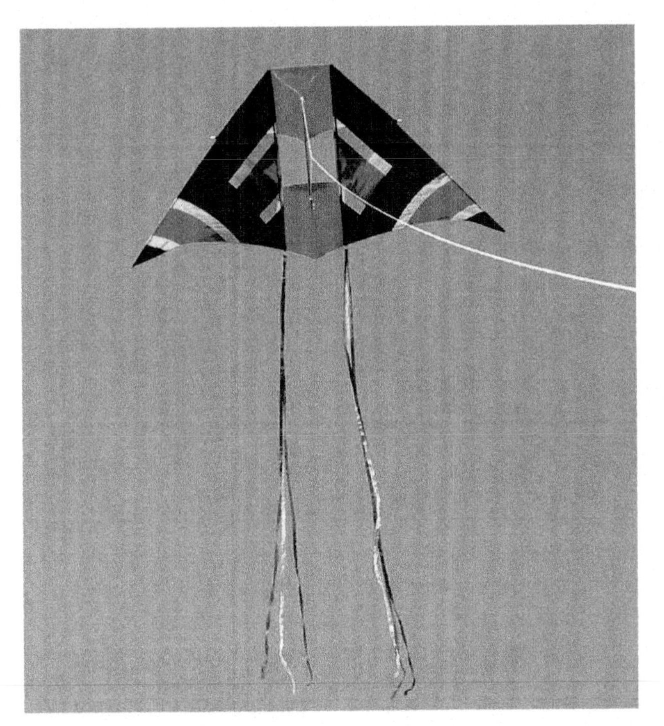

FIGURE 8-26 This *Sun Oak Seminole* delta-conyne is flying with two 4.5-m streamer tails. At 1.4 kg weight, it is sturdy and reliable, but heavy, which reduces its KAP lifting capacity. It folds into a 1.2-m-long case. Photo by JSA

dangerous because of its ability to cut through gloves, skin and clothing like a knife.

Whenever handling the kite line or reel, gloves are highly recommended to protect the hands. Some kite flyers use *Kevlar* gloves, but the authors have found leather to be most practical (Fig. 8-30). For securing kite line, as well as tether lines for blimps and balloons, several knots, bends, and hitches are particularly useful (Fig. 8-31; Pawson, 1998; Budworth, 1999; Jacobson, 1999).

- *Sheet bend*: An excellent knot for joining two lines of equal or unequal diameter. The knot has moderate strength (55%) and is highly resistant to slippage. It is recommended in preference to the square knot. Note both free ends should be on the same side of the knot.
- *Fisherman's (English) knot*: Each line is tied in an overhand knot around the other line, and then the two knots are pulled tight. The fisherman's knot is one of the strongest and most resistant to slippage of all bends for tying two lines together.
- *Lark's head (cow hitch)*: A simple knot created by passing the line through a loop around the anchor. This knot is used to tie a ring into a line. The lark's head is easy to tie and untie, moderately strong, and resistant to slippage.
- *Anchor (fisherman's) bend*: Two half hitches in which the first half hitch is locked by an extra round turn. This is the authors' favorite means to attach kite-flying

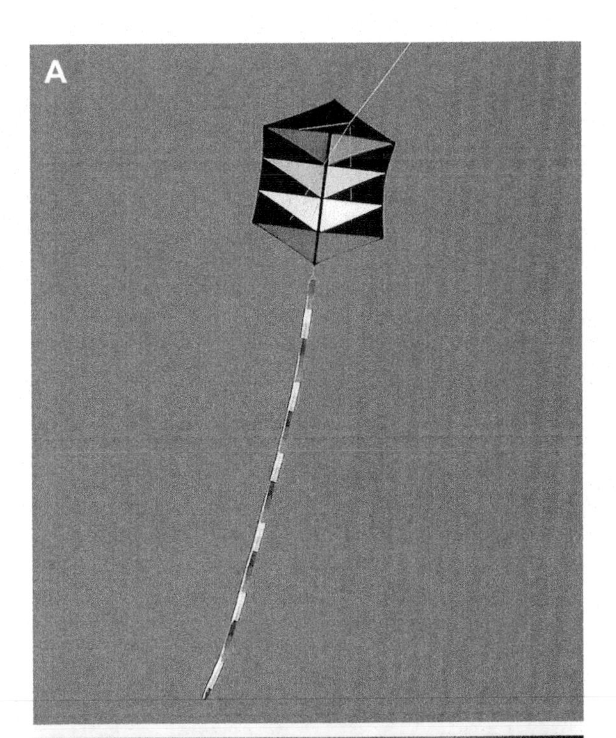

FIGURE 8-27 Rokkaku kites for SFAP. (A) Large rokkaku used most often by JSA. Kite is 2.3 m tall by 1.8 m wide; shown here with a 6-m-long tube tail. This kite weighs <0.8 kg and folds to a compact 0.8 m length. Photo by JSA. (B) Still larger rokkaku of IM and JBR. This kite, designed for lifting an SLR camera with sledge-type rig (see Fig. 7-12) even in lighter winds, is 2.5 m tall by 2.0 m wide. Photo by IM.

FIGURE 8-28 (A) Kite line attached to a wooden handle that is anchored to a fence post set in stone. The extra line is wound on a simple plastic hoop; Kansas, United States. (B) Kite line anchored to the back of a small automobile at a construction site in northern Norway. Photos by JSA.

FIGURE 8-29 Above: large *Strato-spool* reel for handling 300 m of 135-kg dacron kite line. The blue strap locks the handle. Below: disassembled components of the reel. A, main shaft; B, reel with kite line; C, brake lever; D, axle and wrenches. Tape measure is extended ~30 inches (75 cm). Photos by JSA.

line to snaps and rings. Among the strongest knots, its breaking strength is 70–75% of rated line strength.

- *Bowline*: A moderately strong knot for making a loop that will not slip. Its breaking strength is about 60% of the line's rated strength.

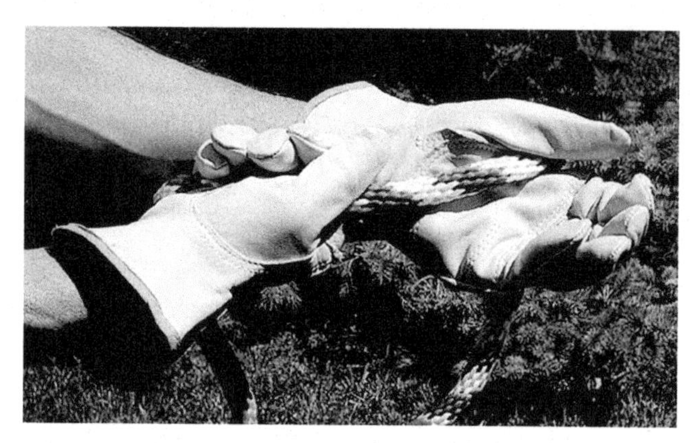

FIGURE 8-30 Gloves for kite aerial photography. These are so-called roper gloves that have an extra patch of leather across the palm of the hand. Photo by SWA.

- *Cleat tiedown*: Proper cleat tiedown involves overlapping the line in a figure-eight pattern. Note the free end (lower right) is tucked under itself on the final wrap to lock the line.

How to attach the camera rig to the kite is a question with many solutions. Nearly all modern KAP is conducted with the camera rig secured to the kite line, some distance down the line from the kite itself (see Fig. 8-22). The purpose of this arrangement is to remove the camera from line vibrations and sudden movements of the kite. Two basic methods are the pendulum and Picavet suspensions (see Chapter 7.3.3).

Finally, a simple gadget made from nylon straps attached to a pulley with pivoting sides may prove extremely helpful for redirecting the kite line (see below) and helping in landing the kite at the end of a survey. Being busy with getting a camera-burdened kite to fly, one tends to forget that

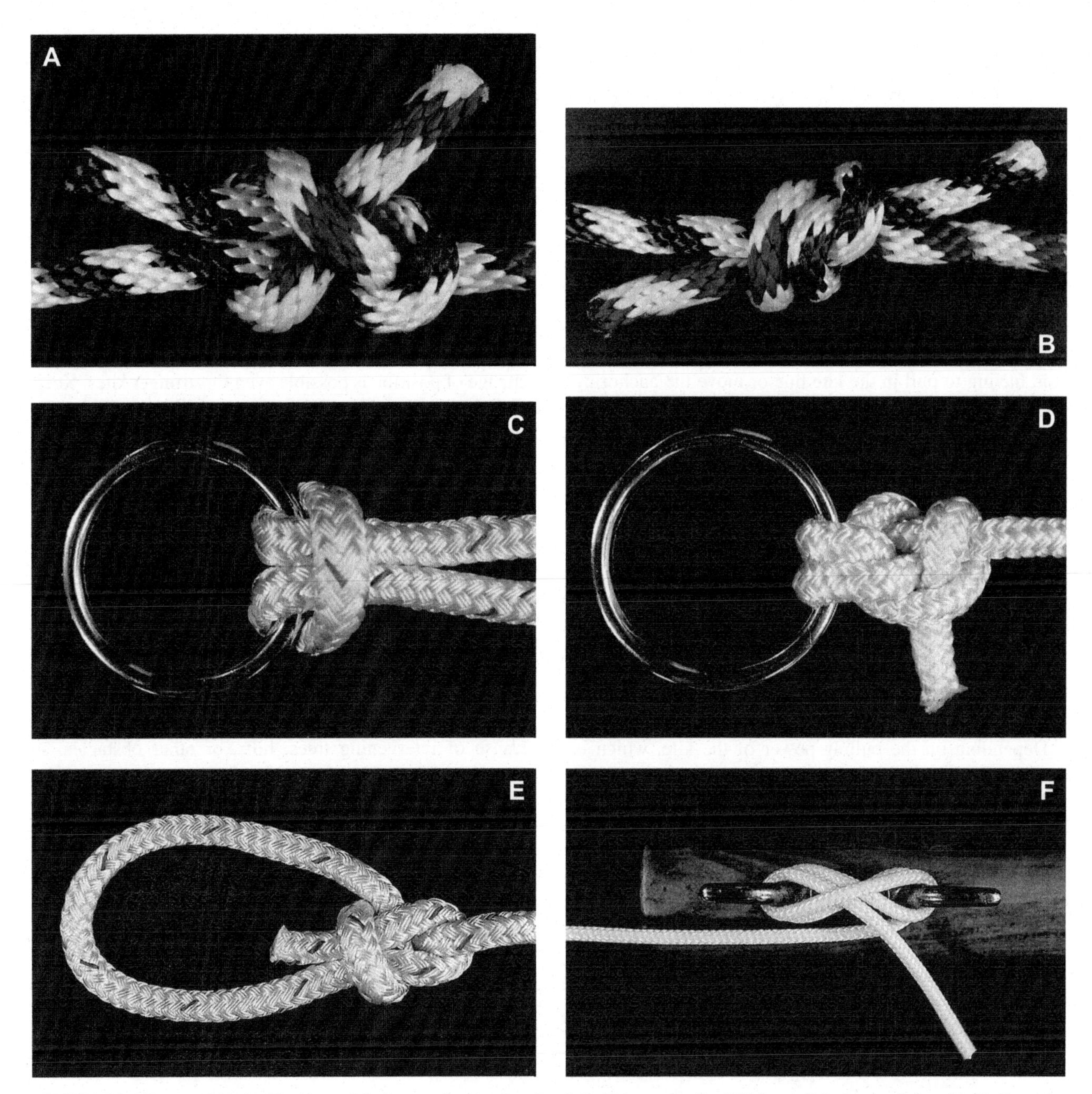

FIGURE 8-31 Knots and hitches for kite aerial photography. A, sheet bend; B, Fisherman's (English) knot; C, Lark's head (cow hitch); D, anchor (fisherman's) bend; E, bowline; and F, cleat tiedown. Knots by JSA.

it might be even more difficult to get it down again, a task that may require several people in strong wind (Fig. 8-32).

8.4.3. Ground Operations

After selecting a suitable open space, free from obstacles and safety hazards (see Chapter 9), the first step is to launch the kite in order to judge wind conditions. Once the kite has reached a suitable flying position, the camera rig is attached and tested. Two approaches may be utilized to lift the camera rig to appropriate height for KAP (see also Chapter 7.3.3). The method used by one of the authors (JSA) involves sending the kite up 30–60 m, then attaching the Picavet suspension and camera rig to the kite line. As the kite line is let out farther, the camera rig then ascends to the desired height (see Fig. 8-22). IM and JBR, however, send the kite to its full height, then pull the camera rig up the kite line using a sledge-and-pulley arrangement similar to a cable car (Fig. 8-33). This minimizes the risk for the camera during positioning maneuvers and enables the operators to take the camera down during the survey for changing lenses, films, or storage cards.

For vertical photography, where the exact camera position is more important than for oblique photography, the position of the launching place relative to the study site requires careful planning, because the wind direction, angle of approach, and desired camera height have to be taken into account. There is never a possibility of changing the camera height only in a vertical ascent. Changing the flying height of the kite by playing out the line (if the rig is directly attached to the kite line) or by pulling the camera up (when using a sledge-and-pulley system) also changes the horizontal location of the camera. Depending on the camera attachment method, there are two different approaches for surveys taken from varying flying heights. Both methods avoid having to pull in the kite line or move the anchoring point backwards against the wind, both of which would mean heavy strain for kite and operators.

- With the fixed-suspension method, start from a launching place closer to the study site and low-flying heights. Then change to increasing flying heights by letting the kite up farther while simultaneously moving the kite anchoring point away from the site.
- With the sledge-and-pulley system, start from a launching place farther away from the study site and high flying heights. Then proceed to lowering the camera sledge while or before moving the kite anchoring point toward the site.

Depending on the pulling power of the kite, which is mainly a function of wind speed, kite size, and camera payload, different methods of anchoring can be chosen. Kites for compact cameras and lightweight suspensions can usually be held by the kite flyer with the great advantage of good mobility for positioning the camera directly above the desired spot on the ground (Fig. 8-34). This should be conducted with a safety harness around the waist and hips; the kite reel and line should never be held only by hand.

Larger kites capable of lifting SLR camera rigs cannot be held by a single person, at least not when the wind is strong enough to keep the kite with camera aloft. They need to be anchored to something heavy or fixed like a fence post, a car, or a suitable rock boulder (see Fig. 8-28). This makes it much more difficult to position the kite or move it during the survey in order to cover larger areas with contiguous images. By slowly driving the anchoring car or redirecting the kite line with the help of one or two people and a pulley system (see Fig. 8-32B) acting as a deflection point, a change of position is possible even for stronger kites. Keep in mind, though, when planning the positioning operation, that it is difficult to pull a large kite in strong wind into a new position toward you—it is much easier to move the kite with the wind.

Owing to the constellations between kite, camera, and anchoring points described above, it may in many cases be difficult for the kite flyer to judge the exact location on the ground directly below the camera. For this purpose a spotter, usually the camera operator, may stand next to the ground target in order to give directions to the kite flyer via a two-way radio or via further team members positioned between the two. The kite flyer and camera operator may be separated by several 100 m and may not be able to see each other because of intervening trees, hills, or other obstacles. A laser range finder could be employed to measure the distance or height of the kite or camera rig.

In some KAP systems, the camera has a video downlink, so the operator can view directly the scene to be photographed. However, this approach adds weight and

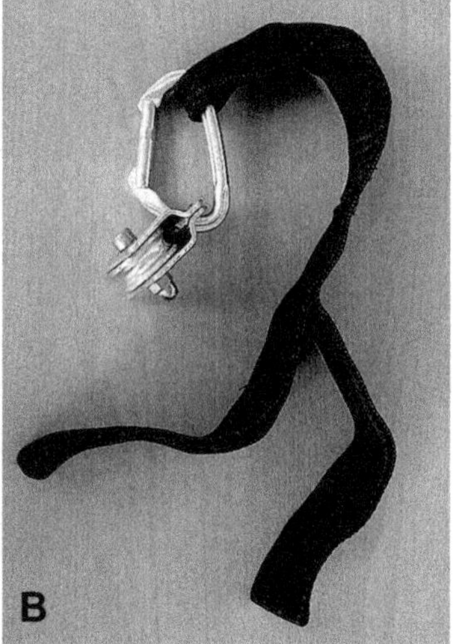

FIGURE 8-32 (A) Pulling down a kite after a photographic survey at an inselberg near Gorom-Gorom, Burkina Faso. The wind was so strong that the team had to wait for an hour until it started to diminish, when four men managed to get the kite back to the ground. (B) Two-man version of the pulley with nylon straps used for the operation in part A. Photos by IM.

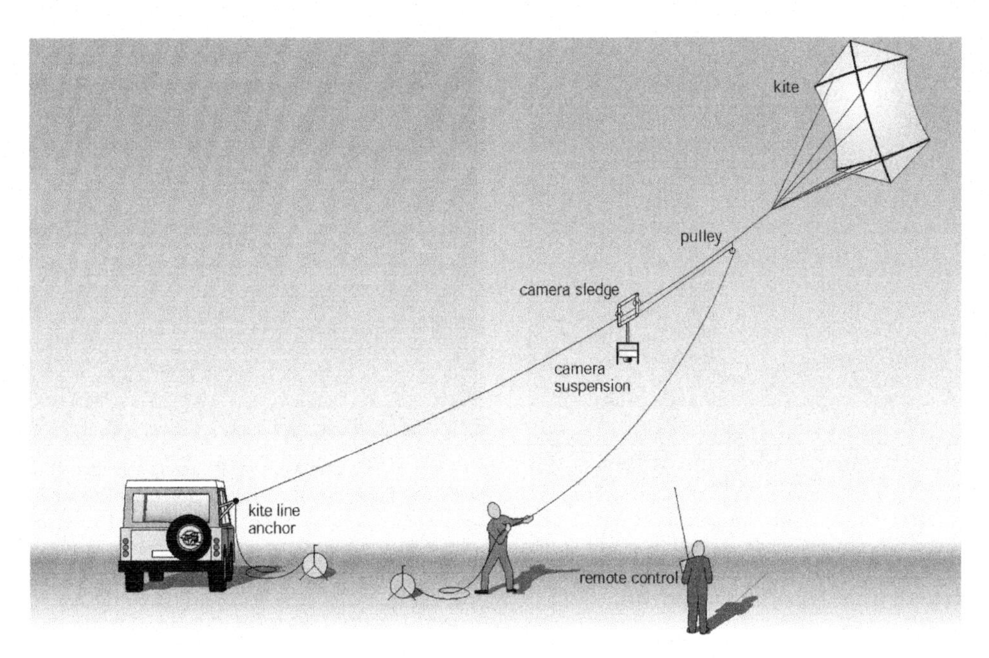

FIGURE 8-33 Method for raising a camera rig on a sledge along a kite line. Taken from Marzolff et al. (2002, fig. 5).

complexity both for the airborne camera as well as ground operation, which may lead to greater chances for equipment failure or procedural mistakes in the field. By the time a target or particular viewing angle is determined in the video viewer, the camera usually has moved on to a new position, so attempting to frame the perfect shot this way is rarely practical. The authors prefer a basic approach without a video downlink. In this case, the position of the camera is judged from the ground, usually by watching the camera with binoculars, and many pictures are taken and stored on large memory cards. These pictures may be downloaded and viewed on a portable image display and storage device or a laptop computer immediately after bringing the camera rig down (Fig. 8-35). Thus, the success of the mission can be evaluated quickly, so that additional KAP could be conducted if necessary.

KAP field logistics involve transporting people and cargo, which are possible in many ways that range from driving an automobile on paved roads directly to the launch site to hiking long distance over rough mountain trails or through wetlands to reach remote locales. In any case, KAP lends itself to efficient packing for easy transportation (Fig. 8-36; see also Fig. 9-2).

FIGURE 8-34 JSA demonstrates a mobile anchor for the kite reel and line. The reel is attached to a harness around the waist and hips of the kite flyer. A small two-way radio (on left arm) allows communication with the camera operator near the ground target. Photo by SWA.

8.5. DRONES FOR SFAP

Drones—also called UAVs or MAVs—are unmanned, powered, free-flying platforms that vary greatly in their technical characteristics and photographic capabilities. They may be controlled by a pilot on the ground, who is in visual contact with the drone, or they may be programmed to fly along predetermined flightlines with pictures taken automatically at specific points. The usual platform is some type of model airplane that is modified to carry a camera. Miniature helicopters, autogyros, powered paragliders, and other platforms also may be utilized as photographic drones (Fig. 8-37). Military surveillance drones, in contrast, are either relatively large or increasingly small, but always highly complex aircraft that are beyond the scope of this book.

FIGURE 8-35 *Digital Foci Picture Porter* image storage and display device. It accepts CF and SD memory cards as well as a USB cable connection for data exchange. This model contains a 75 GB hard drive, and the color screen measures 9 cm diagonally. Photo by JSA.

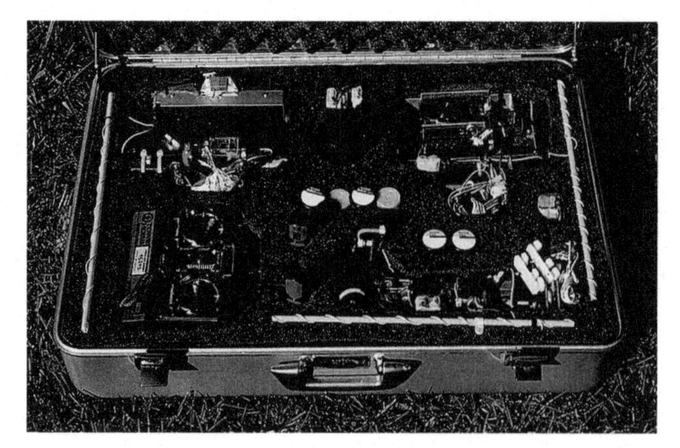

FIGURE 8-36 Case with KAP camera rigs and radio control. The metal case with foam lining contains three complete camera rigs (marked by yellow antenna rods), radio transmitter, and accessories. Case measures ~65 × 50 × 23 cm. Although a bit heavy for long-distance hiking, it may be carried by hand for short distance, conveniently in an automobile, or as checked luggage on an airline. Photo by JSA.

Drones have been used increasingly for SFAP in scientific research during the last two decades (Everaerts, 2008), and the literature reflects the wide range for this type of platform. Most of the earlier works are based on remote-controlled basic model airplanes that are flown manually through visual contact, requiring an operator with considerable experience (e.g., Koo, 1993; Fouché and Booysen, 1994; Walker and de Vore, 1995; Quilter and Anderson, 2000; Hunt et al., 2005). One of the main problems with these platforms is the difficult assessment of the camera's field of view, making it nearly impossible for the operator to navigate the plane to an exact predefined location before triggering the camera. Addressing these complications, the powered paraglider described by Thamm and Judex (2006; see also below) is equipped with a live image transmission system, providing the instantaneous field of view to the operator in real time via goggles or laptop screen.

In recent years, off-the-shelf model airplanes are becoming progressively easier to fly as technical advances enable their upgrading with flight-stabilization systems, autonomous control systems, and GPS/INS (e.g., Hardin and Jackson, 2005; Espinar and Wiese, 2006; Jones et al., 2006; Grenzdörffer et al., 2008), thus making them more independent of continuous manual control and much more precise in following a predesigned flightpath.

Currently, helicopter-type UAVs—including the quadrocopter, hexacopter, or octocopter varieties with four, six, or even eight rotors—are experiencing increased attention by researchers as well as by developers of aerial survey systems (e.g., Lambers et al., 2007; Achtelik, 2008; Eisenbeiss, 2008; Ascending Technologies, 2009; Berni et al., 2009; Rotomotion, 2009; ROTROB, 2009). Such flight devices have the advantage of being capable of hovering above a survey area rather than crossing it in flight. However, the larger ones that are capable of carring the weight of a compact or even SLR camera must be judged rather dangerous aircraft regarding the considerable chopper force exerted by the rotors (Wu, Zhang and Liu, 2008). The examples described in more detail below span the range from quite basic to rather elaborate drones for SFAP.

8.5.1. Basic Model Airplane

A simple, low-cost approach for SFAP was undertaken by Graves (2007). He focused on small, lightweight, highly durable, electrically powered aircraft constructed of expanded polypropylene foam, so-called "foamies" or park-flyer model airplanes. Compared with fuel-powered, piston-engine model airplanes, foamies tend to fly slower, have less vibration, cost less, and require less pilot training, but they have limited payload capacity for a camera. Given the small size of many modern digital cameras, however, this latter limitation was not considered a serious handicap.

FIGURE 8-37 Radio-controlled, gas-powered model helicopter that may be fitted with a compact digital still or video camera. Helicopter is highly maneuverable, but noisy; rotor diameter is ~1.35 m. Photo by JSA.

A further advantage of foamies is that they are quite robust and can be repaired quickly and easily in the event of damage from a crash.

Among many possible models, Graves selected the *Multiplex Easy Star* airplane from Germany, which he modified to hold a small digital camera (Fig. 8-38). The *Easy Star* consists of foam wings, body, and tail; it is powered by a small pusher-prop electric motor mounted above the body just behind the wing. This configuration has two advantages compared with a nose-prop arrangement.

- Propeller and motor are less likely to be damaged in rough landings, which happen frequently during the learning phase.
- Camera could be mounted in the front of the aircraft without the propeller obstructing the view and without having to modify the motor mounting.

Power is provided by a nickel-metal-hydride (NiMH) battery, which can be recharged quickly in the field from a vehicle. The wings have a dihedral design that leads to stable flight, but decreases aerobatic ability. The rudder and elevator control flight direction and turning without the need for ailerons. This simplifies piloting the plane and leaves one channel on the radio control open for operation of the camera shutter microservo. The plane can be broken down and all components can be packed in a storage box, such as a travel golf case, for easy portability.

Pilot training required several weeks of practice flights with a ballast block in place of the camera, during which time situational awareness and wind conditions were emphasized. Graves (2007) noted six key factors for successful model airplane piloting: wind, orientation, speed, altitude, over control, and pre-flight check. After mastering takeoff, flight, and landing maneuvers, a digital camera was placed onboard for further test flights (Fig. 8-39).

Graves and JSA conducted operational testing at Blue Lake, a geothermal spring and marsh complex in the desert basin of western Utah, United States. Three flights were taken, each lasting 12–15 min with about 20 images collected per flight. Between flights the airplane battery was recharged. Excellent overviews of the Blue Lake environment and surroundings were acquired under near-calm flight conditions (Fig. 8-40). Graves was able to achieve his goal of building and operating a basic model airplane for SFAP at low cost—less than $US 500 total.

8.5.2. Autopiloted Model Airplane

Pilot-operated model airplanes, as shown by the previous example, are not the easiest platform for SFAP. They require considerable flying experience of the pilot to start with, and even more experience and technical skills for pin-pointing exact locations with vertical photographs or covering large areas in systematic flightline arrangements. Recent developments in global positioning system (GPS) and inertial navigation system (INS) technologies are now leading to autopiloted model airplanes which can autonomously follow prescribed flightlines.

FIGURE 8-38 Close-up view of the *Multiplex Easy Star* model airplane in the field with small digital camera mounted for oblique views to the forward port (left) side of the plane. The camera and shutter microservo are secured quite simply with rubber bands. Green tape on the nose was used to repair damage from a previous crash. Photo by JSA.

In his Master's thesis, C. Claussen, assisted by M. Niesen, developed an automated navigation system for model airplanes which fully controls flightpaths, flying height, velocity, and airplane landing. Both are among the founders of the German *MAVinci* company (Claussen et al., 2008), which has subsequently expanded and refined the system for controlling unmanned aerial photographic surveys. The autopilot software now integrates the NASA *WorldWind* viewer; the information about the survey area is fed into the software via this viewer or alternatively via GPS coordinates or placemarks digitized in *Google Earth*. The direction and parallel distance of the flightlines, flying height, and flying velocity can be set by the operator or automatically computed by the software for an optimal flight plan.

The target flightpath is transferred to the on-board computer in the plane via radio communication. During the flight, this continuous two-way wireless communication reports the plane's location and orientation, so the actual flightpaths can be visualized in the NASA *WorldWind* viewer or *Google Earth* in real time. The wireless communication also enables to correct deviations from the flightpath or even force the plane to repeat flightlines which have not been met with sufficient precision due to wind drift, etc. The control can be toggled between the autopilot software and a handheld radio controller so that the operator can take charge of the launching and landing phase, but even this is not necessary when the ground conditions allow fully autonomous takeoff and landing. The maximum flying height in Germany, as demanded by national aviation laws, is restricted by the visibility of the plane and amounts to about 500 m.

Claussen and Niesen's autopilot system makes the *MAVinci* planes the most technically advanced and most automated SFAP platform employed by the authors so far.

FIGURE 8-39 B. Graves prepares to launch the *Easy Star* model airplane from his right hand while he holds the radio controller in his left hand. Photo by JSA.

It considerably reduces the role of the operator compared to hand-controlled model airplanes, but some basic flying skills, which could be gained with a similar off-the-shelf model and with a flying simulator, are still indispensable. The plane is launched by hand as shown in Figure 8-41A, and the autopilot takes over immediately, directing the plane in tight circles for gaining height (see Fig. 3-9). However, most ground conditions require the plane to be landed by hand-control to avoid a rough end to the survey (Fig. 8-41B).

So far, the *MAVinci* team has developed two SFAP airplanes meeting different demands, a smaller one with a digital compact camera and a larger one with a calibrated DSLR camera, as prototypes for photogrammetric survey platforms (Fig. 8-42). For the smaller system, a *Multiplex Twinstar II* made from *Elapor* with a wing length of 1.4 m and a weight of 1.7 kg was adapted to accommodate the GPS/INS control system and a *Nikon Coolpix* camera (ca. 200 g). The camera is completely concealed in the body of the plane and points vertically to the ground through an opening in the plane bottom. A single lithium-polymer (LiPo) rechargeable battery (11.1 V, 5000–7000 mAh) provides power for the two propeller motors, the GPS/INS, and camera, allowing a mission time of approximately 40 minutes or mission distance of 35 km. The battery as well as the camera memory card can be reached and removed when the cockpit cover is taken off.

Following the requirements given by two of the authors (IM, JBR) this first system design was modified for a prototype of a larger system capable of conducting stereoscopic surveys with a calibrated DSLR camera suited for photogrammetric analysis. An existing *Canon EOS* 300D camera with 28 mm lens, weighing 1.1 kg in total, was to be installed in the plane. This required a larger model that would be able to carry such high payloads without exceeding the 5 kg limit for authorization-free model aircraft in Germany (see Chapter 9.8.). A *Multiplex Mentor* with a wing length of 1.6 m and a body length of 1.2 m was modified by mounting two propellers to the wings and additional *Elapor* moldings to the bottom of the plane, where the camera lens needed protection (see Fig. 8-41A). The resulting somewhat bulky appearance earned it the nickname "*der Bulle*" (the bull), and its heavy weight of 3.3 kg reduces the mission time to

FIGURE 8-40 Panorama of the Blue Lake hot spring and marsh complex in the desert basin of western Utah, United States. The boardwalk and dock in lower center provide access for scuba divers to enter the deep pool. View toward southeast, October 2006. Photo courtesy of B. Graves.

about 15 minutes, but these limitations will be improved in a future system based on this first prototype. In cooperation with another company, *MAVinci* is currently developing a third plane that will carry a 2.5-kg payload and have a mission time of 50 minutes. This plane will be capable of integrating different types of cameras depending on the application.

The velocity of the plane is an important factor in survey planning (see Chapters 3.3 and 9.6). For the two *Multiplex* models, the nominal airspeed of 45–70 km/h results in a wider range of ground speeds depending on the wind conditions. Usually, the flightlines are arranged parallel to the wind direction, so the airplane is considerably slower out than back, a fact which needs to be taken into account when calculating exposure intervals and image coverage. For stereoscopic coverage from low-flying heights, the slow interval between shots (low frame rate) for continuous exposure series is currently still a problem with many digital cameras. For example, the fastest interval with the *Canon EOS* 300D is 3 seconds for JPG images and 4 seconds for RAW images, which is (assuming a ground speed of 70 km/h and 28 mm lens) too slow for achieving 60% overlaps at flying heights below 240 m. The *Nikon Coolpix* compact camera is not much faster. Recent DSLR camera models and industrial cameras often offer shorter exposure intervals, and this requirement may be satisfied in a future version of the plane. For now, stereoscopic coverage at lower flying heights is simply achieved by repeated overflights on the same flightpaths.

The camera exposure is triggered by the on-board computer in regular pre-set intervals so that a continuous image series is taken from the moment of launching till landing of the plane. In the refined autopilot version of *der*

FIGURE 8-42 The two autopiloted model airplanes developed by the *MAVinci* team. Background: *Multiplex Twinstar II*, carrying a digital compact camera. Foreground: modified *Multiplex Mentor* (dubbed *der Bulle*; see also Fig. 8-41) carrying a DSLR camera. The cockpit cover of both planes is removed, giving access to the batteries. The cameras are mounted beneath the detachable wings. Photo by M. Niesen.

Bulle plane, the triggering of the camera can be constricted to a limited nadir angle of the optical axis. If the plane and thus the axis exceed a tilt of, e.g., 5°, no image is taken; this eliminates the oblique situations at the turning point of each flightline and also temporary tilts by wind pressure. GPS position and tilting angles for each image are logged into a file so that this information can later be added to the EXIF data of the image header. This feature is especially helpful for selecting suitable images for stereoscopic viewing and photogrammetric analysis. For use as exterior orientation parameters in photogrammetric triangulation without additional ground control (see Chapter 3), the recorded data are

FIGURE 8-41 (A) *MAVinci* plane being hand-launched by C. Claussen. The additional *Elapor* moldings mounted to the bottom of the plane serve as a protection for the camera lens and as a skid for softer landings. (B) M. Niesen landing the plane on a plowed field. In spite of the dirt plume the camera lens was barely dusted. Photos by G. Rock, February 2009.

not precise enough, especially as exact synchronization with the camera trigger is difficult owing to the shutter time lag. However, they are sufficient to enable tagging the images with the georeferencing information used by NASA *WorldWind* or *Google Earth*, so the flightpath and whole image block can be visualized in these viewers and checked for completeness and gaps (Fig. 8-43).

The *MAVinci* airplane is currently the only platform in use by the authors which allows the automatic visualization of flightpaths, image exposure stations, and the images themselves by using the exterior orientation parameters in viewers such as *Google Earth*. However, these exciting and useful features, which are enabled by recent GPS/INS technology, also could be incorporated in other platforms. This is rather a question of costs than carrying capacity. In terms of transportability, the autopiloted airplane is quite mobile and can be transported easily by car. The modular assembly design allows uncomplicated replacement of broken or defective components by spare parts.

8.5.3. Powered Paraglider

The powered paragliders belong to the slowly flying, remotely controlled camera platforms that have been developed by the military for simple reconnaissance purposes. Accordingly and in contrast to the positive public perception of kites, balloons, and blimps, the image of these drones in public is rather negatively connotated, as sometimes is the case for model airplanes. The construction is characterized by low weight, robustness, and convenient handling in operation. The development objective of such devices is, after all, the quick glimpse behind the next hill or around the next street corner. The safety of the user and system itself and continuous deployability are not primary attributes of this platform.

Usually a powered paraglider is based on a simple three-wheeled chassis frame with a high-performance gasoline engine and a vertical propeller mounted above the two back wheels. The downward-facing camera is gimbal-mounted to the connection between front and back wheels. Frequently an additional video camera facing in flight direction is attached. By means of a small monitor or life-view video glasses, a realistic flight perspective is given for the pilot, which enables flying the remotely controlled drone without direct visual contact. The whole system is attached to a rectangular parachute (or kite) that is constructed like a parafoil with wing-shaped cross-section, like a manned paraglider. The glider is steered by shortening of the left or right line attached to the trailing wing edges (brake) to turn right or left.

The motor-powered paraglider shown in Figure 8-44 was developed for the French Foreign Legion by *ABS Aerolight*

FIGURE 8-43 The complete set of images taken during a flight of the autopiloted model airplane at Freila, Province of Granada, Spain (see Fig. 3-9) is visualized in *Google Earth* by automatic (approximate) georeferencing using the exterior orientation parameters logged in the EXIF headers. Small airplane symbols indicate the image centers and flying directions. Screenshot from *Google Earth*®.

FIGURE 8-44 Motor-powered paraglider (parafoil not attached) built by *ABS Aerolight Industries* in France. M, 3.7 kW two-stroke motor; T, removable 1-l tank, enabling a flight duration of about 45 minutes if completely filled; C, control box for motor, paraglider, and camera; W, servo-driven winch for height and direction control. The circular lines run from the servos that are fixed next to the back wheels across guide rollers fixed to the upper frame rods; G, circular grate protecting motor, propeller, and operators; B, heavy-weight bandages used for trimming. Photo by G. Rock.

FIGURE 8-45 Start of the motor-powered paraglider. Two persons hold the parafoil aloft and one person controls the motor. If the lines are not held tight they may easily get caught up in the propeller and lead to dangerous situations. Photo by G. Rock.

Industries in Montpellier (France) and can nowadays be purchased for civilian use as well. The frame is made of high-quality aluminum used for aircraft manufacturing. This material is robust and can be bent easily to its original shape after a crash. The aircraft-model wheels have pneumatic tires for shock protection. The 50 cm^3 *Yamaha* chainsaw engine has 3.7 kW and is powered with a two-stroke mixture. The motor drives the wooden 37-cm propeller that creates the necessary propulsion for takeoff. For safety reasons, a simple grate rack is attached to shield the propeller and reduce the risk of injury.

Details of the camera mount are given in Chapter 7.4. The $\sim$4.5 m^2 parachute (Figs. 8-45 and 8-46) is an elaborately stitched parafoil type; it is attached to the chassis with 32 lines merging to 16 and then again to four lines ending in carabiner locks. An additional 12 lines at the rear are used for direction control (Fig. 8-46B). The parafoil is two-colored, yellow at the left and violet at the right side, in order to simplify orientation.

For takeoff, 30 m of agricultural road or smooth field surface are sufficient, but an open flight field of 100–200 m to the left and the right is necessary because even slight irregularities of the surface lead to a lateral breakaway of the three-wheeled frame. Correcting the flightpath by countersteering is only possible at flying height $>$20 m, because it leads to the drone descending. Rough terrain or rocks and small shrubs may cause the drone to overturn.

Flying large loops is the best method for covering the study site and can be performed even by beginners. The flight speed may be as slow as 45 km/h in calm conditions. Because the paraglider is clearly visible over long distances, flying altitudes up to 1000 m can be achieved. In stronger wind, the paraglider may stand stationary relative to the ground or even be driven backwards at medium engine speed. Thus, the drone can be positioned comfortably in the air over the target site to be photographed. However, owing to the lagged reaction of the paraglider to steering and due to the sensitivity to wind influence, the pilot needs considerable practice. A good strategy with climbing, diving, accelerating, and throttling is needed particularly at high wind velocities for exact positioning.

For security reasons the motor is switched off before landing. In gliding flight, the possibilities of maneuvering are limited, so aborting the landing for touch-and-go is not possible. Even on a smooth runway the drone tends to overturn (Fig. 8-47), but with the motor being switched off, severe damages are unusual. The most vulnerable parts are the camera and the propeller, so a replacement propeller and a robust compact camera are recommended.

Altogether, the powered paraglider is impressive by its inertia in flight, the slow possible flight velocities, and its robust landing behavior. However, steering demands much practice even for those already trained in model airplanes. As an aircraft much akin to a flying lawnmower, it is among the more dangerous platforms.

FIGURE 8-46 Powered paraglider in flight. (A) Low-level flight above a slope covered with *Stipa* grass in Andalusia, Spain. (B) Paraglider in sharp left turn. The picture shows how the shortened control lines at the left side of the drone are downfolding the left wing-flaps. During such fast turns the camera cannot be kept vertical by the gimbal mount owing to the intense centrifugal forces. Photos by G. Rock.

FIGURE 8-47 Landing maneuver of the powered paraglider. (A) In landing approach above an abandoned field with *Stipa* grass and *Thymus* shrubs in Andalusia, Spain. (B) Paraglider landed. Despite the spectacular rollover the drone as well as camera remained undamaged; the paraglider took off again a few minutes later. Photos by G. Rock.

8.6. PROS AND CONS OF DIFFERENT PLATFORMS

This chapter has introduced a variety of platforms with different characteristics, all of which have been used by the authors in their work; even more can be found in the literature. A basic distinction between all platforms is the question of tethering or free-flying. Tethered platforms inherently have a lower operating space but are easier to position over fixed points. Free-flying aircraft may cover larger areas and distances but, with the exception of hovering types such as helicopters and multicopters, tend to cross over small areas of interests in the wink of an eye. Autopiloted systems and flightpath planning prior to a survey or live-view transmission systems are therefore highly recommended for such platforms.

The choice of possible platforms for an SFAP project is huge and continues to grow. It is difficult to generalize about advantages or disadvantages for particular platforms, because possible applications and working conditions vary greatly around the world. Cost of equipment and availability of trained personnel must be considered along with necessary logistical support, transport issues, and legal aspects associated with various types of platforms in different countries.

Nonetheless, it is clear that certain platforms would have desirable characteristics for conducting SFAP under particular circumstances, and what might be the ideal SFAP platform in one situation could be ineffective, impractical, or impossible in another. The decision for a particular one depends on numerous aspects which are always specific to the financial means, the technical skills of the personnel, the study area location and size, the required image characteristics, etc. Table 8-1 is an attempt of a more-or-less

TABLE 8-1 Comparison of selected platforms commonly used for SFAP based on authors' experiences and literature reports.

SFAP comparison	Platform cost	Operating energy	Operating cost	Transport-ability	Personnel	Risk	Wind	Flying height	Payload	Technical level	Positional precision	Areal coverage	Stereo coverage	Vertical imaging	Exposure time range
Manned LSA	●●●●●	aviation fuel	●●●●○	●○	1-2	●●●●○	calm to moderate	1000+ m	10 to 50 kg	●●●●○	●●○	●●●●●	●●●●●	●●●	●●○
Helium blimp	●●○	helium	●●○	●●○	2	●○	calm to light	300+ m	5 to 10 kg	●○	●●●●●	●●●○	●●●○	●●●●●	●●●●●
Hot-air blimp	●●●●○	propane	●○○	●●●	4+	●●○	calm	300+ m	5 to 10 kg	●●●○	●●●●●	●●●○	●●●○	●●●●●	●●●●●
Kite	●○	wind only	○	●●●●●	1-4	●○	light to strong	300+ m	<1 to 3 kg	●○	●●●●○	●●○	●●○	●●○	●●○
Basic model airplane	●○	rechargable battery	●○	●●●●	1-2	●○	calm to light	300+ m	≤ 1 kg	●●○	●●●○	●●●●○	●●●●○	●●○	●●○
Autopiloted airplane	●●●●○	rechargable battery	●○	●●●●	1-2	●○○	calm to moderate	1000 m	≤ 2 kg	●●●●○	●●●●○	●●●●●	●●●●●	●●●○	●●●
Powered paraglider	●●●○	two-stroke gas mix	●○○	●●●○	2-3	●●●○	calm to light	1000 m	≤ 2 kg	●●●●○	●●●○	●●●○	●●●○	●●○	●●○
Model helicopter	●●●●○	battery or gas mix	●○○	●●●○	1-3	●●●○	calm to light	300+ m	<1 to 20 kg	●●●●○	●●●●○	●●●●●	●●●●●	●●●○	●○

See text for details of the individual aspects and parameters. Qualitative estimates: five solid dots, highest or best (relative to other platforms); open dots, variable, or uncertain rating. Maximum flying-height regulations vary from country to country and may be more restrictive in some countries.

subjective comparison looking at different parameters and characteristics, resulting from the authors' own experience and assessment combined with reports from colleagues and literature. The following considerations accompany the individual aspects.

- *Platform cost*: including all accessories except camera rig and camera.
- *Operating energy and costs*: electricity, fuel, lifting gas, or other energy sources necessary for platform power, camera, and other equipment.
- *Transportability*: is a function of equipment weight, storage space, and required transport means.
- *Personnel*: minimum number ranges from a single person to more than half a dozen people, as a result from different requirements depending on platform operation and mission objectives.
- *Risk for personnel*: a rough assessment of the risks the platform may present to the personnel (and onlookers) in case of malfunctions, forced landings, crash, etc. Ranging from no risk, to light and serious injuries, to danger of death.
- *Wind conditions*: approximate range of wind velocities in which the platform may be used safely.
- *Flying heights*: approximate realistic flying height at which the platform may be used safely and which may be subject to legal restrictions.
- *Payload*: includes camera, rig, ancillary equipment, and suspension; the ranges result from different carrying capacities depending on the platform size.
- *Technical sophistication*: degree of high-tech components, which may require specialized training to operate, and which could be difficult to adjust or repair in the field.
- *Positioning precision*: refers to the capability for positioning the camera exactly over a given ground target for vertical shots or with a specific look direction and tilt angle for oblique views.
- *Areal coverage*: refers to the ability to cover larger areas with gapless, contiguous vertical images.
- *Stereoscopic coverage*: refers to the ease with which stereopairs with photogrammetrically useful overlaps, exposure angles, and scale similarity may be acquired; in combination with the two preceding parameters for judging the possibility of acquiring individual stereopairs or larger stereoblocks.
- *Vertical images*: reliability with which a vertical image ($<3°$ tilt) may be acquired.
- *Exposure times range*: depends on the platform velocity and vibrations. Low ranges mean short exposure times ($<1/1000$) are required, larger ranges mean longer exposure times (e.g., 1/125) are also tolerated.

8.7. SUMMARY

Much innovation in recent years has led to many kinds of platforms utilized for SFAP. Aircraft include potentially any and all types of flying machines, which may be manned or unmanned, powered or unpowered, heavier or lighter than air, and free flying or tethered to the ground. At a minimum, successful SFAP under various conditions depends upon a platform that is capable of lifting the necessary camera rig and maintaining a position above a desired spot on the ground or flying a line or grid of ground coverage. Size and complexity of these platforms vary enormously, ranging from simple kites to substantial manned aircraft. For the most part, manned platforms are larger, require an experienced pilot, are more expensive to operate, and need greater logistical support compared with most unmanned aircraft. Unmanned platforms have experienced rapid development during the past decade, and many new techniques may be expected in the near future for both manned and unmanned SFAP platforms.

Chapter 9

SFAP Survey Planning and Implementation

Base all your work on the reality of the field; there lies the truth if you are clever enough to extract it.

D. Shearman (1958); quoted by Dewey (2004)

9.1. INTRODUCTION

A successful small-format aerial photography (SFAP) field survey often depends on the ability to react flexibly to a plethora of complications—the more of these that can be anticipated and planned for in advance, the more likely one would return with plenty of good images. SFAP can be undertaken quickly and spontaneously, for example, with a lightweight minimal kite system for an afternoon leisure hour at the beach to capture a bird's-eye view of that spectacular cliff above the churning blue-green surf. But it usually takes more time and effort to prepare and conduct the survey of a specific site for scientific purposes. The considerations related to planning a field survey, which are discussed in more detail in the following, include

- Travel and equipment logistics.
- Accessibility of the site, flight obstacles, wind, and other site characteristics.
- Personnel required (see Chapter 8).
- Ground control.
- Flight planning considerations regarding the desired image area, scale, and resolution.
- Legal issues.

Some of these issues require that the location, size, and characteristics of the survey area are already fairly well known, but this is not necessarily always the case. There are many situations for which not much is known about a site where SFAP is being planned. The authors often have been invited along by other colleagues to take aerial photographs of biological test plots, archaeological remains, or architectural monuments at sites not previously familiar to us. In many cases, our hosts knew little or nothing about the logistical requirements of successful SFAP.

Often the initial information about the site is limited to a rough estimation of size and directions of how to get there. Luckily, in times of *Google Earth* and online map services, it has become much easier to gather information about remote areas around the world. Maps as well as satellite and aerial images can be extremely helpful in SFAP mission planning even for previously known sites, offering possibilities of distance measurements, ground control preparation, access planning, assessment of obstacle problems, and even flightline planning for autopiloted aircraft.

9.2. TRAVEL AND EQUIPMENT LOGISTICS

For a successful mission, the SFAP equipment and personnel must be available to reach the study site at the desirable time of day or year to meet the goals of the project. This invariably requires travel for people and transport of baggage. The logistics could be as simple as packing kite aerial photography equipment in the car for a weekend trip. Still this depends upon good weather conditions—sunshine and favorable wind, which are not always so easy to arrange in practice. At the other extreme, the mission may require extensive pre-flight planning and site preparation on the ground and involve considerable equipment and personnel to operate the SFAP platform. The logistics for this sort of project may take months to organize.

Travel options depend to a large extent on the type of SFAP platform as well as location and conditions of the study site. Nearby sites adjacent to paved roads are the easiest to reach, of course. In such situations, the photographer has few limits on how much equipment and how many field assistants to take to the site. On the other hand are distant or remote locations with poor transportation connections—reached in the final stage perhaps only by footpaths. In these cases, a minimal amount of equipment and personnel would be desirable.

Often the last few kilometers on the way to a specific study site are the most difficult, even in case of rather good connections via roads or tracks. During several field trips in the Bardenas Reales in the Ebro Basin, one of the driest regions in Spain, for example, the research team had severe problems in approaching the study site. Particularly in winter and spring the agricultural roads may be quite muddy and nearly impassable, even for all-terrain vehicles.

For cases of generally or temporarily bad accessibility, the equipment needs to be packed in a way that enables transportation by means of human (or animal) carriers (Fig. 9-1). Waterproof transportation boxes made of plastic or aluminum have shown to be necessary requirements in order to bring flight and camera equipment to the study site facing bad weather and difficult terrain. Oversized cargo boxes as well as cardboard packaging held together with parcel tape should be avoided. Neither has purpose-built packaging with made-to-measure partitions for the individual parts proven to be a sustainable solution, because of continued further development of the equipment and consequent changes in size and shape of camera rigs and other items. Accordingly, the best practice is to use several smaller, stable, dust- and splash-proof cargo or storage boxes with specially fitted, but easily replaceable foam liners for fixing and padding the scientific devices (see Fig. 8-36).

International travel to study sites presents further difficulties, especially if flying commercial airlines, because of customs and security concerns about the purpose of travel and the unusual equipment in baggage (Fig. 9-2). The authors have been questioned often at airports about the nature of our equipment; various cameras, batteries, kites, radios, and other odd-looking gadgets seem quite suspicious to security inspectors.

Certain types of platforms require fuel for power or gas for lift (see Chapter 8). Helium is commonly used for inflating balloons and blimps. However, there are few commercial sources of helium in the world; helium is either quite expensive or simply not available in many countries, and long-distance transport of heavy helium tanks is usually not feasible for most SFAP. Hot-air balloons and blimps are an option where helium is not available, because propane or cooking gas to fire the burner may be found just about everywhere. Decanting gas from commercial gas bottles into fuel tanks modified for flying is not always trouble-free. A special adaptor and tubes kit is necessary, which can be purchased rather cheaply in camping stores, particularly in shops providing motor caravan equipment. Gas-powered model airplanes or drones require fuel that is highly flammable. Transport of such fuels is strictly prohibited on commercial airlines, so a local source would be necessary. Electric-motor power would be more desirable in such cases, as the batteries could be recharged or replaced locally.

FIGURE 9-1 Tackling the 400-m altitude difference between the last possible parking spot (see valley bottom) and a high-mountain survey site in the Cuenca de Izas, High Pyrenees, Spain. JBR (second from left), with the hot-air blimp envelope on a frame-carrier pack, shares the weight of the burner system with one of his students at the beginning of a 1.5-hour ascent. Photo by J. Heckes, October 1995.

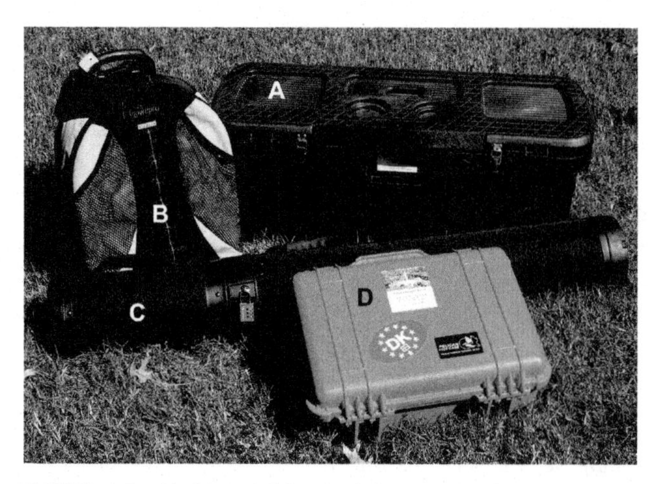

FIGURE 9-2 Airplane-suitable travel baggage for international kite aerial photography. (A) Cargo box with line reel, spare parts, extra batteries, radio controller, etc. (B) Waterproof *Lowepro* pack with one complete camera rig and radio controller. (C) Long kites in golf-club case. (D) Waterproof *Pelican* case with second complete camera rig. (A) and (C) are checked luggage; (B) and (D) are carry-on items. Photo by JSA.

Regarding electricity, it should be noted that two domestic AC power standards are found in different parts of the world: (a) 110–120 V, 60 Hz and (b) 220 V, 50 Hz. In addition, at least half a dozen standard types of electric sockets are utilized in various countries. Most small electronic devices, such as laptop computers, accept either power source, but a selection of socket adapters may be required to "plug in" to the local power grid. In some cases, the authors have modified or customized socket—plug connectors for necessary electricity to power devices or recharge batteries.

Overall, SFAP logistics is usually a matter of common sense and experience. A rule of thumb is to employ a system that is robust, easy to transport, and, most importantly, capable of acquiring the necessary type of imagery. More equipment—platforms, cameras, batteries, fuel, radios, monitors, and other devices—means that more things can go wrong during travel or at the SFAP field site. Careful pre-flight planning and logistical preparations may eliminate many potential problems; still, mishaps cannot be avoided completely and should be expected. Spare parts and backup equipment are often essential for completing a mission.

9.3. SITE ACCESSIBILITY AND CHARACTERISTICS

9.3.1. Local Site Accessibility

Accessibility of a site is an important consideration for unmanned SFAP, as the survey system is controlled by one or more operators on the ground. Depending on the system and size of the area, the operator(s) may need to move around quite a lot, making accessibility and ease of action a rather central issue. Some sites may be impossible or difficult to access owing to their conditions (wetlands) or conservation reasons (nature reserves); others may just be difficult to move or navigate in because of tall weeds, rugged terrain, forest, high mountains, etc. As a general guideline, the more difficult the terrain, the more personnel can be recommended to help with navigating, relaying messages, taking bearings for flight directions, etc.

Direct accessibility by vehicles may be an issue with large and heavy equipment, although most of the systems presented in this book can be transported by foot or small boats with a little goodwill and exertion and have indeed often been so by the authors. Some systems have special requirements for launching places (e.g., size, flatness, local wind conditions, distance to obstacles), which also can mean that the operators have to walk some distance after launching the system if no convenient spot can be found directly at the site.

9.3.2. Flight Obstacles

Reading about Charlie Brown in the comic strip *Peanuts*, if not by personal experience, has taught us that trees are the nemesis of kites. When a kite or its tail becomes caught, it is important not to pull on the kite line, as this would only drag the kite deeper into the tree. Let the line go slack. With luck, wind may lift the kite out of the tree. Otherwise climbing the tree may be the only means to recover the kite (Fig. 9-3). Most such incidents happen because of inattention by the kite flyer; a moment spent adjusting the camera rig or other equipment may be just the instant when the kite makes a loop or takes a dive.

Actually trees and other obstacles present problems for all types of tethered or free-flying platforms. Tangling a blimp tether at the top of a tall tree, for instance, leaves few options for recovering the blimp and camera rig. The most dangerous obstacles are high-voltage power lines, along with telephone wires, railway power lines, radio towers, light poles, and other electrical sources. Dry tether lines normally are not conductive for electricity, but may become conductors when damp, which represents a serious safety issue in cloudy, foggy, or rainy weather. The best solution is simply to keep well away from any such power supply. Other potential obstacles for tethered platforms include buildings, steep cliffs, and high river embankments.

FIGURE 9-3 Rokkaku kite caught in a tree at a cemetery study site. (A) Kite lodged near top of tree; slack kite line to lower left. (B) JSA standing on a stout tree limb to free the kite. The kite and tail were recovered without damage. Photos by SWA, Kansas, United States.

In a safety hierarchy are first and foremost people—both the SFAP operators and bystanders, followed by buildings and vehicles, wildlife and natural features, and lastly the camera rig and platform. This order of priorities should be kept clearly in mind before and during each flight by all members of the survey team. Once the device is in the air and everyone's attention is upward, many dangers may be overlooked on the ground.

9.3.3. Local Wind Conditions

Regarding survey planning, the position of the site within the landscape may be important for the choice of platform or for navigation planning because it may influence local winds or cause windward and leeward situations. For kites, the wind direction always must be taken into account when choosing the launching site, but for most other SFAP platforms, too, wind susceptibility is more or less crucial. Both wind direction and speed may vary at different heights above the ground, especially in alpine terrain, as discussed in the next section.

Usually the wind direction varies during the course of a day. Early in the morning until a short time after sunrise, land or mountain winds, respectively, are dominating. During the day the wind system goes into reverse so that from noon onward sea or valley wind, respectively, may be expected. Due to thermal updrafts, wind velocities are usually higher, so the preconditions are best for the use of the kite as camera platform. Toward evening the wind is likely to calm down in most situations.

Seasonally varying wind systems mainly have to be considered in the subtropics and marginal tropics. In North and West Africa the Harmattan provides excellent flight conditions in the dry period in winter, the same applies to the Cierzo in northern Spain and the Mistral in the western Mediterranean area. In summer these regions are dominated by thermally induced local winds or by calm. Similar seasonal winds are encountered in many parts of the world, for example the Santa Ana, a hot desert wind from the east or northeast in southern California. This should be taken into account if a monitoring project involves conditions in different seasons. In such cases, a combination of camera platforms for different wind velocities including a system for calm conditions has proven of value for the authors. For long-term monitoring purposes it is easier to take the pictures always in the same season and to adapt the camera platform to the wind velocity prevailing at that time.

9.3.4. High-Altitude SFAP

According to Cohen and Small (1998), more than half of all humans live at elevations below 300 m, and nearly 90% live below 1000 m altitude. Thus, it is no surprise that most SFAP has been undertaken at relatively low altitudes. Chorier attempted kite aerial photography at 5600 m at Khardung La pass in the Jammu and Kashmir province of northern India, but was thwarted with strongly gusty wind and temperature of −15 °C. He did succeed nearby in the village of Khardung just below 4000 m altitude (Chorier and Mehta, 2007). This is the highest successful kite aerial photography known to the authors. In many high-mountain regions airfields even for small single-engine microlight aircraft are either not existent or not usable throughout the year. Using pilotless SFAP platforms, higher risks can be taken under difficult flight conditions.

For purpose of this discussion, high altitude is considered to be those regions above ~1000 m in elevation, and tremendous scientific interest exists for all types of environments at these high elevations. Global warming in high-mountain regions and related geomorphological changes provide a great challenge for SFAP. Glacier retreat and glaciofluvial geomorphodynamics in glacier forelands, erosion on moraines, development and outburst of proglacial lakes, changes in permafrost soils (patterned ground) due to permafrost-degradation, landslides, debris flows, and mountain floods are short-term processes whose spatial distribution and change dynamics are of highest interest in the near future. The same applies for vegetation impact by climate change, overgrazing, building operations, and winter-sports tourism.

The force that holds a winged aircraft up is determined by relative wind speed and air density acting on the lifting surface. At high altitude, lower air density means that fewer and lighter molecules (per air volume) flow over the lifting surface at a given wind speed. Air density is governed by three factors—pressure, temperature, and humidity. The combination of these factors determines potential lifting power of a particular winged platform for a given wind speed.

The standard atmosphere is an ideal model of atmospheric conditions (Table 9-1). These values reveal that density decrease is fairly slight up to about 1000 m high. At higher altitudes, however, air density declines significantly. At 3000 m, for example, air density is only about ¾ that of sea-level density. However, this assumes the standard atmospheric temperature of −4.5 °C at that altitude. At

TABLE 9-1 Standard atmospheric conditions for temperature, pressure, density, and density percentage according to altitude.

Altitude (m)	Temp. (°C)	Pressure (hPa)	Density (kg/m^3)	Percent
Sea level	15.0	1013	1.2	100
1000	8.5	900	1.1	92
2000	2.0	800	1.0	83
3000	−4.5	700	0.91	76
4000	−11.0	620	0.82	68
5000	−17.5	540	0.74	62

Based on Williams (2005); taken from Aber et al. (2008).

FIGURE 9-4 Cumulus clouds building over the High Tatra Mountains, while the adjacent foreland remains cloud free in this mid-day, summer view near Strané pod Tatrami, Slovakia. Similar conditions prevail in many mountain systems during the summer monsoon season. Taken from Aber et al. (2008, fig. 3).

FIGURE 9-5 Meteorological observatory atop Kojšovská hol'a at 1246 m altitude in southeastern Slovakia. Kite flyers (*) are positioned on the downwind side of the observatory buildings and towers to avoid any chance of mishap. Access to this site required permission and carrying equipment approximately 1 km up a steep path. Photo by JSA and SWA, August 2007.

higher temperature comfortable for field work and full battery power, say 20 °C, air density is considerably less.

These conditions impact all types of winged platforms, both manned and unmanned. In the case of kites, for example, several components could be adjusted to compensate for decreased air density at high altitude—use a kite with greater intrinsic lift, increase the size of kite, use a train of multiple kites, and reduce the weight of the camera rig (Aber et al., 2008). In general, rigid kites, particularly the rokkaku type, provide the greatest intrinsic lifting power compared with other types of kites (see Chapter 8.4.1). For model aircraft, it has to be considered that lower air density leads to higher flight speed at the same energy input. At the same speed, accordingly, the available weight-bearing capacity would be lower.

Atmospheric conditions also impact lighter-than-air platforms, such as balloons and blimps. Because of the lower air pressure, hot-air systems at high altitudes have lower lifting power at a given air temperature than they would have in lower regions. On the other hand, the lifting power increases with a larger temperature gradient between the air outside and the air within the balloon. As high mountain areas are usually colder than lower regions, this effect compensates to a large extent for the change in lifting

FIGURE 9-6 Panoramic view of Elephant Rocks in the right foreground and fog-covered San Juan Mountains in the background on the western edge of San Luis Valley, Colorado, United States. Elephant Rocks are erosional features on the edge of the San Juan igneous province. Kite aerial photograph at ~2440 m (8000 feet) elevation. Taken from Aber et al. (2008, fig. 7).

FIGURE 9-7 Mount Maestas (left) and Spanish Peaks (right background) as seen from La Veta Pass in south-central Colorado. U.S. highway 160 crosses the bottom of scene. Access to the ground launch site at ~9400 feet (2865 m) elevation was via a jeep trail. Kite aerial photography was conducted with extremely turbulent wind as thunderstorms grew rapidly nearby. Photo by SWA and JSA, August 2008.

capacity. A hot-air balloon operated at 850 hPa pressure and 15 °C air temperature can carry the same payload as it could at 1000 hPa and 25 °C.

High-altitude SFAP often takes place in mountains. Mountain ranges typically create strong local climatic effects, which include colder temperature, enhanced cloud cover (Fig. 9-4), and more precipitation than for adjacent lowlands. Mountain peaks and valleys are well known for rapid weather changes. Swirling wind funnels along valleys and over passes with frequent and abrupt changes in direction and strength; alternating updrafts and downdrafts are routine. Finding a suitable open space for unmanned SFAP can be a challenge in forested mountains, and access to alpine areas above timberline may be quite limited. Areas with good access are often sites with other human structures and activities that could prove risky for tethered platforms (Fig. 9-5). The combination of cloud cover, variable wind, and limited access makes for difficult SFAP in many mountain settings, and small manned aircraft may be particularly dangerous to operate under these conditions.

The authors have conducted considerable high-altitude SFAP with both kites and hot-air blimps, in spite of such limitations. In general, thinner air with increasing elevation is not a serious problem up to ~2500 m. Thin air does become more significant above 2500 m, particularly at usual temperatures (20–30 °C) for typical field work during the growing season. High plains, mountain forelands, and broad intermontane valleys offer excellent SFAP situations with relatively open terrain, stable wind, and abundant sunshine (Fig. 9-6). Mountains are more difficult, however, because of frequent cloud cover, gusty wind, and limited ground access (Fig. 9-7).

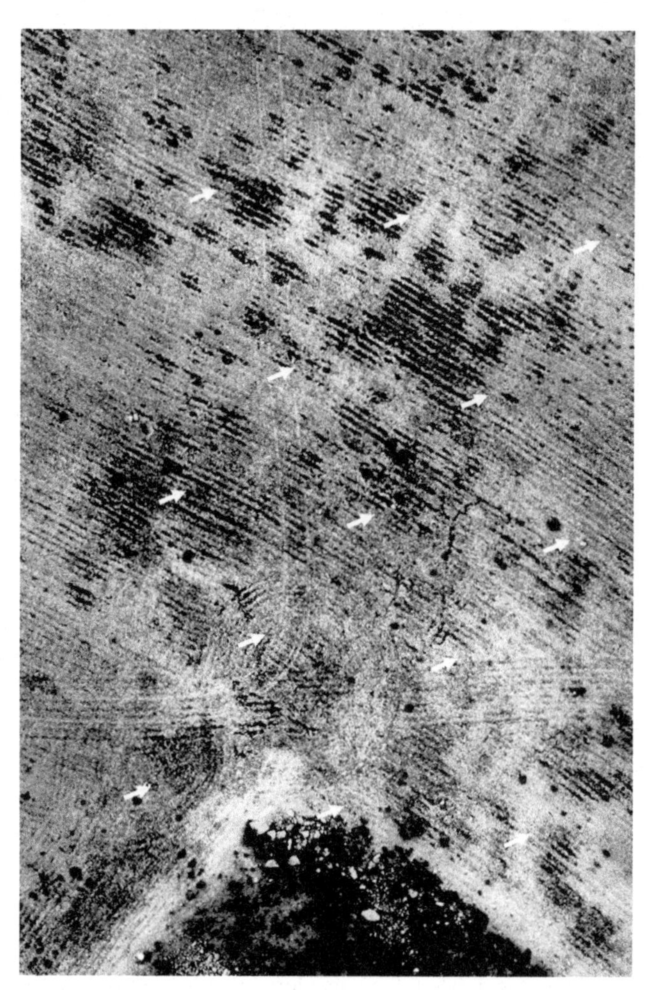

FIGURE 9-8 Regular pattern of ground control points (arrows) on a 24 m × 36 m, slightly sloping site for vegetation and erosion monitoring near María de Huerva, central Ebro Basin, Spain. Red target signals (larger in corners and center) are placed around permanent markers made from metal pipes to allow easy identification in the image. Hot-air blimp aerial photograph by IM and JBR, April 1996.

9.4. GROUND CONTROL

For all applications that involve measuring and mapping, ground control is necessary for georeferencing and geometric correction of the images (see Chapter 3). Ground control points (GCPs) are features that appear on the photographs and whose locations in a reference system are known (Warner et al., 1996; Chandler, 1999; Wolf and Dewitt, 2000). Given the typically large scales of SFAP images and small sizes of the covered areas, GCPs are usually small but well-defined natural features or pre-marked artificial features whose coordinates have to be determined in the field using total stations or global positioning system (GPS).

The geometric accuracy of a map derived from the photographs would depend strongly on the accuracy and precision with which GCPs were measured in the reference system and can be identified in the image. Therefore, care has to be taken that the quality of ground control matches the purpose of the survey and resolution of the images. Establishing ground control can be a time-consuming and costly part of the survey and must not be underestimated.

9.4.1. GCP Installation

As discussed in Chapter 3, a theoretical minimum of three GCPs is necessary for adjusting an image to a reference coordinate system. In reality, more GCPs are necessary to cover an area with differing elevations well enough for georeferencing and to ensure that sufficient GCPs appear on each photograph (see Chapter 11.2). For more advanced photogrammetric analysis, a surplus of GCPs is advisable which can be used as check points and enable error assessment of digital elevation models (DEMs) derived from the images (Chandler, 1999). The points should be distributed evenly over the area so they represent the different terrain heights (Figs. 9-8 and 9-9A) and can be seen easily from above (i.e., not too close to trees, bushes, or walls so they are not concealed by shadows or objects tilted due to relief displacement). In order to avoid unnecessary work, it is also advisable to check during installation of the GCPs if they can be arranged so they are easily visible from one common vantage point for coordinate measurements (see below).

Especially when establishing permanent ground control for monitoring purposes, a detailed plan or sketch of the site and the control-point positions (Fig. 9-9B) is extremely valuable for avoiding GCP confusion during image processing and for re-discovering the points in the field in future missions. Although the images themselves would help to locate the GCPs, the first survey might not have covered the complete marked area and patterns on the ground might have changed, so additional information could be quite helpful, for example, "on highest point of small bump" or "2 m south of dead pine tree."

The position of the control point has to be fixed with some pinpoint object like a metal pole. Pieces of construction steel or water pipe are useful and cheap options and are easily available around the world from plumbing or hardware stores in the local industrial park. During the SFAP survey, the points have to be marked additionally with target signals for easy identification and accurate location in the image, preferably with eye-catching shapes and colors contrasting well with the natural background. If the ground cover is

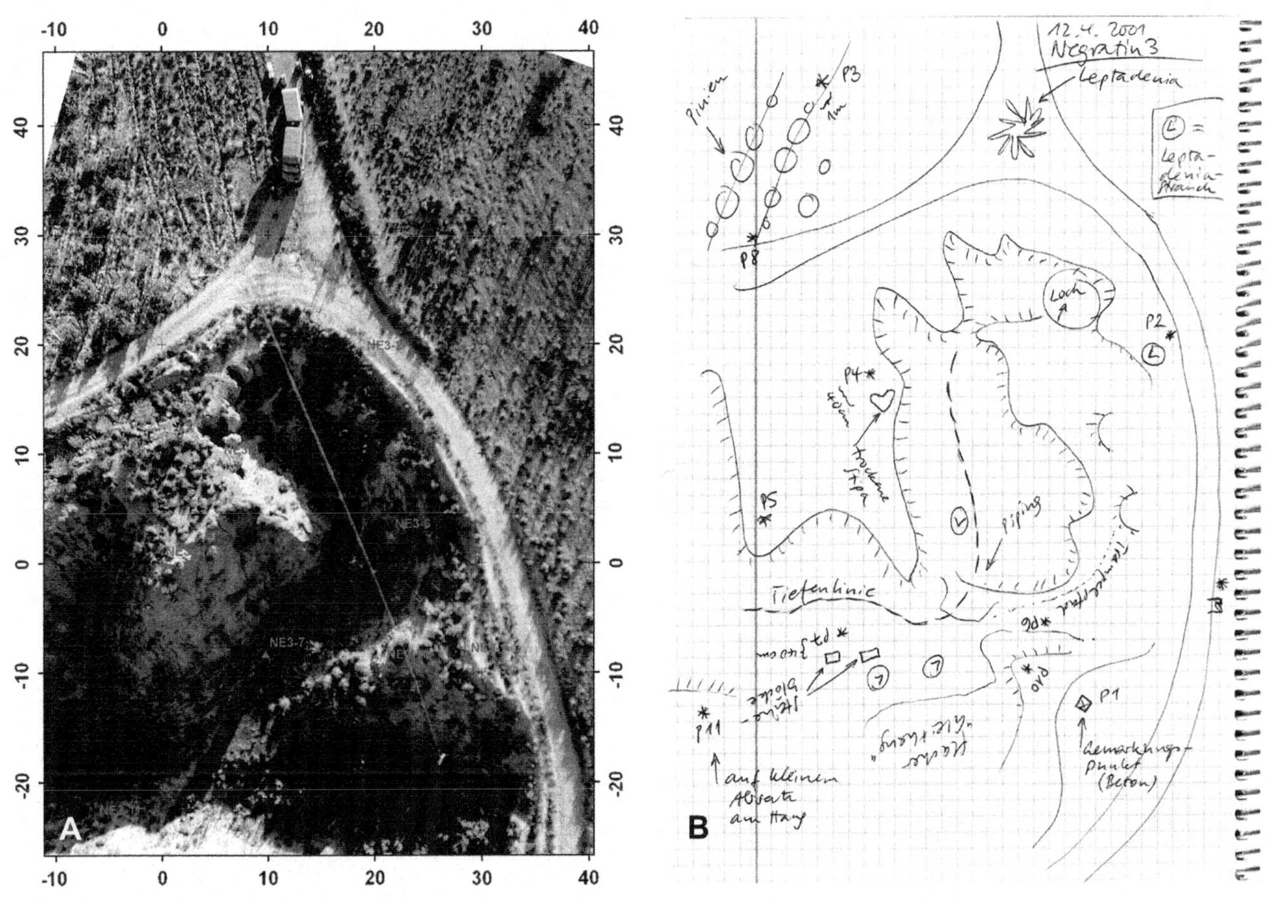

FIGURE 9-9 Gully Negratín 3 near Baza, Granada Province, Spain. (A) Image map with ground control points. GCPs are distributed in and around the gully at varying terrain heights. Hot-air blimp aerial photograph by IM and JBR, March 2006. (B) Field-book sketch indicating the positions of GCPs and landmarks for orientation in the field. Note the internal distortions of the sketch map compared to the corresponding image map (A), which result from the difficulties of overseeing a complex terrain while mapping from different vantage points. Sketch taken from IM's field book 2001.

uniform, e.g. grassland, and without features that allow an unambiguous allocation of the images, it might be necessary to use different markers in order to distinguish the individual points. In addition, a large north-arrow marker can be placed in proper position in the study site (see Fig. 2-3).

The size and design of the target signals influence their visibility in the digital image—they should be large enough to be identified easily in the image and yet enable the user to locate the actual point position accurately. Consider the example given in Chapter 2 with a vertical photograph acquired at a height of 100 m with 35 mm lens and 0.009 mm pixel size: the ground sample distance (*GSD*) is 2.6 cm. In order to be recognized as a distinct object, the target signal has to be at least 3–5 times larger than the *GSD* (8–13 cm). Depending on the "background" of the surface cover, this object could still be completely lost between patterns of vegetation, stones, and clumps of soil, so it would be best to surround this pinpoint signal centered over the GCP with something conspicuous, again 3–5 times larger, e.g. approximately 40 cm in diameter.

Depending on the nature of the study, temporary or permanent ground control could be desirable—for long-term monitoring, permanent control points that endure years or even decades are best, while for a one-time survey, a few hours for measuring the GCPs and taking the aerial photography might be enough. For long-term monitoring of various study sites in Spain, permanent GCPs were installed by hammering 40 cm (15-inch) long pieces of 2-cm metal water pipe into the ground so they protrude a few centimeters. A 30 cm × 30 cm red cardboard square with central hole is fitted over the pipe before the survey and secured by a long carpenter's nail on windy days. For improved definition of the point location, a white ring is painted around the edge of the pipe-hole (Fig. 9-10A). An even better definition of the exact location and height of the GCP (important for photogrammetric accuracy) could be achieved by fixing a circular target made from a CD and metal washer with a long nail and drop of glue to the pipe's top end (Fig. 9-10B). Both types of target make the GCP clearly definable in the aerial image (Fig. 9-11).

After the survey, the pipe should be covered with something protective and conspicuous like big stones to warn animals and people from stumbling and to make it easier to retrieve a year or two later. Keep in mind that it might be quite difficult to find something as small and inconspicuous as a piece of metal pipe unless well marked with something intentionally recognizable. When covered by vegetation or buried by soil, such markers also may be found again with a metal detector. However, the utilization of these devices is not allowed in all countries because they are often applied by tomb raiders and hobby archaeologists for illegal purposes. In Spain, for example, even carrying along a metal detector in a vehicle is indictable. The reader is therefore advised to seek information about the local regulations before employing a metal detector.

In many countries, metal pipes have a high value; they will be dug out by the local population and utilized for a more sensible purpose. Color-marked stones as well are eye-catching and seem too valuable to let them lie around in the field, so children tend to gather and carry them away. On a wasteland site in the periphery of El Houmer village in South Morocco, the installation of permanent GCPs seemed useless considering the continuing destruction by erosion and the many uses a piece of water pipe can have for the village youths. As a quick and easy method of GCP marking, red rings with 15–20 cm diameter were painted directly on the hard bare ground with spray paint (Fig. 9-12). Both SFAP and coordinate measurements were taken on the same

FIGURE 9-10 Permanent GCPs made from a piece of metal water pipe. (A) GCP marked with a 30-cm cardboard target signal for the survey. (B) Top-precision GCP design. The exact top end of GCP is additionally marked with a circular target made from a CD labelled with target-rings. A metal washer is mounted in the center of the CD to hold a large nail inserted in the GCP pipe, and the CD is temporarily fixed with a few drops of glue to avoid displacement by wind. Photos by IM.

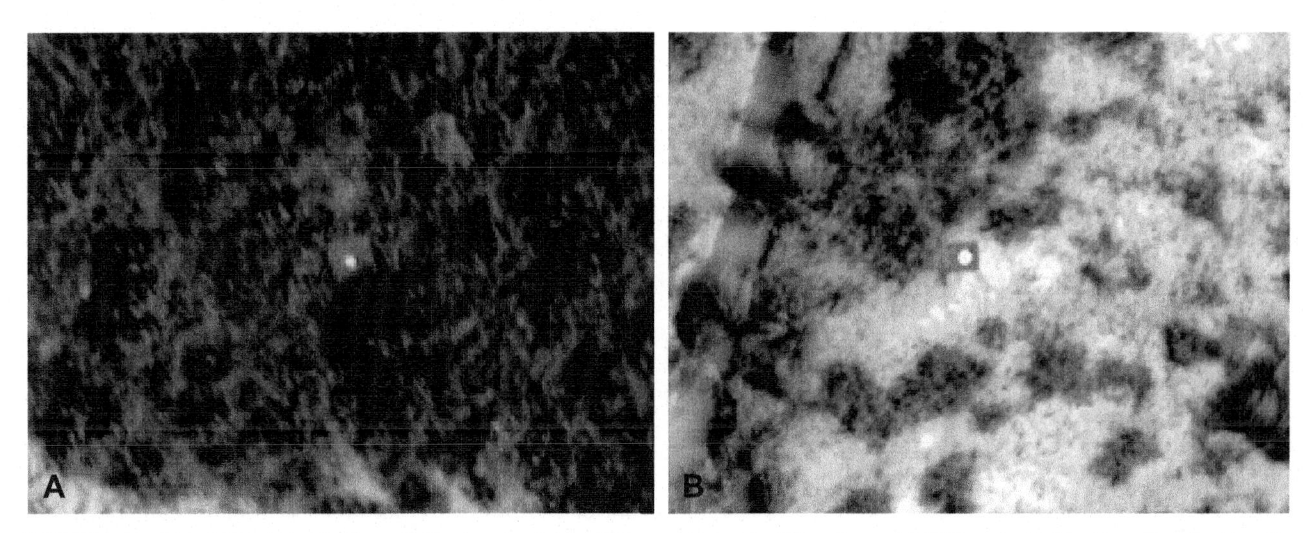

FIGURE 9-11 The GCPs shown in Figure 9-10 appearing in aerial photographs with *GSD* ~3 cm. (A) GCP marked with red cardboard with white center ring. (B) GPC additionally marked with CD target. Kite aerial photographs (subsets, in original resolution) by IM and JBR, February 2009.

afternoon. Signals like these are quite sufficient if a local reference system for scaling and georeferencing photographs is needed in a one-time only survey.

9.4.2. GCP Coordinate Measurement

Given the large image scales for the SFAP emphasized in this book, coordinates for the GCPs usually have to be determined by terrestrial survey with a total station (Chandler, 1999), providing a measurement precision of ±10 mm. Only expensive differential GPS offers similar precisions which can correspond to the high pixel resolution of SFAP imagery. Unlike GPS survey, however, terrestrial survey with a total station for measurement within a particular reference system requires pre-existing trigonometrical points with known coordinates—a feature that is bound to be absent at many SFAP sites. In this case, an arbitrary local coordinate system has to be used, which can be transposed later by offsetting and rotating the local system in order to adjust it to the desired reference system (see Chapter 11.2). However, some sort of link to the latter is still required, e.g., two points in the local system that also have been measured by GPS and which are visible in other, georeferenced, remote-sensing images like those provided by *Google Earth*.

If a global or national reference system is not required, using a self-defined local system throughout also can have advantages. This approach enables the surveyor to align the axes of the system to match with the site of interest, e.g., line up the *y*-axis with the direction of a river bed or oblong test plot. Apart from aesthetic aspects—largely empty maps with diagonal stretches of stitched images just do not look as pleasing as a perfect-fit layout—an optimal alignment of study area and coordinate system also saves much file size in raster data.

When setting up the total station for measurement in the field, it is advisable to choose a spot for the tripod that has a clear view of all GCPs. This is best checked, if in doubt, with the help of the person holding the reflector staff. If the GCP distribution does not permit a single measurement station and a local coordinate system is used, at least two points measured during the first round need to be measured again in the second; this would enable combining the two measurement sets into the same coordinate system by a mathematical transposition.

The authors can report overwhelmingly positive experiences with most local populations during SFAP activities in various countries around the world. The ascent of blimps and kites is a spectacle that often attracts an audience (Fig. 9-13), and most people are interested and supportive

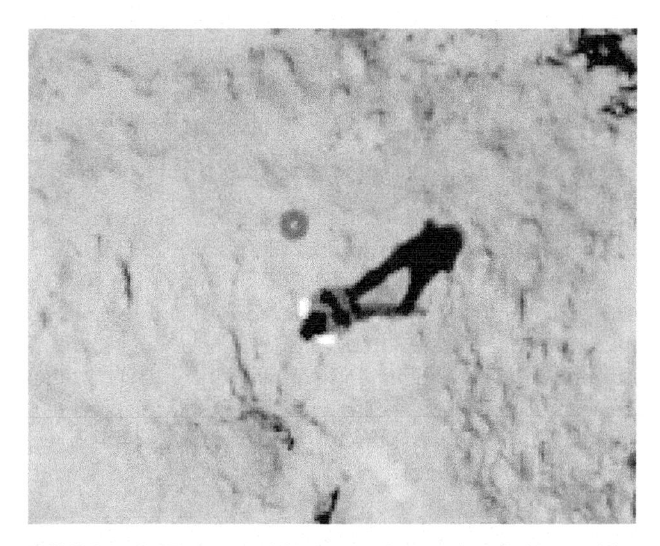

FIGURE 9-12 Temporary GCP for one-time use installed by applying spray paint to the bare soil surface. El Houmer rill erosion test site, Souss Valley, Morocco. Kite aerial photograph (subset, in original resolution) by IM and JBR, March 2006.

FIGURE 9-13 Because most SFAP platforms can be seen from a distance, visitors from the local population or administration during a survey are not rare. Bear in mind, when unpacking the survey equipment in the field, that it might be useful to delegate one of the team as a visitor supervisor, because it might be necessary to enforce the look-but-do-not-touch principle. Here, JBR is keeping the village youth away from dangerous kite ropes near Gangaol, Burkina Faso. Photo by IM, December 2001.

without having any objections against the flying machines or the activities. Nevertheless, this attitude can change quickly, regardless of the country, when surveying markers are set out and equipment, such as a total station, is unpacked or employed. People become apprehensive that some type of building or development is about to take place without their prior knowledge and which might damage their land, economic stability, or privacy. Even parking a vehicle in an unexpected location could raise concerns by land owners, who may call the local police to investigate. Sensitive explanation and authorization papers (see below) are necessary, but still may not clear up all reservations. One of the most important points is providing frank information to the local population concerning the purpose of the flight and survey activities, preferably before a confrontation happens in the field.

9.5. FLIGHT PLANNING CONSIDERATIONS

Flight planning depends on the nature of the study site and its surrounding terrain as well as the type of photographs that are required for the project. A basic distinction can be made between (a) general scene coverage and (b) mapping or cartographic applications. The former may involve a combination of oblique and vertical shots; whereas, the latter would necessitate vertical or even overlapping, stereo imagery. For oblique SFAP, sun position and shadows relative to the camera location are key factors for achieving desirable lighting (see below). The remainder of this section focuses on vertical SFAP for cartographic purposes.

Collection of vertical imagery over a study site ideally includes complete coverage by individual photographs that overlap each other in a predetermined pattern. For some platforms, such as autopiloted model airplanes, pre-designed mission plans can be followed fairly accurately. But for many SFAP platforms, the practical implementation of a survey may be rather far off a preconceived mission plan. Kites do not often obediently comply with the flying-height wishes of their handlers; blimps prefer to turn their nose to the wind rather than following a pattern of straight flightlines. Even so, and regardless of the platform and its maneuverability, air survey calculations should be an integral part of any mission planning in order to ensure the best possible imagery for the intended application.

9.5.1. Image Scale and Resolution

One of the following questions is usually the first to be addressed in this context:

- What is the size of ground details to be distinguished, i.e., what *GSD* is desired?
- What is the size (width and length) of the area to be covered?

The answers to these questions determine at the least the target flying height and minimum number of images required, but they might also affect the choice of camera lens or even camera model and the platform to be used.

If the image *GSD* is of primary concern, the flying height above ground that is necessary for the target *GSD* can be calculated for a given lens focal length by transformation of Equation 3-4. This flying height and focal length in turn determine the area covered by the camera image format, either film or electronic detector (Eq. 3-3). If the calculated area turns out smaller than desired, there are now three possibilities:

- Change lens to a shorter focal length or increase flying height. Both would result in a larger area being covered, but also would proportionally increase *GSD*.
- Cover study area with multiple overlapping and adjacent photographs rather than a single image. This would result in additional field effort and expenses during image processing as the individual images have to be georeferenced, mosaicked, and color-balanced (see Chapter 11.2).
- Change image pixel size by choosing a camera with smaller CCD element size. For analog photography, use film with higher photographic resolution and scan the photograph with higher spatial resolution.

If the size of the area covered by a single photograph is of primary concern, the width and length of the area are usually of greater interest than its extension in units of area. Consider a study site or object with a certain shape: its width and length would have to fit onto the image format completely if it is to be captured by a single image. Unless the width/length ratio is identical with that of the camera sensor or film negative, care has to be taken that the target flying height is calculated using the side of the area that is longer in comparison to the image. This would result in the image covering more than the required area in one direction, rather than cropping the area in the other direction (see below). Remember that all these calculations only apply for the ideal case of the camera being in exactly the right position at exactly the right height; depending on the controllability of the SFAP platform and camera alignment, a generous amount of error margin should always be allowed.

When deciding on the area to be covered by the photographs for a monitoring project, the expected changes—e.g. the upslope movement of gully headcuts, the shifting of river meanders, or the retreat of glaciers—have to be taken into account from the beginning. The image area should be chosen to provide enough room for the expected changes, so the time series would in the earlier images also show the yet unchanged areas and allow for comparison and interpretation of the changes documented in the later images.

9.5.2. Stereoscopic and Large-Area Coverage

Stereoscopic coverage of a study area is necessary for photogrammetric analysis of terrain heights and orthophoto correction, but even if no such advanced analysis is intended, stereophotos may help enormously with photointerpretation (see Chapter 10). There are two possibilities for achieving stereoscopic coverage: in-flight simultaneous image acquisition with two cameras or along-flight consecutive image acquisition with a single camera. There are advantages and disadvantages associated with both stereoscopy methods.

- In-flight simultaneous image acquisition is achieved with two cameras mounted at a fixed distance, e.g. a twin-camera boom. This has the advantage that the stereo images always have the same scale and relative alignment. The amount of image overlap and stereoscopic parallax, however, differ with flying height, as the photogrammetric base (see Chapter 3.3) is predetermined. In order to achieve accurate stereoscopic measurements, the base–height ratio should be at least 0.25 (Warner et al., 1996). For the stereo boom presented earlier (see Fig. 7-8), this would result in optimal flying heights below 4 m! Although much smaller base–height ratios are sufficient for 3D viewing and photointerpretation, the stereo-boom solution is not recommended if photogrammetric measurements are intended.

  Another difficulty is that few SFAP platforms are suitable for supporting a stereo-boom. For kites, the boom represents a considerable challenge as an entanglement with the kite line and Picavet lines become more likely, and stronger microservos would be required for pan and tilt of the cameras. Also, movement and vibration of the boom may impair image quality. Stable constructions require a considerably higher weight-bearing capacity. In case of a blimp, this capacity may be at hand but, nevertheless, the maneuverability is restricted as well by the boom, making take-off and landing operations more vulnerable as compared to flights with only one camera.
- Along-flight consecutive images can achieve better base–height ratios as their distance can be controlled by the operator during the flight. Depending on the platform, however, it may be difficult to ensure image pairs with the same flying height and scale and with regular alignment and spacing (see Fig. 3-10). In this case, more is definitely better: the more images are taken, the better is the choice for selecting those images that provide optimal stereo coverage for analysis.

Tips for navigating individual platforms for stereo-coverage are presented in Chapter 8. Similar considerations apply to non-stereoscopic coverage of larger areas with multiple images for creating seamless image mosaics.

Generally, covering an area with a regular pattern of overlapping images is easier with free-flying platforms, which can be navigated more easily along straight lines than can tethered platforms. The exposure calculations for stereo-survey flightlines in the following are useful for planning SFAP missions with free-flying aircraft with near-steady speed. An ideal flightline pattern, as in Figure 3-8, requires a sophisticated combination of flying height, focal length, image extent, aircraft velocity, and exposure interval.

The first difficulty, therefore, is where to start with the calculations—which of these variables are dependent and which are independent? With a given aircraft and camera, the most easily adaptable variable probably would be the flying height. But the flying height directly controls image extent and *GSD*, and their predefinition is normally of primary concern. This leaves the exposure interval as potentially the most independent variable, suggesting the following sequence of calculations:

- Calculate image scale *S*, image ground cover *A*, *GSD*, and flying height H_g as shown earlier (Eqs. 3-1 to 3-4).
- Decide for the orientation of the rectangular image format with respect to the flight direction.
- Calculate the length of base *B* for the desired forward overlap *PE* (Eq. 3-5).
- Calculate the exposure interval ΔT (Eq. 3-6).

9.6. FLIGHT PLANNING EXAMPLE

The following example for planning an SFAP survey is drawn from a real situation resulting from long-term cooperation between two of the authors (IM, JBR) and Prof. Jean Poesen, Leuven University, Belgium. For several years during the 1990s, Poesen's research group has monitored the advancement of a large gully near Darro, Province of Granada, Spain. The aim of this consecutive study is to resume and expand the monitoring using SFAP and 3D analysis.

FIGURE 9-14 Ground photograph of Gully Belerda, near Darro, Spain, taken during a field trip in spring 2008; view to southeast (Sierra Nevada in the background). The gully cuts several meters deep into the sedimentary valley filling used for cereal and olive cultivation. Photo by IM, March 2008.

A brief field visit in 2008 proved the general suitability of the site for SFAP: good access by car, sufficiently open surroundings for aircraft launching and navigation, and no critical obstacles. Of primary interest is the upper part of the gully (Fig. 9-14), including the orthogonal tributary branch, so the initial question was which flying height would be necessary to cover this area in one image. The prospective camera was a *Canon* EOS 350D (Digital Rebel XT) with 20-mm focal length and a 22.2 mm × 14.8 mm image sensor (3456 × 2304 pixels).

9.6.1. Initial Calculations: Complete Coverage with Single Image

The online GIS *SIGPAC* of the Spanish Ministry of Agriculture, Fisheries and Food was used as a convenient tool for the required measurements and for providing an airphoto basemap in the field. The length and width of the minimum area of interest are 128 m by 120 m (Fig. 9-15, red box). As this oblong is proportionally wider in comparison to the image sensor, the target scale needs to be calculated by inserting the width values into Equation 3-1:

$$S = 1/s = d_W/D_W = 14.8\,\text{mm}/120\,\text{m} = 1/8108$$

with d_W being the image width and D_W, ground area width. This scale results in a longer area covered than necessary (area length $D_L = 22.2\,\text{mm} \times 8108 = 180\,\text{m}$, see green outline in Fig. 9-15). To double check, consider the image scale resulting from the length values (1/5766). This larger scale would result in too small a ground coverage width (14.8 mm × 5766 = 85.34 m) and is, therefore, insufficient for covering the whole area of interest.

From Equation 3-2, the target flying height can be determined as:

$$H_g = f \times s = 20\,\text{mm} \times 8108 = 162.16\,\text{m}$$

and the ground size of a pixel as:

$$\begin{aligned} GSD &= (14.8\,\text{mm}/2304) \times 8108 \\ &= 5.2\,\text{cm or} \sim 2\,\text{inches} \end{aligned}$$

Thus, the whole area of interest would fit into a single image if the camera was positioned and oriented as depicted in Figure 9-15 at a flying height of 162 m, yielding a resolution of 5.2 cm on the ground.

9.6.2. Revised Calculations: Optimal Image Resolution

These initial calculations revealed that the idea of covering the whole area of interest in a single shot had to be abandoned because the *GSD* was larger than desired for detailed gully monitoring. Therefore, the survey design was recalculated

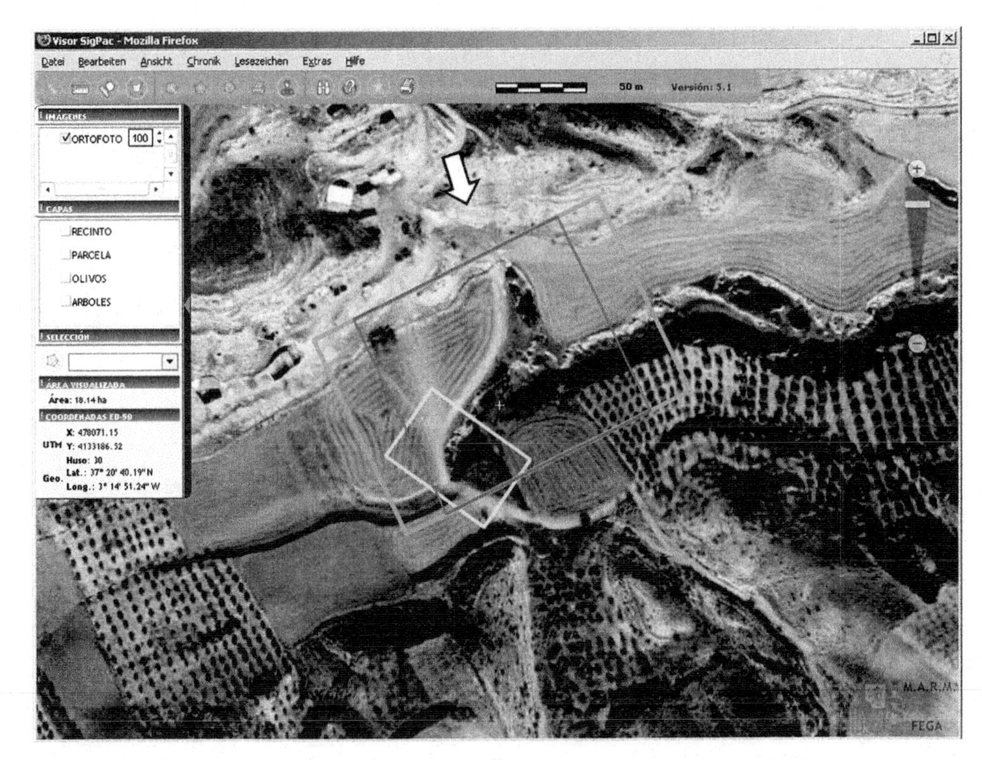

FIGURE 9-15 Orthophoto of the Gully Belerda area in the SIGPAC viewer (©Ministerio de Agricultura, Pesca y Alimentación de España, 2008). The screenshot was used in graphic software for compiling a flight-planning map. Minimum area of interest for monitoring indicated in red, corresponding ideal image footprint in green. Exemplary footprint for optimal image resolution ($GSD = 2$ cm) in yellow. White arrow shows position and viewing direction of Figure 9-14.

for a target GSD of 2 cm using Equation 3-4 (resolved for H_g) and Equation 3-1 (resolved for D):

$$\begin{aligned} H_g &= f \times GSD/(\text{pixel size}) \\ &= 20\text{ mm} \times 2\text{ cm}/(14.8\text{ mm}/2304) = 62.27\text{ m} \end{aligned}$$

which resulted in a much smaller area coverage of approximately 69 m by 46 m (see Fig. 9-15, yellow outline):

$$\begin{aligned} D_L &= s \times d_L = \left(62.27\text{ m}/20\text{ mm}\right) \times 22.2\text{ mm} \\ &= 69.12\text{ m} \end{aligned}$$

and

$$\begin{aligned} D_W &= s \times d_W = (62.27\text{ m}/20\text{ mm}) \times 14.8\text{ mm} \\ &= 46.08\text{ m} \end{aligned}$$

9.6.3. Consequences for Aerial Survey Design

From these results and their visualization in Figure 9-15, the following conclusions could be drawn:

- In order to cover the gully with the desired small GSD, at least 6–8 carefully oriented images would be necessary, accordingly more so for stereoscopic coverage.
- The images should be oriented in transverse direction to the gully's longitudinal course in order to grant enough navigation tolerance and space for covering the surrounding surface.
- Due to the elongated shape of the gully, an easily navigable platform (i.e., model airplane) for following the gully course is preferable to a more static tethered system.
- GCPs would have to be installed rather close to the gully edges to ensure their appearance on the photographs.
- As the gully is surrounded by arable land, GCPs cannot be expected to stay put for years, so we need to be prepared for renewing them for future surveys. At least four GCPs should be placed in potentially permanent positions to allow for restoring the reference system.

9.6.4. Ideal Flightline Calculation

Following these conclusions, it remains to be calculated which exposure interval would be necessary if a free-flying platform, specifically the autopiloted model airplane presented in Chapter 8.5.2, is used for continuous flightline coverage, and if this interval is feasible with the chosen camera. Taking into account that in windy conditions the airplane might not yield a perfectly aligned image series, a 70% forward overlap is considered as minimum.

FIGURE 9-16 Aerial photograph of Gully Belerda taken with a digital compact *Nikon Coolpix* S600 from an autopiloted model airplane at 400 m height. Photo by C. Claussen, M. Niesen, and JBR, September 2008.

Using Equation 3-5, the image base is calculated as:

$$B = D_W \times (1 - PE/100) = 46\,\text{m} \times (1 - 0.7) = 13.8\,\text{m}$$

From this image base and the minimum possible ground speed of the airplane—25 km/h or 7 m/s—the required exposure interval (Eq. 3-6) can be determined as:

$$\Delta T = B/Vg = 13.8\,\text{m}/(7\,\text{m/s}) = 1.97\,\text{s}$$

The results show that this survey design is critical for two reasons:

1. The exposure interval of ~2 s between image frames is quite short for a digital camera of this generation. The survey was conducted later with a compact camera (*Nikon Coolpix* S600) set to 3 s, the fastest possible interval.
2. The base–height ratio resulting from this survey design amounts to 0.22, which is at the lowest limit deemed useful by photogrammetrists with regard to height accuracy. Although the area of interest, the gully, has high relief variations and thus potentially large stereo parallaxes, this base–height ratio is still lower than desirable for the exact height measurements required for monitoring purposes.

The image base in this flightline calculation could be increased if the images were oriented *along* the gully (resulting in $B = D_L \times (1 - PE/100) = 20.7$ m and thus a base–height ratio of 0.33). This also would result in a lower exposure interval (3 s). However, taking into account the width of the gully and area covered, this fit is too close for granting sufficient coverage of the gully.

The above example of a give-and-take compromise situation is quite typical for SFAP applications. The survey actually was conducted in September 2008 with the mentioned model airplane (but different camera for technical reasons) at various flying heights: the coverage area of the lowest flight corresponded to the yellow footprint in Figure 9-15, and the coverage for the highest flight produced images 530 m by 400 m in extent (Fig. 9-16). For the low flight, stereoscopic coverage could not be achieved along-flight with consecutive images, as foreseen, owing to the exposure interval, but for the high flight, dense stereoscopic overlaps (>85%) resulted.

9.7. FLIGHT PLANNING FOR OBLIQUE SFAP

Flight planning for oblique aerial photography is rarely done to achieve overlapping images for mosaics or stereoimagery, as detailed in previous sections. Rather the oblique vantage is chosen to depict the spatial relationships of objects and features within the landscape in a more pictorial fashion. The geometric distortions of single-point perspective, relief displacement, and scale variations are accepted, and the

FIGURE 9-17 High-oblique view northward over the Elk River Wind Farm in the Flint Hills of south-central Kansas, United States. Each tower is 262 feet (~80 m) tall and blades are 125 feet (~38 m) long. Kite aerial photography with the kite positioned upwind from the turbines (note positions of blades) and camera ~120 m above ground. Long shadows show the sun position in this late autumn afternoon view. Photo by JSA and SWA, November 2009.

emphasis is given to best lighting, shadow placement, tilt angle, and other factors that impact the visual scene. In general, views looking away from the sun without excessive shadowing and without the hot spot are preferred (see Chapter 4).

The type of flying platform, terrain conditions, wind direction, and other factors influence what type of oblique imagery may be acquired under favorable conditions. Consider, for example, the challenge of acquiring aerial photographs of huge turbines in a wind-energy farm. Individual turbines typically stand 90–120 m tall, and they are arrayed in fields with many dozens spaced a few 100 m apart. In vertical view, the turbines may be almost impossible to see; the towers have small "footprints" looking straight down. Their shadows may be distinct, depending on lighting conditions, and they may appear in side profile toward the edges of vertical photographs because of relief displacement. Nonetheless, oblique views might well be more desirable (see Fig. 10-24).

Flying any type of manned aircraft in close proximity to such turbines would be at least foolhardy, in view of turbulent wind currents, if not prohibited. Various types of unmanned SFAP platforms could be positioned for overviews and closeup oblique images. Tethered platforms would give the greatest control from the ground. Balloons and blimps would be safe only under calm to light wind (turbines not operating), which leaves kites as the safest platform when turbines are in motion (Figs. 9-17 and 9-18). Ideally the kite should be positioned well upwind or downwind from the turbines with the sun behind the camera, but such arrangement may not be possible depending on ground access. A free-flying unmanned platform would be subject to the same turbulent wind currents, which definitely would require a human pilot to control the aircraft. Flight planning should include considerable flexibility and quick response under these conditions.

FIGURE 9-18 Close-up view of wind turbines at the Spearville Wind Energy Facility, southwestern Kansas, United States. The tip of the upright blade of the near turbine is ~390 feet (120 m) above the ground. The camera is positioned ~250 feet (75 m) high for this side view of the turbines. View toward the north; kite aerial photo by SWA and JSA, November 2008.

9.8. LEGAL ISSUES

Having worked through all the logistical and technical issues for an SFAP mission, the photographer also must be cognizant of local rules and regulations for flying and taking aerial photographs. Just as with manned aircraft, the operation of free-flying as well as tethered unmanned aircraft—balloons, blimps, model airplanes, and kites—comes under the legal constraints of aviation laws in most countries. Also, restrictions may exist for taking photographs from the air of certain objects or areas, e.g., public buildings, military grounds, power plants, nature reserves, or private houses. The regulations may be quite specific regarding the type of aircraft, flying height, survey area, and mission purpose, and restrictions may range from none to special requirements of aircraft signaling, insurance obligations, or absolute prohibition. Usually, third-party liability insurance coverage and overflight permission from the site's owner or the local municipal administration are prerequisites for obtaining flying authorizations from the aviation authorities.

Keep in mind, when planning a mission, that investigating and obtaining flying authorizations may take some time, as one authority's permit may depend on those of other agencies. Unfortunately it is not always obvious who is responsible for the permissions. If a civil authority grants the flight permission, it is still possible that involved police or military authority may recognize other problems. Long-term experience has shown that especially with the latter institutions, personal contact and demonstration of the survey devices as well as the presentation of examples of the aerial photographs add to a positive solution. However, it should be considered that aerial photographs are frequently made for espionage purposes, both military and commercial, therefore various officials may be suspicious. This applies especially to free-flying, remotely controlled model aircraft, as opposed to tethered systems; the latter usually remain within a restricted area and may be maneuvered and observed more closely.

All necessary permits and insurance policies should be carried personally into the field and ready to present immediately if required. Nothing is more annoying than having the aircraft positioned directly above the study site in perfect lighting conditions, and then be forced to break off the survey before taking the first photo by the local police or civil guard because the necessary papers with the right seals are lacking.

Considering the differing national regulations and the wealth of individual rules, it is not possible in this book to give an accurate and up-to-date list of legal issues. The reader should refer to the individual country's body of laws. Information about current regulations can be gathered from aeronautical authorities, insurance companies, and model flying associations. Nevertheless, the following sections aim to give an introduction to some of the diverse legal conditions that apply in selected countries. The authors would like to point out explicitly that the following is given for information only and without responsibility for any eventual errors or omissions.

9.8.1. German Regulations

In Germany, several laws and orders are in force concerning the use of unmanned model aircraft. The current versions of these statutes are published by the German Federal Aviation Administration (Luftfahrtbundesamt, 2009), and an application form for flight clearance may be downloaded from its homepage (search term *Fluggenehmigung*). § 16 of the German Air Traffic Order (*LuftVO*) regulates the permission for use of the German airspace. It is summarized in the following without claim to completeness and warranty of correctness, particularly regarding the fact that the legal framework changes occasionally owing to technical development and changing hazards.

§ 16 Airspace use requiring a permission (*Erlaubnisbedürftige Nutzung des Luftraums*)

(1) The following kinds of airspace use (*Nutzung des Luftraums*) require a permission:
   1. The ascent of model aircraft (*Aufstieg von Flugmodellen*)
      (a) with more than 5 kg overall mass,
      (b) with rocket propulsion, provided that the rocket propelling-charge exceeds 20 g,
      (c) with combustion engine (*Verbrennungsmotor*) in a distance of more than 1.5 km from residential areas,
      (d) of all kinds within a distance of less than 1.5 km from the zones of airfields (*Begrenzung von Flugplätzen*); on airfields the operation of model aircraft additionally requires the accordance of the aviation authority or of the air traffic control (*Luftaufsichtsstelle oder Flugleitung*),
   2. the flying of kites and parachute kites if they are tethered with a line longer than 100 m, [...]
   4. the ascent of captive balloons (*Fesselballone*) if they are fastened with a rope of more than 30 m length,
   5. the operation of remotely controlled and uncontrolled automatic missile (*fern- oder ungesteuerten Flugkörpern mit Eigenantrieb*), [...]

(2) The tether line of unmanned captive balloons (*unbemannte Fesselballone*) as well as kites has to be marked every 100 m with red and white flags at daytime and with red and white lights at night, in order to be visible for the pilots of other aircraft.

(3) The authority for the granting of permission according to Section 1 is the locally responsible authority of the federal state, provided that the representative according to § 31c of the Air Traffic Act (*Luftverkehrsgesetz, LuftVG*) is not responsible.

(4) The permission is granted if the intended uses cannot lead to a danger for the safety of air traffic

(*Sicherheit des Luftverkehrs*) or the public safety and order (*öffentliche Sicherheit oder Ordnung*). The permission may be provided with collateral clauses, and it may be granted to individuals or associations of individuals for particular cases or generally. The authorities determine according to their dutiful assessment (*nach ihrem pflichtgemäßen Ermessen*), which documents the application for the grant of permission needs to contain. In particular, an expertise concerning the suitability of the landscape and the airspace in which the aircraft operations are planned to take place may be required.

(5) The grant of permission may be conditioned upon evidence of agreement of the property owner or another holder of rights of use. (End of citation of the law.)

For the SFAP user, the cited law indicates that the ascent of electrically driven model aircraft with an overall weight of less than 5 kg in a distance of more than 1.5 km to the next residential area and in a distance of more than 1.5 km to the next airfield is possible without any permission, although § 16 (1) 5 does leave some room for interpretation of this statement (see also Müller and Wöhlecke, 2002). Kites and parachute kites are not allowed to ascend higher than 100 m and blimps not higher than 30 m without permission. According to the interpretation of the law by German authorities, these measures refer to the total length of the lines on the reel and not to the amount of uncoiled line during the flight only.

The permission has to be obtained from the responsible aviation authority, which differs between the federal states. As the authors have experienced, the grant of permission for scientific purposes is usually given. Particular security regulations are in force next to civil and military airfields; these may even apply for considerable distances. In the case of an SFAP survey 60 km north of Frankfurt International Airport, for example, the permission was granted with the constraint to inform the responsible aviation authority 24 h in advance of the ascent, providing information on the exact coordinates, time, and aircraft in written form, and calling the authority again immediately before the survey.

9.8.2. Regulations in the United States and Other Countries

The United States represents a quite liberal set of conditions for SFAP, as regulated by the Federal Aviation Administration (FAA). Tethered platforms, such as balloons, blimps, and kites, may be flown up to 500 feet (150 m) above the ground without a special permit in most circumstances. Flying such platforms higher requires filing a flight plan with the nearest airport. On the other hand, manned aircraft should not be flown below 500 feet height in the countryside and not below 1000 feet (300 m) in urban areas.

The airspace is available for public access, except where restricted for military, security, safety, environmental, or wildlife-habitat reasons (see Fig. 14-2). No private land owner may limit overflights for the purpose of taking aerial photographs; however, land owners may restrict access to their properties on the ground. "Posted, no trespassing" signs are commonplace. For all ground operations, it is recommended to obtain prior approval from local authorities or land owners for launching and landing SFAP platforms, as well as approved vehicle parking, equipment setup location, etc.

Flying-height regulations differ considerably in individual countries elsewhere. In the United Kingdom, for example, normal flying ceiling for tethered kites is only 60 m, which is not enough height for many SFAP purposes. Special application to the Civil Aviation Authority (CAA) is required to fly higher, but the application procedure may take three weeks for approval. In Spain, where model aircraft come under the regulation of a royal decree to article 151 of the aviation law (*artículo 151 de la Ley sobre Navegación Aérea*), unmanned aircraft—balloons, model airplanes, and rockets—may ascend up to 300 m without authorization, provided they are more than 10 km away from airports and feature some sort of self-destruction or deflation mode in case of remote-control failure or tether line rupture. For all exceptions, permission by the AENA (*Aeropuertos Españoles y Navegación Aérea, Dto. Coordinación Operativa del Espacio Aéreo*) is required, and additional approval is necessary by the military if the study site is located in militarily sensitive areas.

India is a country well-known for strict security; in fact, taking any kind of aerial photograph is prohibited under nearly all circumstances. In order to conduct kite aerial photography, Chorier had to obtain permissions from various governmental agencies, and he had to carry permits with him or risk arrest by local police which happened several times (Chorier and Mehta, 2007). The Khumb Mela, the largest religious gathering in the world, takes place every 12 years at the confluence of the Ganges, Yamuna, and mythical Saraswati rivers. He was authorized to take kite aerial photographs of preparations for the celebration, but not the actual Kumbh Mela, whereas all other types of aerial photography, even military, were ruled out by officials. In other cases known to the authors, people who attempted SFAP met with stiff resistance from Indian authorities.

Land ownership rights vary substantially in other countries. In Slovakia, for example, the public is allowed into harvested or fallow agricultural fields, as long as the ground is not disturbed. Farmers may not restrict access; such fields are often convenient places for SFAP operation. However, the Tatra Mountains of Slovakia and

Tatrzański Park Narodowy, ul. Chałubińskiego 42a, 34-500 Zakopane, tel/fax: 0-prefix-18 2063203 e-mail: sekretariat@tpn.pl

Zn.spr.: **NB-056/71/07** Symbol: **Geo** Numer **582** Zakopane, dnia **2007-04-20**

ZEZWOLENIE

na realizację badań na obszarze Tatrzańskiego Parku Narodowego

Kierownik tematu: Tytuł: prof. Imię: James Nazwisko: Aber

Instytucja: Emporia State University Ośrodek: USA Termin badań: od 2007-07-15 do 2007-08-15

Temat: Pilotażowe wykonanie zdjęć latawcowych w TPN dla zastosowań geologicznych

Uprawniony: **prof. J. Aber** Osób towarzyszących: **1**

Obszar badań: Cały obszar TPN. Z prawem: Zastosowania latawców do wykonania zdjęć fotograficznych.

Warunki realizacji zezwolenia: Z badań wyłącza się rejon Kominiarskiego Wierchu(oddz.275-280, 295-298). O każdorazowym wejściu należy powiadomić właściwe służby terenowe TPN.

Zezwolenie upoważnia do poruszania w wyznaczonym terenie TPN poza znakowanymi szlakami turystycznymi. Korzystający z zezwolenia mają obowiązek przestrzegania przepisów obowiązujących w strefie granicznej i w parku narodowym. Posiadacz zezwolenia zwolniony jest z opłat za wstęp do TPN. Prowadzący badania jest zobowiązany złożyć w Dyrekcji Parku pisemną informację z realizacji prac terenowych do końca grudnia każdego roku.

FIGURE 9-19 Kite aerial photography in Tatra National Park, Poland. Left: zezwolenie (permission) to conduct kite aerial photography for geological applications in Tatrzański Park Narodowy (TPN). Permit was approved in April 2007 for a limited period of KAP, July 15–August 15, 2007. Below: panoramic view toward northwest over Skupniów Uplaz ridge (lower center) in the Polish Tatra Mountains (Aber et al., 2008).

Poland are protected in national parks; all public, commercial, and scientific activities within these parks are strictly regulated. Obtaining permission requires the assistance of local colleagues and may take considerable time, even months, to arrange in advance. Necessary permits must be carried in the field at all times while conducting SFAP in these circumstances (Fig. 9-19), and it is best to have local colleagues along to explain the SFAP mission.

9.8.3. Insurance

Generally, third-party liability insurance is a requirement for obtaining flight clearance, and also would be in the best interest of SFAP operators. From the example of insurance issues in Germany, it can be seen that the matter is not a trivial one. As per August 11, 2005, each model aircraft in Germany has to be insured independent of its weight. Private indemnity insurances cover the risks of model flights only in exceptional cases. Some older contracts merely include an insurance protection for model aircraft up to a maximum weight of 5 kg. This protection in most cases only applies for models without self-propulsion, and insurance policies should be checked carefully for details.

A summary of insurance offered for model builders and flyers by various organizations (*Deutscher Aero Club DAeC, Verband Deutscher Modellflieger DMFV, Deutsche*

Modellsport Organisation DMO) with identification of possible loopholes has been compiled painstakingly by Steenblock (2006). These insurance packages are included in the membership fee of the organizations (ca. 50–80 €/a) and customized to the specific activities of the members.

From the point of view of a worldwide active German SFAP user, membership in the DAeC appears the most sensible option, because it is the biggest organization for pilots of all kinds in Europe and the insurance is globally valid. Model aircraft are insured up to a total starting weight of 25 kg, and no co-payment is required except in case of double-channel occupancy. Supplementary agreements also enable insurance of models heavier than 25 kg. For blimps and balloons, it has to be considered that the flight weight is not calculated just from the summed weight of the components but rather from the addition of component weight and the weight of the covered volume within the balloon, in other words the gas (specific weight of air: 1.225 kg/m^3, helium: 0.17 kg/m^3, all assuming ICAO standard atmosphere). This complicated calculation may result in surprisingly large flight weights that exceed the 25 kg insurance limit, although all components together may weigh much less than 25 kg.

Another option is separate insurance policies that can be taken out by individuals or institutions with *Allianz* or in special cases also with the *Gerling* Insurance Group. In this case, a specific model aircraft or blimp may be insured with the obligatory specification of its serial identification number. However, the insurance rates are significantly higher and are in the experience of the authors only suitable for heavier and larger aircraft. Also, according to Steenblock (2006), these individual insurance packages show some loopholes in contrast to the organizational insurance offers, and consequently the authors recommend to check their conditions in detail before conclusion.

In other countries, individuals may seek insurance through hobby or professional organizations. In the United States, for example, the American Kitefliers Association provides limited liability insurance at officially sanctioned kite competitions and tournaments. Another option is a professional rider that may be added to an individual's homeowner insurance policy. When acting in professional capacity, for example, teaching a field-methods course in SFAP for a university, the individuals and activity may be covered by the institution. But the conditions and exclusions vary greatly with each type or source of insurance, so the SFAP operator needs to pay some attention to appropriate coverage.

9.9. SUMMARY

A successful small-format aerial photography (SFAP) survey requires planning of various issues, of which some may need to be prepared well ahead. This process is aided by good knowledge of the study site gained from personal experience, maps, satellite images and airphotos, and local colleagues.

For a given SFAP mission, both equipment and personnel must be available at the appropriate time of year under suitable weather conditions. This usually requires some travel for the involved people and shipment of necessary equipment, which may prove difficult for international missions. The accessibility of the site by automobile or boat as well as for the survey personnel on foot can be an important consideration when choosing the survey system, as some sites like wetlands, nature reserves, or rugged terrain can be impossible or difficult to enter and navigate on the ground. Site characteristics—obstacles, topography, lighting, wind—also may influence field procedures. High-altitude SFAP presents special circumstances because of thin air and alpine weather effects.

For all applications that involve measuring and mapping, ground control for georeferencing and geometric correction of the images is necessary. Given the large scale of SFAP images, measuring coordinates of the ground control points (GCPs) would usually require the high accuracies of a terrestric survey by total station. Temporary GCPs are sufficient for one-time surveys; permanent GCPs need to be installed for long-term monitoring purposes. Careful targeting with colors that contrast well with the surrounding area helps to identify the GCPs in the photographs.

Although most SFAP platforms are not navigable precisely, some basic air survey calculations should be an integral part of any mission planning in order to ensure the best possible imagery for the intended application. Target flying heights can be calculated prior to the survey for the desired area coverage or ground sample distance (pixel resolution), and the possibilities of achieving stereoscopic coverage can be assessed for a given platform and camera system. Such calculations can help to assess if the desired image characteristics can be realized with the existing equipment and which compromises or adaptations could be necessary.

Legal constraints of aviation laws may apply for SFAP depending on the country, the platform used, and the mission purpose. Permission to conduct SFAP may require authorizations from various governmental agencies as well as insurance policies, and careful review of the individual country's body of laws and the details of insurance policies is highly recommended. The help of local informants is most useful in situations that may be unfamiliar to the photographer. The best approach is to utilize an SFAP system that is robust, easy to transport, supported by available personnel and supplies, and capable of acquiring suitable imagery within the local legal framework.

Chapter 10

Image Interpretation

One picture is worth ten thousand words.

F.R. Barnard (1927)

10.1. INTRODUCTION

Visual interpretation of aerial photographs is based on recognition of objects, as seen from above, in pictorial format. This visual recognition often takes place without any conscious effort by the interpreter. Nonetheless, several basic features aid in the examination and interpretation of airphotos (based on Avery and Berlin, 1992; Teng et al., 1997; Jensen, 2007).

- *Tone or color*: gray tone (b/w) and color are how we recognize and distinguish objects. The tone or color of an object helps to separate it from other features in the scene, especially for features with high contrast. Color may be an important clue to the object's identity—water, soil, vegetation, rocks, etc.
- *Shape*: natural shapes tend to follow the lay of the land. Cultural (human) shapes, on the other hand, are often geometric in nature with straight lines, sharp angles, and regular forms. In monoscopic images only 2D shape can be appreciated fully, although an object's height or relief may be detected to some extent by shadow or relief displacement. In stereoscopic images, the full three dimensions become apparent.
- *Size and height*: absolute size and height are important clues that depend on the scale of the photograph. Some object sizes are common experience, for example, a car, tree, and house. But many objects seen on airphotos are not so obvious, and height may be difficult to judge in vertical airphotos. Always check photo scale for a guide to object size.
- *Shadow*: shadows can be useful clues for identifying objects, as seen from above. Trees, buildings, animals, bridges, and towers are examples of features that cast distinctive shadows. Shadows on the landscape also help with depth perception. Photographs without shadows often seem "blank" and difficult to interpret; however, too much shadowing may obscure features of interest.
- *Pattern*: the spatial arrangement of discrete objects may create a distinctive pattern. This is most apparent for cultural features, for example, city street grids, airport runways, and agricultural fields. Patterns in the natural environment also may be noticeable in some cases, for example, bedrock fractures, drainage networks, and sand dunes.
- *Texture*: this refers to grouped objects that are too small or too close together to create distinctive patterns. Examples include tree crowns in a forest canopy, individual plants in an agricultural field, ripples on a water surface, gravel sheets of different grain size in a river bed, etc. The difference between texture and pattern is largely determined by photo scale.
- *Context*: the association and site location of objects are often important for aiding interpretation. Note land cover and land use as clues to help identify related features in the scene, and refer to existing maps or census data for ancillary information. This is often a matter of common sense and experience for image interpretation.

Each of these basic features is seldom recognized on its own. Rather it is the combination of visual elements that allows interpretation of objects depicted in an aerial photograph. These elements are explored in further detail with selected small-format aerial photography (SFAP) examples later in this chapter and in the case studies. The issue of stereoscopic viewing is covered together with stereoscopic mapping in Chapter 11.

10.2. IMAGE INTERPRETABILITY

Visual identification of objects in vertical airphotos normally requires ground sample distance (*GSD*) 3–5 times smaller than the object itself (Hall, 1997), as noted before (Chapter 2). For example, to positively identify a house (~ 10 m $\times$ 10 m), the camera/image system needs to achieve a *GSD* in the range of 2–3 m. Factors other than spatial resolution often affect the ability to recognize objects, however. Contrasts in color and brightness are important clues as are size, shape, context, shadows, and other factors noted above. Keep in mind that with increasing height above the ground, atmospheric effects begin to degrade image quality regardless of spatial resolution.

With these limitations in mind, a perceptual measure of image quality or interpretability has been developed. This measure is the U.S. National Imagery Interpretability Rating Scale—NIIRS (Leachtenauer et al., 1997). An image rating depends on the most difficult interpretation task that can be performed, which indicates the level of interpretability that can be achieved (Table 10-1).

TABLE 10-1 **The civilian U.S. National Imagery Interpretability Rating Scale levels (0–9) and examples of interpretation tasks (based on Leachtenauer et al., 1997).**

Rating Level	Interpretation tasks
0	Interpretability precluded by obscuration, degradation, or poor resolution
1	Distinguish between major landuse classes-urban, agricultural, forest, etc. Detect medium-sized port facility Distinguish between runways and taxiways at a large airport Identify regional drainage patterns-dendritic, trellis, etc.
2	Identify large (800 m diameter) center-pivot irrigated fields Detect large buildings-hospital, factory, etc. Identify road pattern-highway interchange, road network, etc. Detect ice-breaker tracks in sea ice; detect wake of large (100 m) ship
3	Detect large area (800 m square) contour plowing Detect individual houses in residential neighborhoods Detect trains or strings of railroad cars on tracks Identify inland waterways navigable by barges Distinguish between natural forest and orchards
4	Identify farm buildings as barns, silos, or residences Count tracks along railroad right-of-way or in train yards Detect basketball court, tennis court, volleyball court in urban area Identify individual tracks, control towers or switching points in railroad yard Detect jeep trails through grassland
5	Identify Christmas-tree plantation Recognize individual train cars by type-box, flat, tank, etc. Detect open bay doors of vehicle storage buildings Identify large tents within camping areas Distinguish between deciduous and coniferous forest during leaf-off season Detect large animals (elephant, rhinoceros, bison) in grassland
6	Detect narcotics intercropping based on texture Distinguish between row crops and small grains Identify passenger vehicles as sedans, station wagons, etc. Identify individual utility poles in residential neighborhoods Detect foot trails in barren areas
7	Identify individual mature plants in a field of known row crops Recognize individual railroad ties Detect individual steps on stairways Detect stumps and rocks in forest clearings or meadows
8	Count individual baby farm animals-pigs, sheep, etc. Identify a survey benchmark set in a paved surface Identify body details or license plate on passenger car or truck Recognize individual pine seedlings or individual water lilies on a pond Detect individual bricks in a building
9	Identify individual seed heads on grain crop-wheat, oats, barley, etc. Recognize individual barbs on a barbed-wire fence Detect individual spikes in railroad ties Identify individual leaves on a tree Identify an ear tag on large animals-deer, elk, cattle, etc.

FIGURE 10-1 *Ikonos* satellite image showing a portion of Fort Leavenworth military base, Kansas, United States. Vehicles are clearly visible on roads and in parking lots; parking pattern – diagonal or parallel – is evident. The smallest features that can be identified (under enlargement) are individual parking stalls, which give an interpretability rating of 6. The slightly fuzzy appearance demonstrates the spatial limitation of such imagery. Panchromatic band (green + red + near-infrared) with 1 m spatial resolution, August 2000. Image processing by JSA.

Manned-space small-format photography, as practiced on space-shuttle and space-station missions, provides images of the Earth's surface for rating levels 1–3. Rating levels of 4–6 are typically attained by conventional aerial photography and high-resolution commercial satellite imagery (Fig. 10-1), and sometimes level 7 is possible. However, the highest rating levels (8–9) are generally not available for civilian use. These highest NIIRS rating levels may be acquired, in many cases, with SFAP from low-height platforms (Fig. 10-2).

10.3. SFAP INTERPRETATION

The basic visual aspects of airphoto interpretation are illustrated with small-format aerial photographs representing selected common subjects. Typically, several of the basic features noted above are combined for interpretation of the subjects discussed below. This review is not intended to be exhaustive, but rather to introduce the general concepts of interpretation as applied to SFAP. For in-depth discussion of airphoto interpretation, readers should consult Avery and Berlin (1992), Philipson (1997), Jensen (2007), Lillesand et al. (2008) and similar texts on this subject. Classical textbooks with emphasis on geoscientific subjects are Lueder (1959), Schneider (1974), and Kronberg (1995).

10.3.1. Water and Drainage

Water is the most widespread feature on the surface of the Earth, and water bodies exist in many forms—seas, lakes, rivers, ponds, estuaries, bayous, lagoons, etc. In aerial

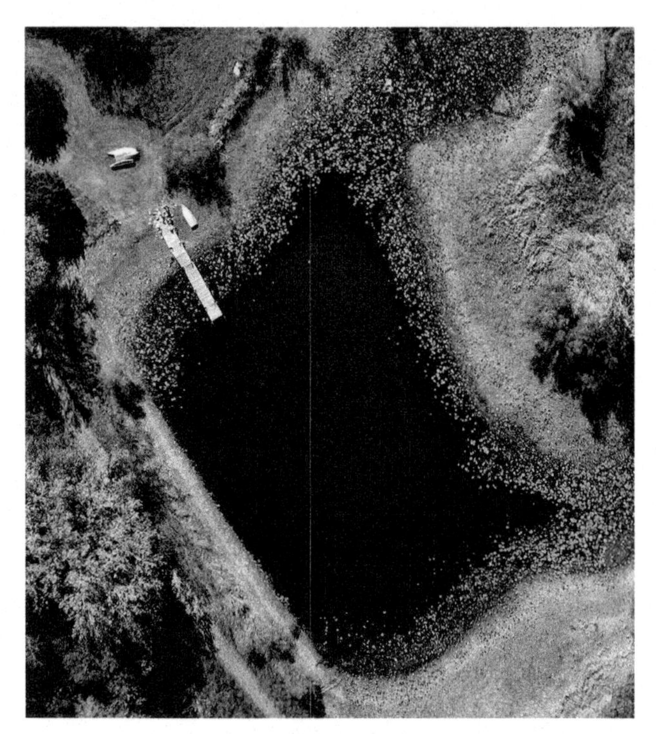

FIGURE 10-2 Vertical SFAP of a pond with emergent leaves of the American lotus (*Nelumbo lutea*). Individual leaves are clearly visible around the margin of the pond; these leaves are typically about one foot (30 cm) in diameter. The dock is ~15.5 m long. *GSD* for the original image is ~5 cm, and the NIIRS rating is 8. Kite aerial photo by JSA, October 2006.

FIGURE 10-3 The man-made pond in the foreground trapped recent runoff and has a high content of yellowish-brown suspended sediment. The next pond downstream did not receive this sediment and has a clean, dark blue appearance. Central Kansas, United States. Kite aerial photo by SWA and JSA, May 2007.

photography water color is most obvious, and the color of water bodies is a good indication of suspended sediment. Clean water reflects blue light weakly, but reflectance drops off sharply for green and red light and is essentially zero for near-infrared radiation (see Fig. 4-17). Thus, clean water typically looks dark blue or sky colored in visible imagery.

FIGURE 10-4 Low-oblique view over a small, meandering stream channel (left) flowing into Tar Creek (right). Sun glint highlights water in the smaller channel; however, muddy water in the larger stream is barely visible through deciduous trees with sparse leaves in early spring, Miami, Oklahoma, United States. Helium-blimp aerial photo by JSA and SWA, April 2008.

FIGURE 10-5 Sun glint highlights tiny islands in the shallow Väinameri Sea south of Vormsi, Estonia. Kite aerial photo by JSA and SWA, August 2000.

Suspended sediment impacts the color of water, which depends on the sediment composition and turbidity (Fig. 10-3; see also Fig. 14-20).

Aerial photography of water bodies often displays sun glint and glitter (see Fig. 4-5). Although usually considered undesirable, under some conditions sun glint is advantageous for identifying small water bodies that would otherwise be difficult to distinguish (Fig. 10-4). Conversely, sun glint may aid recognition of small, emergent islands within larger water bodies (Fig. 10-5), and sun glitter may highlight the pattern of wind-driven waves on water surfaces (see Fig. 5-18). Water flooding also may assist in interpreting subtle relief variations in an area, revealing shallow depressions and drainage structures that otherwise would not be identifiable. Figure 10-6, for example, contributed to recognizing the importance of

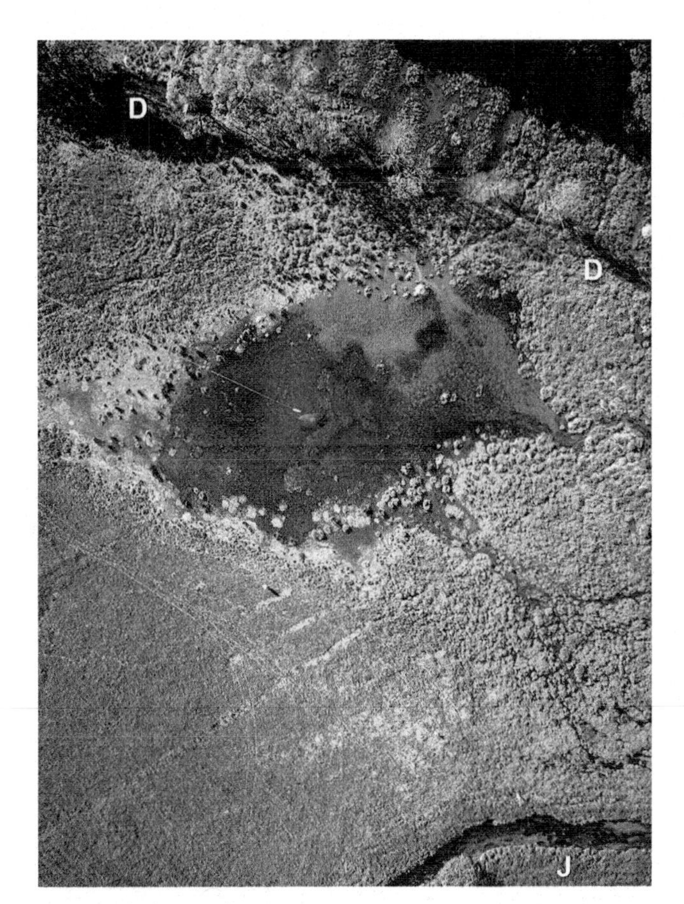

FIGURE 10-6 Flooding caused by a beaver dam in the Jossa Valley near Mernes (Spessart), Germany. On the inundated floodplain between the Jossa River (J) and the main irrigation ditch crossing the upper part of the image (D), a multi-channel drainage network and small lake have developed. The water table also traces a series of parallel ditches running at right angles to the main ditch; they are the remains of "Rückenwiesen", a nineteenth century irrigation system. More recent drainage ditches are revealed by water and yellow-flowering buttercups in the lower image half. Field of view ~150 m across. Hot-air blimp photograph by JBR, IM and A. Fengler, June 2001.

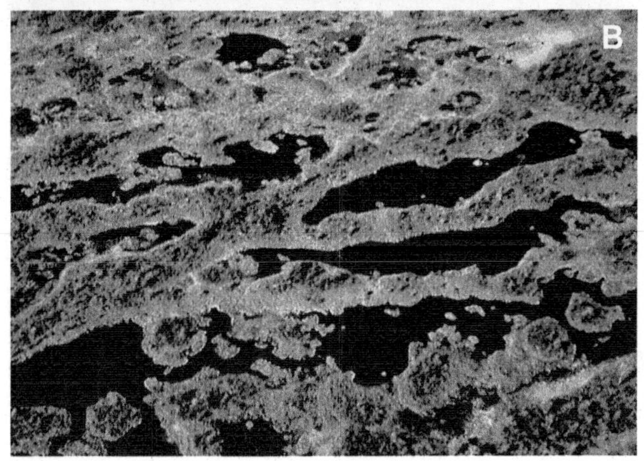

FIGURE 10-7 Color-visible (A) and color-infrared (B) views over Mannikjärv Bog, eastern Estonia. The water pools are quite shallow (<1 m) and clear. Submerged aquatic vegetation is visible in (A), but water pools are completely black in (B). Kite aerial photos by JSA and SWA, September 2001.

beaver activity in floodplain evolution in a study by John and Klein (2004). Beavers in large numbers have inhabited the river systems of central Europe and North America until the middle ages and nineteenth century, and beaver-induced biogeomorphological processes can be expected to have influenced the floodplain sediments and alluvial soils significantly. This raises many exciting questions in fluvial geomorphology.

In color-infrared imagery, for which blue light is excluded, water bodies are usually dark blue to black regardless of suspended sediment or water depth (Fig. 10-7). This fact can be used to advantage for distinguishing muddy water from wet, bare soil of similar visible color (Aber et al., 2009). However, color-infrared images over water are subject to strong sun glint and opposition effects that degrade image interpretation (Fig. 10-8).

Stream, delta, and tidal channels of various sizes and types are the products of erosion and deposition by concentrated water flow. Such channels are integrated into patterns found in nearly all land and coastal regions of the world. Individual channels are recognized in airphotos by the presence of water, vegetation, shadows, and other characteristic features (see Fig. 5-8). Networks of channels define drainage patterns that are useful clues for interpreting the geology and geomorphology of the landscape (Fig. 10-9). Such patterns are developed at all scales and spatial resolutions from huge river systems covering subcontinental areas to tiny gully networks (Fig. 10-10).

10.3.2. Geomorphology

Geomorphological forms and processes are often highly influenced by water and drainage, especially on the spatial scales captured by SFAP. The high resolutions of SFAP here

FIGURE 10-8 Sun glint (A) and opposition effect (B) in oblique color-infrared images. Brightness contrast is so high that few features can be discerned in detail. (A) Nigula Bog and (B) Teosaare Bog, Estonia. Taken from Aber et al. (2002, figs. 11 and 12).

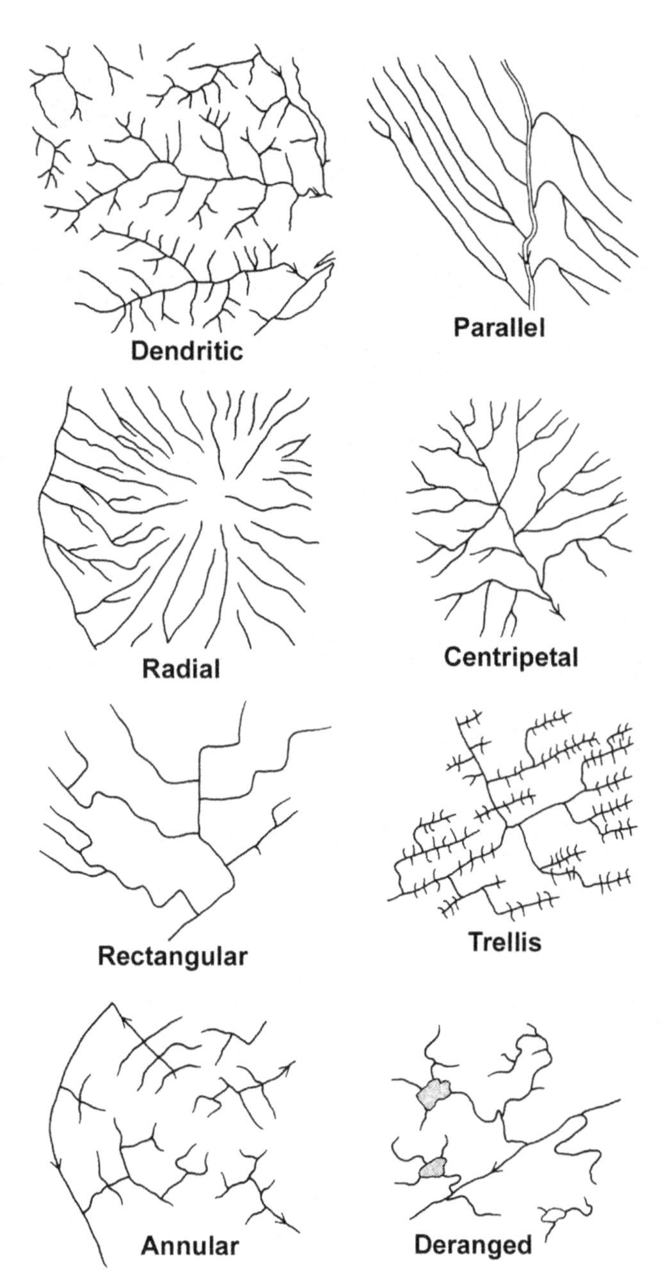

FIGURE 10-9 Schematic illustrations of drainage patterns that reflect underlying bedrock structure or sediment accumulation. Adapted from Drury (1987, fig. 4.4).

offer interpretation possibilities not given with conventional airphotos. While geomorphological processes forming the landscape cannot be seen directly in an aerial image, and rarely may be observed in the field, the correlative forms and deposits may become apparent to the skilled observer by characteristic shapes, patterns, and textures (Marzolff, 1999; Fig. 10-11). Smooth, homogeneous surfaces tell of soil sealed by splash erosion and crusting; dense networks of parallel rills and residual enrichment of stones on the surface may be the result of strong sheet-wash processes. Sediment fans of accumulated material patterned with desiccation cracks at the outlet of larger rills testify to their geomorphodynamic activity.

Beneath the edges of a wide gully cutting into a nearly flat *glacis d'accumulation* in Burkina Faso's Sahel (Fig. 10-12A), soil debris broken off the walls creates a distinctive pattern on the ground where it has not yet been washed away by runoff water. Clearly visible drainage lines have been carved within the gully by the first storms at the beginning of the rainy season. Some islands and peninsulas are left of the former surface, mostly where tree roots strengthen the ground. In spite of the obviously heavy sheet-wash processes, the gully rim itself is sharply edged above the washed-out hollows in the more erodible subhorizon (Fig. 10-12B). An explanation is offered by the polygonal pattern of the emerging grass cover on the surrounding glacis: the vertic character of the soil leads to the development of fine-webbed desiccation cracks. It is along these tension cracks that clumps of soil break off along the gully edge when the soil dries up again. Also, when water infiltrates through the cracks, not only vegetation development

FIGURE 10-10 Multiple ravines with dendritic channel patterns on the North Sea coastal cliff at Bovbjerg Klit, western Denmark. Walking path across grassy cliff top is ~1.5 m wide. Kite aerial photo by IM, SWA, and JSA, September 2005.

FIGURE 10-11 Rill erosion on fallow land in the Spanish Ebro Basin near María de Huerva, Province of Zaragoza. Hot-air blimp photography by IM (lower right corner, with remote control) and JBR, October 1995. White measuring stick in lower right corner is 2 m long. S, strong sheet wash with small rills; H, headcut of large rill; K, knickpoint where small rills merge and incise to form larger rill; F, sediment fan of accumulated material eroded from the rill; C, soil crusted by splash erosion and week sheet wash.

FIGURE 10-12 Gully erosion near Gorom-Gorom, Province of Oudalan, Burkina Faso. (A) Piping processes (arrow) and desiccation cracks, traced by vegetation in the vertic soil of the *glacis d'accumulation*, play an important role in developing and shaping the gully. Field of view ~30 m across. Kite aerial photograph taken by IM, JBR and K.-D. Albert, July 2000. (B) Seen from the ground, the undercuts carved by runoff water flowing over the gully's edge become apparent. Soil clumps below the wall have broken off the rim along the desiccation cracks. Photo by IM.

FIGURE 10-13 Rill and gully erosion on an abandoned agricultural field near Taroudant, South Morocco. The main drainage lines have a dendritic pattern, while the small rill tributaries follow the parallel patterns carved by a bulldozer during land levelling some years previously. Field of view ~75 m across. Kite aerial photograph by IM, JBR, and A. Kalisch, March 2006.

is encouraged but also subsurface erosion, leading to overhanging walls and piping processes (Albert, 2002).

On the abandoned agricultural field heavily affected by soil erosion in south Morocco (Fig. 10-13), a natural dendritic pattern is linked to an anthropogenic parallel pattern. In a vain endeavor to stop gully erosion, the field was levelled with a bulldozer some years before. The resulting grooves now channel the surface runoff into thousands of small parallel rills, encouraging incision and subsurface erosion. The rills continue to merge into larger rills and gullies that retrace the former, pre-levelling drainage network. The statistical analysis of the drainage network shows that more than one-third of the total rill length is directly predetermined by the mechanical intervention (Kalisch, 2009).

A further example for the influence of agricultural treatment patterns on rill erosion is shown in Figure 10-14. The field in the lower right, situated in the Mesa landscape of the Spanish Bardenas Reales, had been plowed one year before the picture was taken. Rills of ~10–30 cm depth have developed due to inflowing water from the adjoining cuesta slope. The tillage direction has a clear influence on the course, depth, and distribution pattern of the rills. Such features are clearly related to the few torrential rainstorm events and usually are destroyed with the next tillage operation. However, the barely visible and measurable depression that the material loss has created in the terrain surface makes it likely that a similar rill system would develop in the same places during the next heavy rainstorm event.

On a neighboring field that has lain fallow for three years, a gully up to 1 m deep has been cut by the concentrated overland flow from the old furrows (Fig. 10-15; see also Giménez et al., 2009). The super position of several tillage operations leads to a complex course of this linear erosion form that cannot be understood from the ground perspective (Plegnière, 2009). This example demonstrates the difficulties that may arise for integrating the factor of tillage direction into GIS-based erosion models.

Some 1800 m higher in elevation, the vertical view onto a 40° high-mountain slope (Fig. 10-16) presents an altogether different aspect of geomorphology. Here, vegetation patterns can be seen reflecting periglacial geomorphic processes. The freeze–thaw cycles in this alpine area have created solifluction lobes in the form of minute terraces. While the step's surface, which is stretched by soil creep, is without cover, the front of the each lobe is bordered by thick turf that retards the gelisolifluction process.

10.3.3. Vegetation and Agriculture

After water, vegetation is the next most common land cover, including natural, agricultural, and other human-modified plantings and ranging from formal gardens, to tropical forests, to tundra. Given that most active, emergent vegetation is green, visible color is often less important for interpretation than are other visual clues such as size (height), shape, pattern, texture, and context. These factors are especially important for interpretation of wetland vegetation (Fig. 10-17).

In the case of forest canopies, for example, the size, shape, and pattern of tree crowns are most noteworthy in addition to seasonal phenological characteristics (Murtha et al., 1997). Furthermore, it should be understood that nearly all vegetation is managed, manipulated, or altered by human activities in various obvious or subtle ways, which impart additional distinctive aspects for photo interpretation.

Consider this view of mixed conifer and deciduous forest in north-central Poland (Fig. 10-18). Both conifer and deciduous trees are present in distinct stands. The conifers are dark green to blue-green in color and conical in shape. Deciduous trees, in contrast, are light green to yellow-green in color and have billowy or tuffed crowns; some display autumn colors (right background). In addition, linear and rectangular boundaries mark some stands, which represent artificial plantings. The same view taken in winter would show even greater contrast between the conifers and bare deciduous trees.

As another example, regard this view over bottomland forest of the Missouri River floodplain taken in early spring (Fig. 10-19). Cottonwood (*Populus deltoides*) trees display green flowers, but leaves have not yet sprouted. The dark appearance of the forest floor is because the understory was intentionally burned a few days before the picture was taken. The cottonwood trees are arranged in curved, linear patterns that reflect natural morphology, soils, and drainage on the floodplain surface. To the right, an open, green strip marks an abandoned river channel that is too wet for trees to grow. Knowing the vegetation phenology, human activity, and geomorphic context are all essential for interpretation of this image.

The same applies for the interpretation of the vegetation patterns in Figure 10-20, where shrubs can be seen encircling a grassland area in wavy formations. The image was taken in a small catchment in the mountains of the Spanish Pyrenees, which was arable land until the second half of the twentieth century. This particular field, which is bordered by grove-hemmed brooks and stone walls, was abandoned in the 1970s and only occasionally used for sheep grazing since. Subsequently, a spiny broom common in Spanish matorral (*Genista scorpius*) began invading the enclosure from the side, divulging its ballistic dispersal mechanism in a wavy encroachment front.

Agricultural fields are human-managed vegetation plots, ranging from fruit orchards to rice paddies, that typically have sizes, shapes and patterns determined by property boundaries as well as drainage and contour of the land. Where soil is fertile and land is relatively flat, people tend to impose various geometric field shapes—square, rectangle, circle, etc. (Fig. 10-21). Where land is steeper, terraces and

FIGURE 10-14 Rill erosion on a field tilled one year previously in the Bardenas Reales, Province of Navarra, Spain (field of view ~60 m across). The tillage pattern of ridges and furrows curving around the sloping edge of the field is dissected by dendritic rills that merge into deeper rills following the circular pattern left by the tractor. In the corner of the field, the two rills upslope of the crawler transporter (loaded with water cans for soil erosion experiments) are redirected diagonally to the gradient by the furrows. These rills are significantly deeper and more linear. In the lower left corner, sheep trails can be seen between the patchy vegetation cover. The trails converge at the field border where the sheep have to cross a trench. Kite aerial photograph by IM, JBR, and S. Plegnière, February 2009.

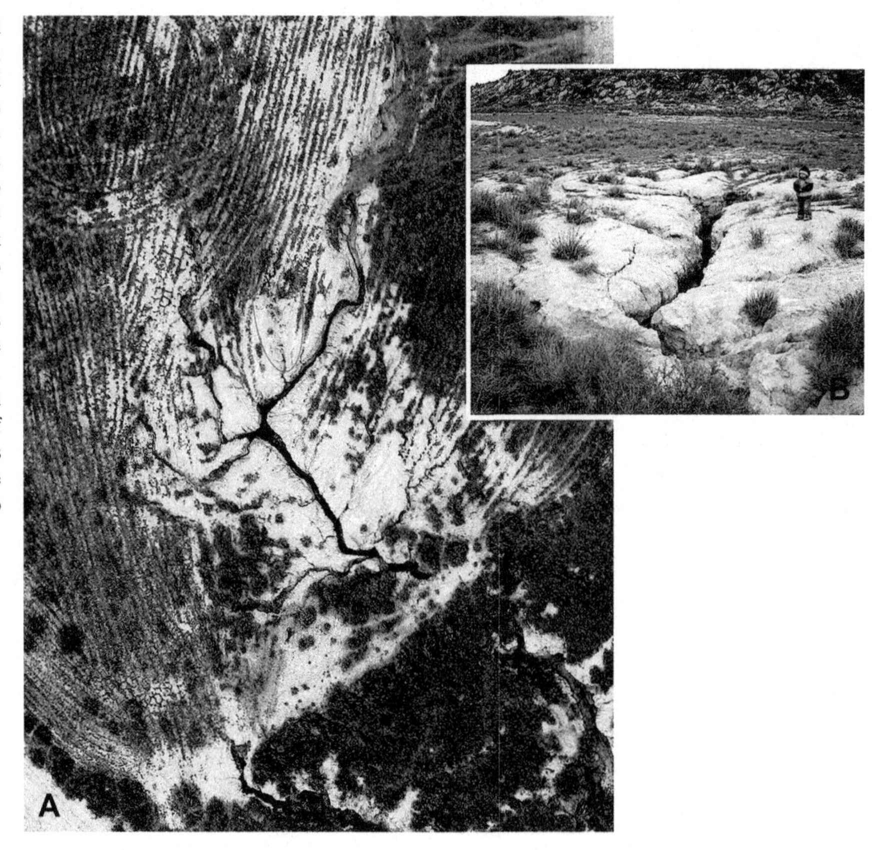

FIGURE 10-15 Rill erosion on the field adjacent to Figure 10-14. (A) Complex pattern of tillage furrows resulting from repeated crossings by the tractor. The field slopes slightly from top to bottom in the image; the upper part of the gully and the tributary rills from the right all run more or less oblique to the inclination and follow the furrows of the last tillage operation. The lower gully part not only follows the slope direction but is also predetermined by an underlying tillage pattern. A chain of piping holes (circled in red) indicates the subsurface rill course in this older tillage direction. Kite aerial photograph by JBR, M. Seeger, and S. Plegnière, March 2007. (B) From the ground perspective, the destructive effect of gully erosion becomes clear. Such erosion forms are already too deep to be leveled by plowing. The junior gully researcher measures ~70 cm. Photo by JBR.

FIGURE 10-16 Close-up vertical shot of steep mountain slope with turf-banked solifluction terracettes in the Spanish Pyrenees near Tramacastilla de Tena, Province of Huesca, Spain. Slope descends from right to left; blimp flyer near scene center for scale. Hot-air blimp photograph by IM, JBR, and J. Heckes, October 1995.

FIGURE 10-17 Plant heights, patterns, textures, and color variations are visual clues for recognizing distinct vegetation zones in these coastal wetlands. (A) Mississippi Sound, United States. (B) Island of Vormsi, Estonia. Kite aerial photos by JSA and SWA, March 2004 and August 2000.

FIGURE 10-18 Low-oblique, autumn view over mixed forest near Mława, north-central Poland. Kite aerial photo by JSA and D. Gałązka, October 1998.

FIGURE 10-19 High-oblique, early-spring view across the Missouri River valley floodplain forest at Fort Leavenworth, Kansas, United States. Kite aerial photo by JSA, April 2000.

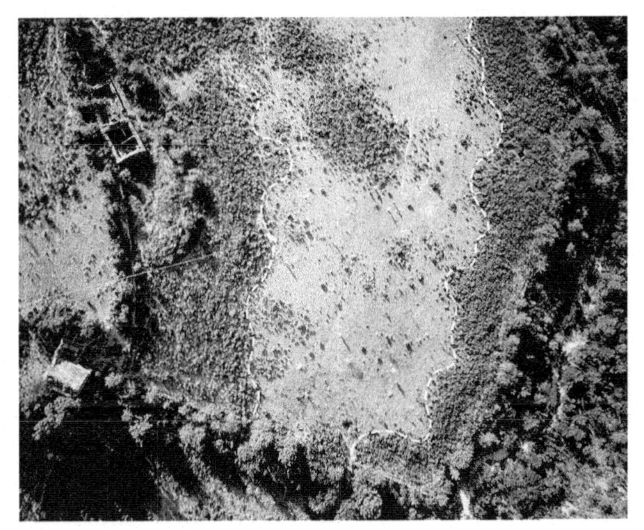

FIGURE 10-20 Abandoned farmland in the Arnás Catchment in the Spanish Pyrenees near Jaca, Province of Huesca, Spain. Note wavy vegetation front marked by dashed white lines. Hot-air blimp photograph by IM, JBR, and M. Seeger, August 1998.

FIGURE 10-21 Rectangular shapes and patterns are most common in agricultural fields. (A) Blue Hills of west-central Kansas, United States, May 2006. Rectangular fields cover 80 acres (~0.2 km^2) each. (B) Foreland of the Tatra Mountains, Slovakia, August 2007. Kite aerial photos by JSA and SWA.

FIGURE 10-23 Grazing lands in the High Plains of western Kansas, United States. (A) Dry rangeland in a drought year, June 2006. (B) Sand hills terrain in a wet year, May 2007. Kite aerial photos by SWA and JSA.

contour plowing follow hill slopes to minimize erosion and retain water (Fig. 10-22). Rocky, sandy, steep, or erosion-prone lands, as well as lands too dry or wet for crops, are normally left in pasture or range for hay or livestock grazing (Fig. 10-23).

Most agricultural crops are annual monocultures, for which knowledge of crop phenology and local agricultural practices are essential for successful interpretation (Ryerson et al., 1997). As an example, consider winter wheat, a major crop in the High Plains of southwestern Kansas.

FIGURE 10-22 Terraced fields on sloping land under fallow conditions. Dark stripes represent water and wet soil behind terraces following heavy rain. Smoky Hills, central Kansas, United States. Kite aerial photo by SWA and JSA, May 2006.

The wheat is planted and begins growing in early autumn, overwinters, grows rapidly in the spring, and is harvested in early summer. From autumn until late spring, it is the only active crop in the region. Furthermore, the semi-arid climate limits winter wheat growth in non-irrigated fields to every other year. In other words, each field is used for growing wheat one year and stands fallow the next year in order to accumulate soil moisture. This practice results in a patchwork of fallow fields intermingled with crop fields, which alternate every other year (Fig. 10-24).

Active, green vegetation has quite distinct spectral characteristics (see Fig. 4-16), which are most obvious in color-infrared photography (Fig. 10-25). The strong near-infrared (NIR) reflectivity of active "green" leaves was discovered a century ago, and is known as the Wood Effect after Prof. R.W. Wood who first photographed this phenomenon in 1910 (Finney, 2007). NIR radiation (~0.7–1.3 μm wavelength) is strongly scattered by leaf cell walls, and some NIR energy passes through an individual leaf and may interact with adjacent leaves or soil (Fig. 10-26). Red and blue light are absorbed by chloroplasts for photosynthesis. Thus the ratio of NIR to red is an indicator for

FIGURE 10-24 Winter wheat fields, fallow fields, and wind turbines on the High Plains of southwestern Kansas, United States. Kite aerial photo by JSA and SWA, April 2006.

photosynthetically active vegetation. This gives rise to several vegetation indices, such as the normalized difference vegetation index (NDVI), which are important for ecological studies of biomass, leaf area index, and photosynthetic activity (Tucker, 1979; Murtha et al., 1997).

Aerial photography is widely employed for interpreting and monitoring vegetation conditions (e.g., Baker et al., 2004; Imeson and Prinsen, 2004; Lesschen et al., 2008) and making estimates of crop yield, forest production, and similar purposes (e.g., Grenzdörffer, 2004; Jensen et al., 2007; Berni et al., 2009). SFAP is used increasingly by farmers to help increase crop yields in conjunction with precision agriculture. Impacts of drought and flooding

FIGURE 10-25 Color-infrared photograph of the Arkansas River floodplain and recreation area near Pueblo, Colorado. Active vegetation appears in bright red, pink, and magenta colors. Kite aerial photo by JSA and SWA, May 2003.

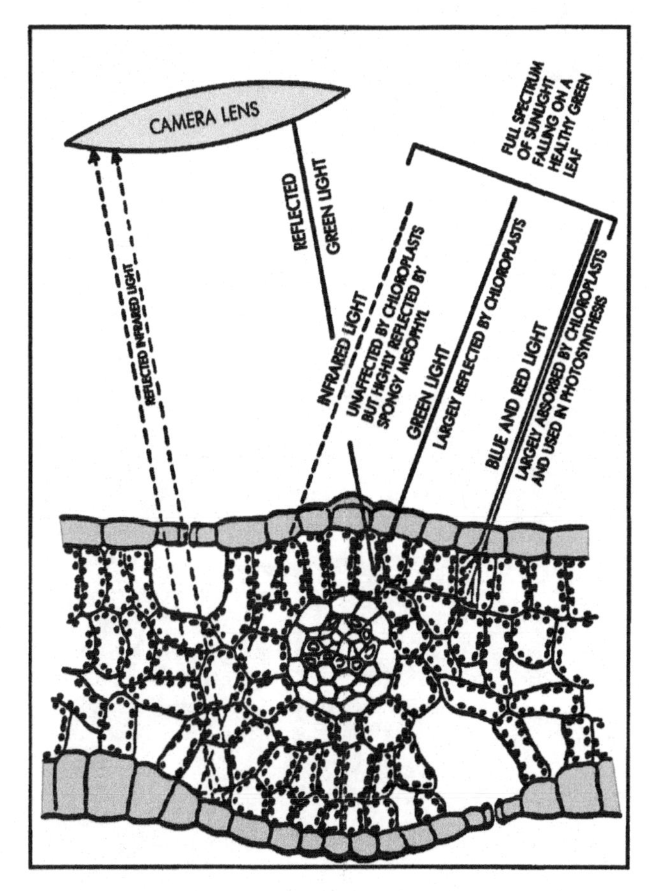

FIGURE 10-26 Schematic illustration of sunlight interaction with a healthy green plant leaf. Most near-infrared is reflected from leaf cell walls, and is not affected by chlorophyll. Blue and red light are absorbed by chloroplasts (black spots). Adapted from Murtha et al. (1997, fig. 5.7) and based on Colwell (1956).

FIGURE 10-27 (A) Floodplain of the Hornád River. Overflow channels (c) are revealed clearly in this fallow field, which was flooded one year before. The white building on the floodplain is a water well for the city of Košice. Southeastern Slovakia with Hungary across the river in right background. Kite aerial photo by JSA, SWA, and J. Janočko, August 2007. (B) Ground view of Hornád River flood in 2006; photo courtesy of J. Janočko.

(Fig. 10-27), disease and insect infestation, and many other factors can be evaluated to various degrees in all types of SFAP—panchromatic, color-visible, and color-infrared.

10.3.4. Cultural Heritage and Archaeology

This section deals with historic and prehistoric human-made structures including houses, churches, canals, roads, mills, monuments, graveyards, and other constructions. Size, shape, pattern, shadow, context, and other basic visual clues are important for recognizing and identifying such human structures, whether modern (Fig. 10-28) or ancient (Fig. 10-29).

Long after the above-ground structure is removed, covered over, or destroyed by the passage of time, traces of the foundations, alterations of the terrain, or disturbances in the soil are preserved. Consider the Amache Japanese Internment Camp that operated briefly during World War II near Granada, Colorado. It had a peak population of ~7500 Japanese Americans; today it is a ghost town with almost no surviving structures above ground. Nonetheless, the street network and building foundations are well preserved and clearly visible from above (Fig. 10-30).

Aerial photography has long been utilized to find, document, and map archaeological sites. Human structures tend to display regular, linear, geometric shapes and patterns that contrast with natural objects. These patterns may be recognized in airphotos by subtle variations in vegetation cover, soil color, and shadows—crop, soil, and shadow marks (Avery and Berlin, 1992). A good example is the Knife River Indian Villages in central North Dakota. This U.S. National Historical Site consists of Hidatsa and Mandan villages that were settled around A.D. 1300 and occupied until the mid-nineteenth century. The villages were the focus of a trading network that encompassed

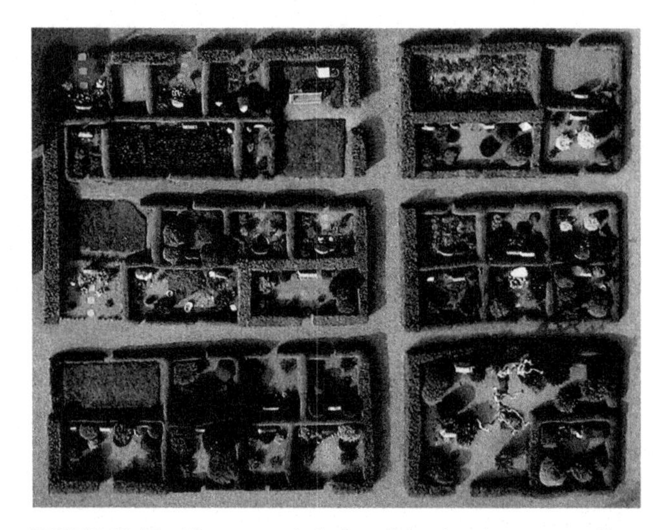

FIGURE 10-28 Close-up vertical view of churchyard cemetery at Gammelsogn Kirke (old parish church), near Ringkøbing, western Denmark. Tombstones stand within carefully manicured burial plots marked by hedges. Field of view ~16 m across. Kite aerial photo by JSA, SWA, and IM, September 2005. See also Figure 2-2.

FIGURE 10-29 Stone Age and Bronze Age burial mounds at Ramme Dige, a protected site in western Denmark. Overview of multiple mounds with sheep grazing in between. Shadows and vegetation patterns help to define the mounds. Kite aerial photo by JSA, SWA, and IM, September 2005.

FIGURE 10-30 Amache Japanese Internment Camp near Granada, Colorado, United States. (A) Overview looking toward the south. Building foundations are arranged in a uniform grid pattern. (B) Each rectangular foundation represents a barrack that housed a family. (C) Vertical view over the remains of the high school building. Kite aerial photos by SWA and JSA, May 2007.

FIGURE 10-31 Bird's-eye view of the Mandan Village, 1800 miles above St. Louis. George Catlin, 1837–39, oil, ~62 cm × 73 cm. Adapted from Dippie et al. (2002, p. 155).

FIGURE 10-32 Superwide-angle view of the Big Hidatsa Village site showing several dozen earth-lodge remains in the mowed portion. Kite flyers are standing in upper right corner of the scene. Note: some viewers may see raised bumps instead of depressions; this is an optical trick (see Fig. 5-3). Kite airphoto by JSA, October 2003, North Dakota, United States.

a vast region between the Great Lakes and northern Rocky Mountains.

Each village contained several dozen earth lodges (Fig. 10-31). The villages moved and were rebuilt over the centuries, such that remains of several hundred earth lodges are preserved as circular depressions, each about ~10–12 m in diameter. The earth lodges are still quite distinct in spite of former agricultural land use at the site in the late nineteenth and early twentieth centuries (Fig. 10-32). Depressions mark the former interior floors of lodges, which were built at ground level—not dug out. Surrounding each depression is a raised rim composed of material that fell off the lodge wall and roof. Some portions of the site are mowed, so the lodge remains can be seen easily on the

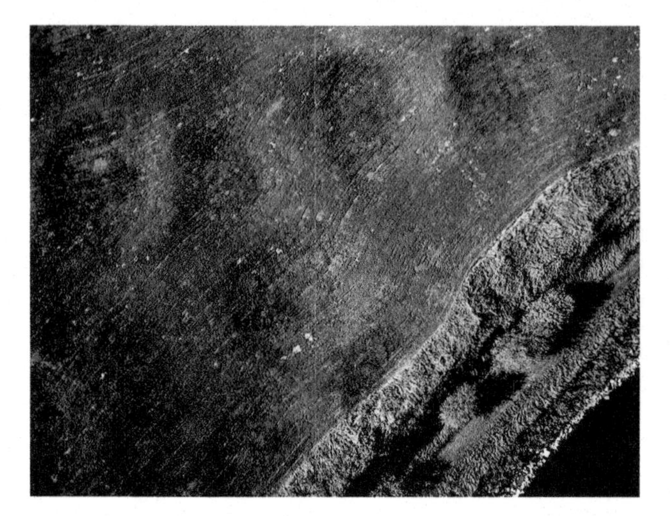

FIGURE 10-33 Close-up vertical view of earth-lodge remains in the Awatixa Village. Green circles (~ 10 m across) mark floors of former lodges. A distinctive mowing pattern crosses the lodge remains, and small white patches are animal burrows (gophers and squirrels). Erosion by the Knife River (lower right) has removed part of this village site. Kite airphoto by JSA, October 2003, North Dakota, United States.

FIGURE 10-34 Vertical view of an unmowed portion of the Big Hidatsa Village site. Circular patches and patterns in nearly dormant vegetation give clear indications of earth-lodge remains. Kite airphoto by JSA, October 2003, North Dakota, United States.

ground (Fig. 10-33). Other portions are covered by prairie grass, shrubs, and brush. Earth lodges in these unmowed portions are not obvious from the ground, but are quite evident from above because of variations in vegetation cover (Fig. 10-34).

Several settlement mounds of Medieval origin are distinctly visible today as mounds of pottery shards and stones in this oblique view of the dunes of Oursi in Burkina Faso (Fig. 10-35). These settlement mounds have extremely high runoff rates, up to 90%, as measured by rainfall simulations, and limited infiltration capacity, thus encouraging linear erosion in the sandy Pleistocene dune deposits downslope. The gullies are now already cutting into the

FIGURE 10-35 Medieval settlement mounds (West African Iron Age) at the Oursi Dunes, Province of Oudalan, Burkina Faso. The high runoff rates of the mounds' shard-and-stone composition cause gully erosion between the individual sites. Oblique kite aerial photograph by IM, JBR, and K-D. Albert, July 2000. Taken from Marzolff et al. (2003, fig. 8).

FIGURE 10-36 Early spring vertical view of cemetery and adjacent suburban housing. Two rows of unmarked graves (>) are visible in the upper left portion, as shown by greener grass cover. Each grave is ~1 m × 2 m in size, oriented with long dimension E–W. The green circles are fairy rings created by fungus in the soil and visible at this time of year. North arrow is 4 m long. Kite aerial photo by JSA and SWA, late March 2003. Emporia, Kansas, United States.

archaeological layers, and it is of high interest to both geomorphologists and archaeologists researching this site if the hilly surface relief seen today actually represents the historical settlements or if it has developed by erosion only fairly recently (Albert, 2002).

Subtle variations in crops and lawns often reveal buried archaeological features that may be nearly invisible on the ground. In this regard, drought conditions may accentuate variations in vegetation cover that depict archaeological remains (Eriksen and Olesen, 2002). Likewise, seasonal growth of vegetation, particularly in spring or autumn, may be critical for recognizing buried features. Consider, for example, the problem of locating unmarked graves in cemeteries. Past disturbance of soil may be revealed by slight changes in vegetation, soil moisture, or microtopography (Fig. 10-36).

Sea Song Plantation was the home of Andrew Jackson, Jr., the nephew and adopted son of U.S. President Andrew

FIGURE 10-37 Ground shot of Sea Song Plantation as seen from the pier ca. 1904, near Waveland, Mississippi, United States. Photo courtesy of M. Giardino.

FIGURE 10-38 Overview of Sea Song Plantation site, now a public park with a picnic area and playground. Kite aerial photo by JSA, SWA, and M. Giardino, March 2004.

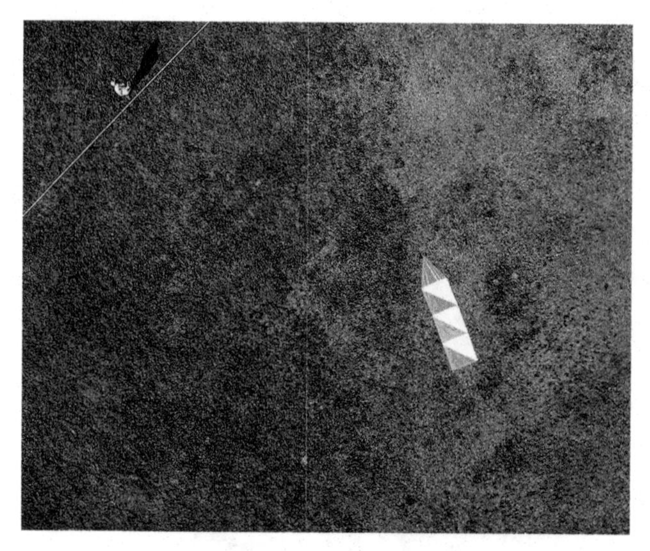
FIGURE 10-39 Close-up vertical view of Sea Song Plantation site. Note the square shape revealed in the grass cover around the scale arrow (4 m long). The nature and identity of this feature are unknown, but it could represent one of the smaller outbuildings of the former plantation. Kite aerial photo by JSA, SWA, and M. Giardino, March 2004.

FIGURE 10-40 Late-Neolithic moat fortification near Ottmaring-Niendorf in Lower Bavaria, Germany. The fort was built for reasons of defense before 5500 BP. The course of the palisade trench can be detected easily from the dark lines in the image center. Its left lower edge is missing or not visible anymore following the erosion of ~50 cm of soil since the abandonment of the settlement. The central rise where the fort is situated and the ridge running uphill to the left show considerably lighter tones than the rest of the field. Here, topsoil rich in humus has been eroded, ringed by a narrow band of the Luvisols' dark-colored Bt horizon. Farther downslope in direction of the trench, the medium-brown colluvium built up from the material eroded upslope covers everything else. Small-format aerial photograph taken from a light airplane by R. Christlein in 1982. Taken from Christlein and Braasch (1998, plate 18).

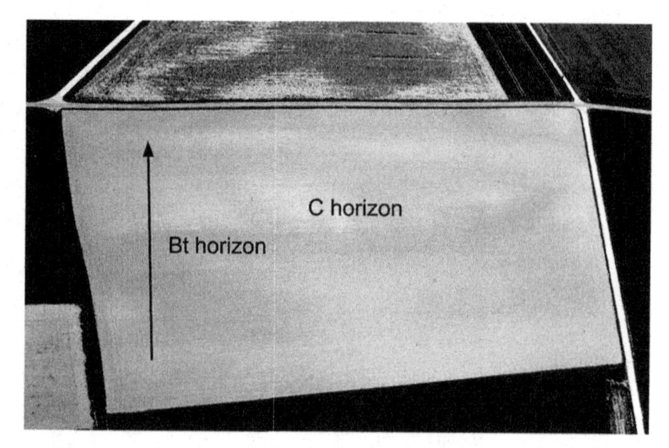

FIGURE 10-41 Considerable differences in color on a gently sloping field near Nieder-Weisel (Wetterau), Germany indicate areas with intense (C horizon) and less intense (Bt horizon) erosion. Slope direction indicated by arrow. Helicopter aerial photograph taken by A. Fengler, May 1998.

Jackson in the mid-nineteenth century. Located on the Mississippi Sound coast, the house survived into the early twentieth century before it was destroyed by fire (Fig. 10-37). The site is now a small park with a picnic area and playground (Fig. 10-38). Although the general locations of the main house and outbuildings are known, the exact positions and identities of various structures are obscured by the passage of time. These buildings had no foundations, so the most noticeable traces are driplines around the former roofs. Close-up vertical SFAP revealed square features of unknown origin (Fig. 10-39), but hurricane Katrina interrupted further archaeological investigations at the site.

10.3.5. Soils

Soil discolorations in aerial photography are helpful aids for archaeology as well as soil science. From the small-scale distribution of soil units, conclusions about soil thickness and degradation state of the soil may be drawn. Not only old postholes, pits, or walls can be detected from soil color changes, but also individual soil horizons distinguished by their colors (Fig. 10-40). Topsoils (Ah horizon) rich in humus are usually darker than the parent material (C horizon) in which the soils have developed. Subsoil horizons that are situated in between usually show their own distinctive colors, mainly brown or reddish-brown (B horizon). This is typical at least for the loess landscapes in Europe, Asia, and northern North America with their Luvisols, Chernozems and Kastanozems. When the soils are eroded, truncated profiles result, exposing the underlying soil horizons or even the parent material (Fig. 10-41). Such erosion areas may be detected easily from the air by their lighter colors.

Truncated horizons are particularly frequent in the landscapes of central and southern Europe and of the Near East that have been cultivated for millennia. In the United States and southern Canada, soils that were still largely intact until two centuries ago have been affected severely also by mechanized agriculture. The intensity and small-scale distribution of soil impairment is today particularly important for precision farming techniques. Soil mapping from remotely sensed images requires a soil surface largely devoid of vegetation and a homogeneous, preferably fine-aggregated condition of a dry soil surface. In central Europe, such conditions may be found for only a few days between the end of February and mid-May. Thus, flexible timing of aerial surveys, such as offered by SFAP, is needed for soil mapping applications.

10.4. SUMMARY

Interpretation of aerial photographs begins with basic aspects—color, size, shape, pattern, texture, and shadows—combined with scene context and experience of the interpreter. Visual recognition of objects in vertical images normally requires a ground sample distance 3–5 times smaller than the object itself, although other non-spatial factors also may impact interpretability. Atmospheric effects tend to degrade image quality with increasing height regardless of spatial resolution. SFAP from low height may provide interpretability ratings in the range of 7–9, noticeably higher than conventional airphotos or high-resolution satellite imagery.

Visible color is diagnostic for suspended sediment in water bodies, whereas color-infrared imagery is most appropriate for vegetation. Cultural features tend to display linear and geometric shapes and patterns that contrast with natural objects for both modern urban and rural features as well as prehistoric remains. Subtle crop, soil, and shadow marks may reveal the presence of buried archaeological features. Often, the most interesting information in an image does not reveal itself immediately but rather indirectly by visual clues to the conditions and processes that induce a certain pattern, shape, or color.

For the presented and other applications of SFAP, knowledge of conditions on the ground and familiarity with local human activities are invaluable for improving image interpretation. The scientific benefit arises from the combination of knowledge achieved by conventional field work and the visual information gleaned from SFAP. In monitoring studies, temporal changes of forms and patterns give additional clues. The coupling of ground evidence and interpretation of SFAP may lead to a deeper understanding of the geoscientific processes.

Chapter 11

Image Processing and Analysis

Οἱ δ' ἀριθμοὶ πάσης τῆς φύσεως πρῶτοι.

Numbers are the ultimate things in the whole physical universe.

Pythagoras (quoted by Aristotle in *Metaphysics*, 1.985b, Fourth century BC)

11.1. INTRODUCTION

Basically, a digital image is just an array of numbers arranged in rows and columns. Processing an image by enhancing its contrast, correcting its geometry, or classifying it into various categories means processing numbers with mathematical and statistical algorithms. The same image-processing techniques that are used for analyzing remotely sensed images from multispectral satellite sensors may be applied to small-format aerial photography (SFAP) (e.g., Kasser and Egels, 2002; Mather, 2004; Bähr, 2005; Richards and Jia, 2006; Jensen, 2005, 2007; Lillesand et al., 2008). Again, this chapter is not intended to be exhaustive, but meant to cover selected aspects that the authors consider specific or particularly useful for digital analysis of SFAP. It begins with aspects of geometric corrections and registration, continues to image enhancement techniques, image transformation and classification, and concludes with stereoscopic viewing and photogrammetric analysis. The final section gives a brief overview of software suitable for SFAP processing.

11.2. GEOMETRIC CORRECTION AND GEOREFERENCING

Correction of geometric distortions and georeferencing to a common object coordinate system are necessary for many SFAP applications, e.g., quantitative analysis of areas and distances, monitoring projects, or integration of various data sources in a geographical information system (GIS). The first decision to be made concerns the reference coordinate system to be used; the second concerns the image rectification method. Finally, covering large areas with high-resolution SFAP might require mosaicking several images together.

11.2.1. Reference Coordinate Systems

Two basic types of coordinate systems exist for geographic data: geodetic coordinate systems based on map projections and geographic coordinate systems based on latitude and longitude (for details, see for example Hake et al., 2002; Longley et al., 2006). The main difference is that projected, geodetic coordinates are Cartesian coordinates with two equally scaled orthogonal axes. Distances and areas calculated in these reference units are comparable across the globe. Geographic (unprojected) coordinates, on the other hand, are polar coordinates defined by a distance (the radius of the Earth) and two angles (between a given location and the equator and between this location and the prime meridian). Because the spacing between lines of longitude decreases from the equator to the poles, they are not useful for comparing distances or areas across the globe. However, they are useful as a comprehensive global system unaffected by distortion issues associated with map projections. The following coordinate systems may be considered for georeferencing SFAP.

- The national map projection system locally used for topographic mapping, e.g., the Gauß-Krüger or ETRS89/UTM system in Germany or the state plane systems in the United States. This is most useful when the images are to be combined with other, existing data such as topographic maps or digital cadastral data, etc. It requires the connection to the national system by nearby trigonometric points or highly precise differential global positioning system (DGPS) when measuring ground control points.
- An arbitrary local coordinate system (see Chapter 9.5). This is useful when correct scale, area, and distances are required but absolute position of the study site within a national or global system is not a necessity, because no data with other spatial references have to be overlaid. Local coordinate systems subsequently may be transposed into national reference systems by rotating and shifting them, either by adjusting them visually over other referenced image data in a GIS or by mathematical transformation of the coordinates in a spreadsheet or conversion program (e.g. for Helmert transformations).

- No pre-defined coordination system at all but image file coordinates. For some applications, not even scale, areas, or distances are of interest because relative values like percentage cover or mean normalized difference vegetation index (NDVI) value are the only output information required. If the image is not significantly distorted, it does not need to be geometrically transformed. Even time series and change analysis may be carried out when subsequent images are relatively referenced to the first by image-to-image registration.
- Geographic coordinates in decimal degrees or degrees, minutes, and seconds. This worldwide reference system (see above) is useful because of its universality and global availability, but has the already mentioned disadvantage of angular coordinates. Unless the study site is situated near the equator, images referenced in lat/lon should be projected into a geodetic coordinate system for displaying or printing, or they would appear badly stretched. Also, a coordinate precision adequate for the small ground sample distances (*GSDs*) of SFAP requires many decimal places: Consider that one degree of latitude equals ~111 km; thus, a centimeter precision for a point on Frankfurt University's Riedberg Campus requires recording a latitude coordinate as precise as 50.1794072° N.

11.2.2. Image Rectification

Basically, three different types of distortion may be present in an SFAP image: lens distortion, image tilt, and relief displacement (see Chapter 3.2). Two methods of correcting these errors, polynomial rectification by ground control points (GCPs) and orthorectification, have already been introduced in Chapter 3.4.1. Also, the precision and survey methods for GCPs have been discussed in Chapters 3 and 9. The following sections look at some exemplary questions and cases related to geocorrecting small-format aerial photographs.

A highly exact geometric correction requires time, effort, a digital elevation model, and excellent ground control. Many SFAP applications may not really need such efforts, and depending on the image and relief characteristics, simpler solutions might be quite sufficient. So how does one judge how correct is good enough? This is always a question of relating the residual error to the desired spatial information detail. The following concerns may play a role in this decision.

- Maximum location accuracy to be expected with a given *GSD*.
- Desired accuracy for locating or delineating individual, well-defined features.
- Required accuracy for measuring changes in monitoring projects.
- Achievable precision of object delineation.

Consider a project where the invasion by a non-native plant species in an estuarine wetland reserve is to be assessed by quantifying the percentage of affected vegetation cover from aerial photographs. Here, precise orthorectification with <2 cm error would certainly be a wasted effort for several reasons. The images would be affected little by relief displacement because of the flat terrain; a measurement accuracy in the centimeter range is not required for the purpose; and indeed the delineation precision of something as fuzzy as vegetation patches would be much inferior to 2 cm. Another application might be monitoring the development of sharp-edged ephemeral rills on a hillslope during one season and relating their retreat rates to precipitation and runoff measurements. In this case, the measured changes may be grossly inaccurate if the image time series were rectified using a few GCPs only, with check-point RMS errors in the decimeter range.

The differences in positional accuracy achievable with various rectification techniques for a situation with variable terrain are illustrated with Figure 11-1. This vertical photograph of a long gully cutting into a shallow depression between two agricultural fields was taken from 82 m height above field level with an original image scale on the sensor of 1:4100. The altitudinal differences present in the scene amount to ~4.5 m (top of field-bordering ridge to bottom of gully). Forty-one GCPs were measured with a total station; 22 of those were used for georectifying the image with different methods in *ERDAS IMAGINE* and *Leica Photogrammetry Suite*, while 19 were reserved as check points for error assessment.

Results show that the abrupt relief changes (see cross-profile in Fig. 11-2) cause relief displacement that cannot be corrected sufficiently with polynomial transformations. Although a large number of GCPs was used, the third-order transformation shown in Figure 11-1A leaves considerable residuals at both control and check points, with a maximum displacement of 43 cm. Rubbersheeting—a piecewise non-linear polynomial transformation—results in no errors at the control points, but even larger errors at the check points (Fig. 11-1B). This confirms that the GCPs, in spite of their good distribution and abundance, are not able to represent the variations in terrain height adequately. The image is now geometrically correct at the control points, but still distorted in the areas between, with a total RMS error of nearly 30 cm at the check points.

By far the best results are achieved with orthorectification (Fig. 11-1C), where the image is differentially corrected using a digital elevation model. This was realized by setting up a photogrammetric block file with both interior and exterior orientation information (see Chapter 3) and triangulating several overlapping images. A digital elevation model (DEM) was extracted from the image shown in Figure 11-1 and its stereopartner, and this DEM provided the necessary height information for computing and correcting relief distortion. The residual errors, which amount

GCP position in reference system

× used as control point in rectification

× used as check point in rectification

RMS errors in reference units

Radius proportional to linear RMS, exaggeration 5 x

● error at control point

● error at check point

75 cm
50 cm
25 cm

N

0 5 m

		A 3rd order polynom [cm]	B Rubber-sheeting [cm]	C Ortho-rectification [cm]
GCPs	Min RMS	1.08	0.00	1.05
	Max RMS	42.89	0.00	11.19
	Total RMS	17.90	0.00	3.73
Checks	Min RMS	1.64	2.07	0.63
	Max RMS	38.44	71.98	6.57
	Total RMS	21.83	29.25	3.09

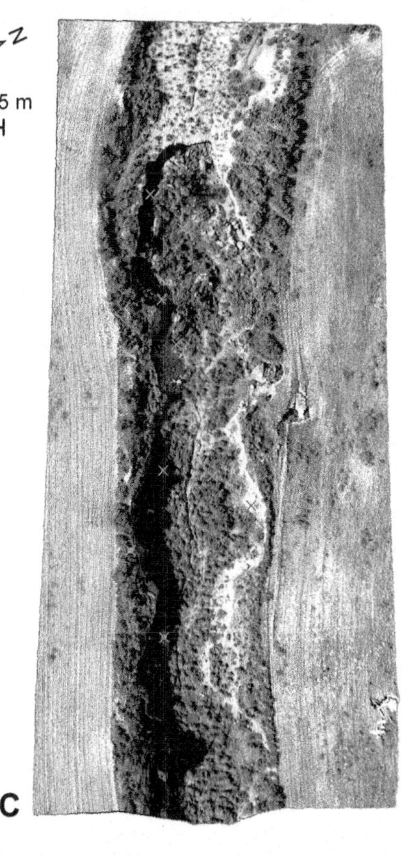

FIGURE 11-1 Georectified images of gully Bardenas 1, Bardenas Reales, Province of Navarra, Spain. Kite aerial photograph by JBR, IM, and S. Plegnière, February 2009; image processing by IM. (A) Rectified by third-order polynomial transformation. (B) Rectified by non-linear rubbersheeting. (C) Orthorectified using a DEM derived from the same image and overlapping stereopair. Location and RMS errors of 22 control points and 19 check points shown by diagram radius with fivefold exaggeration. *GSD* of all rectified images is 2.5 cm.

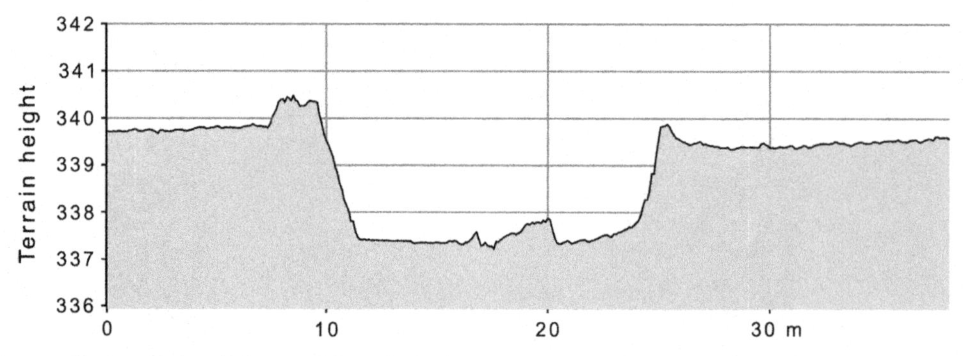

FIGURE 11-2 Cross-profile through the middle part of the gully in Figure 11-1, derived from the DEM used for orthorectification.

to only about 3–4 cm of total RMS, are due to the small remaining errors in triangulation and extraction of the DEM, which was not corrected for vegetation errors.

There is no general solution as to how many GCPs are needed for a good georectification. For flat areas with no relief, a theoretical minimum of three is enough, but not recommended because of the lack of error control. A second-order transformation, which is able to take simple smooth relief undulations into account, requires a minimum of six GCPs. As a rule of thumb, twice the minimum number required is a good measure for reducing residual errors.

Because establishing ground control every time a survey is conducted is costly and time-consuming (see Chapter 9.4), it can be a useful method to supplement or even substitute control points with image-to-image rectification in monitoring projects. Non-referenced images may be warped onto a first, thoroughly rectified image by image-to-image registration, picking homologous points in both the reference and input images. This can be done manually or with software for automatic point matching (e.g. *IMAGINE AutoSync*).

11.2.3. Image Mosaics

SFAP survey planning often requires compromising about small *GSD* of the images and large area coverage (see Chapter 9). In order to achieve a desired degree of detail, it might be necessary to take multiple images for a given area that subsequently need to be assembled into an image mosaic. In this way, even unmanned SFAP may cover considerable areas with high resolution, provided a platform capable of travelling longer distances is used, e.g. a model airplane (see Chapter 8.5.2).

The term mosaicking is commonly used in remote sensing and geoinformation sciences when contiguous images are joined to form a single image file. However, this does not necessarily require that the images be georeferenced—rather than using GCPs for positioning the individual images in a common reference coordinate system it is also possible to use any homologous image objects for a relative spatial adjustment of the photographs (uncontrolled mosaic). This can be done either in remote sensing and GIS software or with dedicated photo-editing software, with varying degrees of automation.

Most digital camera software now includes so-called stitching functions, and there is a large choice of free and commercial panorama software for fully automatic merging of multiple photographs into larger composites. These tools can produce results that are visually highly appealing (see Fig. 11-3), but they are not geometrically correct. Note the angular skew and straight edges of the left image tiles in Figure 11-3. If the mosaic were georectified, the edges would be bent and warped following the changing relief heights of the hillslopes, similar to the bottom edge of Figure 11-1C. Remote sensing or GIS software is required for georeferenced (controlled) mosaics, which can be constructed from image tiles previously geocorrected with polynomial transformations or orthorectification.

Image mosaics mercilessly reveal exposure differences across or between photographs. Regardless of controlled or uncontrolled and no matter which software used, a good mosaic requires some sort of color matching between the stitched images in order to conceal the seams. Not all software capable of stitching adjacent images together may provide color-correction tools, and matching the image tiles for a balanced overall impression may be a difficult task especially for large mosaics with strongly varying brightness distributions (Fig. 11-4).

11.3. IMAGE ENHANCEMENT

SFAP images for scientific applications are taken usually with the intention to show certain information about a site or an area; thus, optimizing the visual interpretability of an image with image enhancement techniques is often the primary concern. As discussed in Chapter 6.4.6, photographs taken with small-format cameras are not without radiometric deficiencies. Image-degrading effects like chromatic aberration (CA), vignetting, and noise can be corrected partly or completely with camera software or professional image-editing software. This may minimize or

FIGURE 11-3 Uncontrolled image mosaic created with *Autopan Pro* panorama software from 17 aerial photographs taken from 150 m height by an autopiloted model airplane. The mosaic covers a 600 m stretch of the Rambla de Casablanca near Guadix, Province of Granada, Spain (northeast is up). The hillslopes at the top and bottom are dotted with small terraces recently installed for a reforestation project; two patches with pine trees left of the image center are the remains of an earlier, failed reforestation attempt. Photos by C. Claussen, M. Niesen, and JBR, October 2008; image processing by C. Claussen.

remove radiometric flaws determined by lens and camera characteristics. However, the image objects themselves also may be responsible for visual shortcomings, especially regarding color contrast. Image enhancements may range from fine adjustments for improving the general visual appearance to considerable modifications aimed at emphasizing certain information. The following sections cover some of the enhancement methods the authors most frequently use on their SFAP images.

11.3.1. Correcting Lens-Dependent Aberrations

Lens-dependent aberrations are circular effects centered on the optical center of the image. Therefore, it is important to correct them before cropping or warping the image. Many photo-editing software packages and some modules of professional remote sensing software include tools for correcting vignetting, CA, and sometimes even lens distortion.

The degree of vignetting depends on lens characteristics, aperture, and exposure and is thus not an invariable effect—it is difficult to correct automatically and its correction settings usually need to be judged visually for each image. Using sliders for the amount and progression radius of the lightening, the brightness of the image is modified until the light falloff toward the corners is compensated. Correcting for vignetting, which is most noticeable in homogeneous, light portions like the sky or bare soil, can be sensible not only for aesthetic reasons but also prior to image classification in order to balance the spectral characteristics of the scene.

The correction of CA and lens distortion involves not brightness or color modifications but geometric transformations of the image. Latitudinal CA is rectified by slight adjustments of the image channel sizes, usually by radially scaling the red and blue channels individually to match the size of the green channel. Radial lens distortions (see Figs. 6-7 and 6-14) are lens-specific geometric distortions that are best corrected with lens-specific and focal-length-specific settings using dedicated software like *PTLens*. It is also possible to correct radial lens distortion visually, but only for scenes where its degree can be judged from the deformation of straight lines or grids—something that is usually not the case for SFAP. Correcting radial distortions prior to quantitative image analysis might be especially useful if the image is to be georectified using simple polynomial equations. Some recent camera models automatically apply CA and lens-distortion correction in-camera to JPG or even RAW files, so the user does not even get to see the original distortions any more.

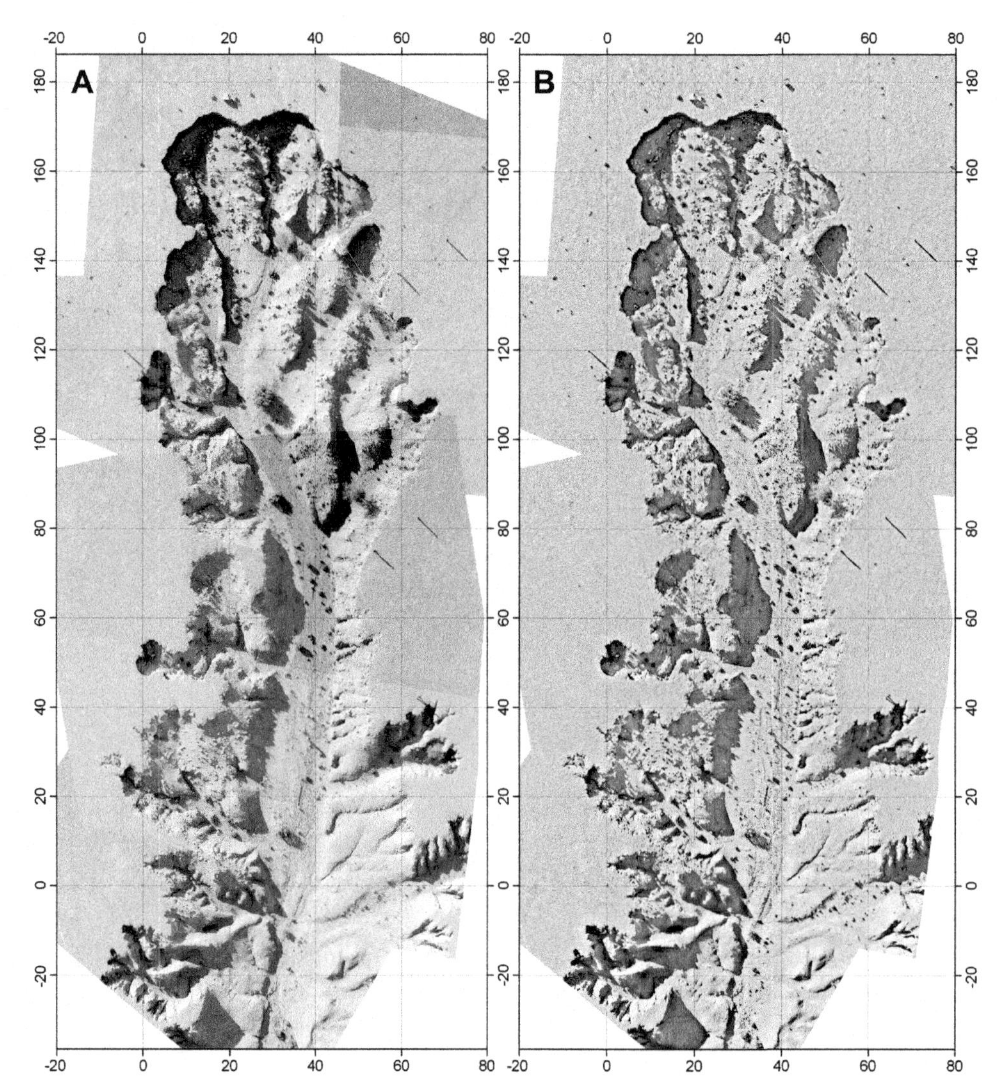

FIGURE 11-4 Controlled image mosaic of gully erosion near Icht, South Morocco, created with *ERDAS Imagine* 9.3 from nine orthorectified aerial photographs. (A) Original images mosaicked without color corrections. (B) Original images mosaicked with color dodging corrections. Contrast enhancement (see Fig. 11-8) applied subsequently to both mosaics. Note the perfect color-adaptation of the surrounding desert surface but still different tints of shadowed areas in version B. Kite aerial photographs by IM, JBR, and M. Seeger, March 2006; image processing by IM.

11.3.2. Contrast Enhancement

Most aerial photography covered in this book shows landscapes of all sorts, and few of them abound in striking colors with lively contrasts. Even more than high-oblique images, vertical aerial photographs for geoscientific applications tend to be restricted to shades of green, brown, and beige. Yet exactly the subtle differences in color shades may be interesting for distinguishing certain soil properties or vegetation types.

Various ways exist for manipulating the distribution of tones across the brightness range between 0 and 255 (for an 8-bit image), all of which modify the image histogram that shows the relative frequency of occurrence plotted against pixel values. Depending on the software used, the following methods might be available (e.g., Richards and Jia, 2006; Langford and Bilissi, 2007).

- Simple global brightness and contrast controls.
- Advanced controls for specific color and brightness ranges.
- Histogram adjustments using levels.
- Histogram adjustments using curves/graphic functions.

Color enhancements may be done in any of the available color spaces (see Chapter 5.2.5). For example, instead of manipulating the three primary colors red, green, and blue, the saturation component of the IHS color space may be used for slight enhancement of dull colors due to backlit situations. Using the intensity and saturation components of the IHS system would not change the hues in the image, while enhancements in the RGB space are able to produce complex color shifts.

Consider the view of a network of erosion rills on a river terrace (Fig. 11-5). This is what the original image looks

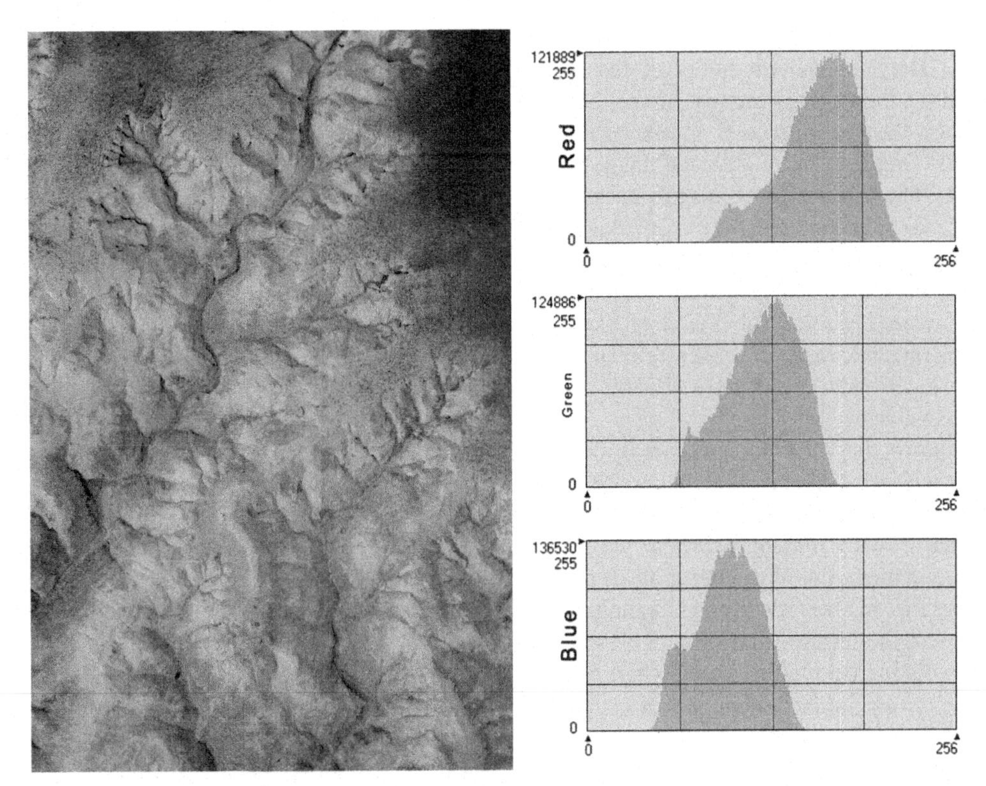

FIGURE 11-5 Vertical kite aerial photograph (with image histograms) of rill and gully erosion on a sedimentary river terrace near Foum el Hassane, South Morocco. Taken with a *Canon* EOS 350D (Digital Rebel XT) with 28 mm *Sigma* lens, no image processing applied. Field of view ~70 m across; photo by IM, JBR, and M. Seeger, March 2006.

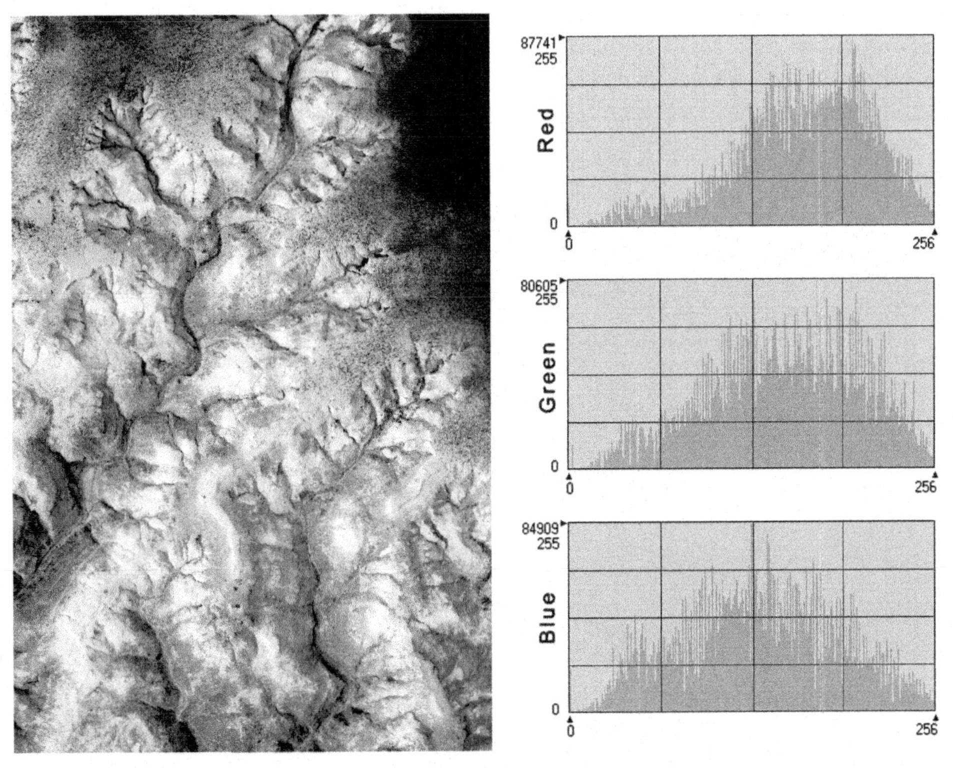

FIGURE 11-6 Image in previous figure after histogram contrast enhancement. Image processing by IM.

like, shot as RAW and converted to TIF without any additional processing. The monotonous colors of the loamy sediments differ little from the darker hue of the terrace's rock-strewn surface in the upper and lower right. The indirect light of the overcast day renders the image nearly devoid of shadows. The histograms of image values reflect the homogeneous visual impression by a unimodal distribution in the mid-ranges. The slight broadening and shift toward the higher values from blue to red results in the overall brownish image hues. It is obvious that the scene does not exploit the dynamic range of the sensor (and screen or printer) at all; each band covers only 40–50% of the possible brightness values.

Figure 11-6 shows the same image after a contrast enhancement that expanded the histograms between their extremes, slightly saturating the high ranges. The scene is now a bit unnatural with respect to what we would actually see with our own eyes in the field, but it reveals remarkable variations in the seemingly homogeneous sedimentary layers of the terrace. The smooth crusted top surface now has a distinctive pinkish shade where bare of rocks, while the layer immediately underneath and another one farther below appear bright with a turquoise tinge. Where the two main gullies have cut deeper, the sediment is much darker and reddish-brown again. The color differences result from differences in pedogenetic processes, proportions of organic and inorganic material, and grain sizes in the individual layers, which have been deposited during the Late Glacial and Early Holocene (Andres, 1972).

A similar contrast enhancement was applied to this view of sand-sage prairie vegetation (Fig. 11-7). In the original image, only subtle variations in colors and shadows can be seen. Disregarding the bright gravel road and dark shadows of the trees, the brightness values present for the vegetated areas of the scene only covered 25–30% of the 8-bit dynamic range. By stretching them to the full 256-value range, great variations of color, texture, and shape in the vegetation cover become apparent. Individual sagebrush (*Artemisia* sp.) bushes become quite distinct as pale-green clumps on subtle sand dunes. Between dunes, circular swales with dark green prairie grass are apparent. Orange spots are patches of prairie wildflowers.

In the examples discussed above, the camera sensor's dynamic range was wider than the scene required. Starting with simple linear contrast stretches (min–max or about ±2.5 std. dev.) often yields good results with such images. A more difficult case is the presence of deep shadow in a scene (see also Chapter 4.7). A contrast stretch as applied in Figure 11-6 would deepen them even further, obscuring the shadowed objects still more and increasing their radiometric difference to the sunlit parts of the image. This might impede further image processing like classification or stereoscopic feature extraction.

In Figure 11-8, the original image of a heavily shadowed scene is compared with two contrast-enhanced versions. Note the strongly bimodal distribution of the original image values shown in Figure 11-8D (gray histogram) and the generally dark appearance of the desert scene. In real life, the scenery was much brighter with dazzling light, but the automatic, non-corrected exposure settings of the camera have resulted in comparative underexposure (see Chapter 6.4.5). The surfaces and objects obscured by the shadow are of the same type as those illuminated by the sun, so the shadows would decrease or even disappear if the two peaks could be moved or even merged together. The colored

FIGURE 11-7 Sand-sage prairie vegetation in sand hills of western Kansas, United States. Kite flyers next to abandoned farmstead at upper right edge of view. Photo taken by JSA and SWA with a *Canon* PowerShot S70, May 2007; image processing by IM. (A) Original image without processing. (B) Contrast-enhanced version emphasizing variations in vegetation cover. Small pale-green clumps in lower and left sides are sagebrush on dune crests; dark green prairie grass in swales and around farmstead; orange indicates wildflowers in bloom.

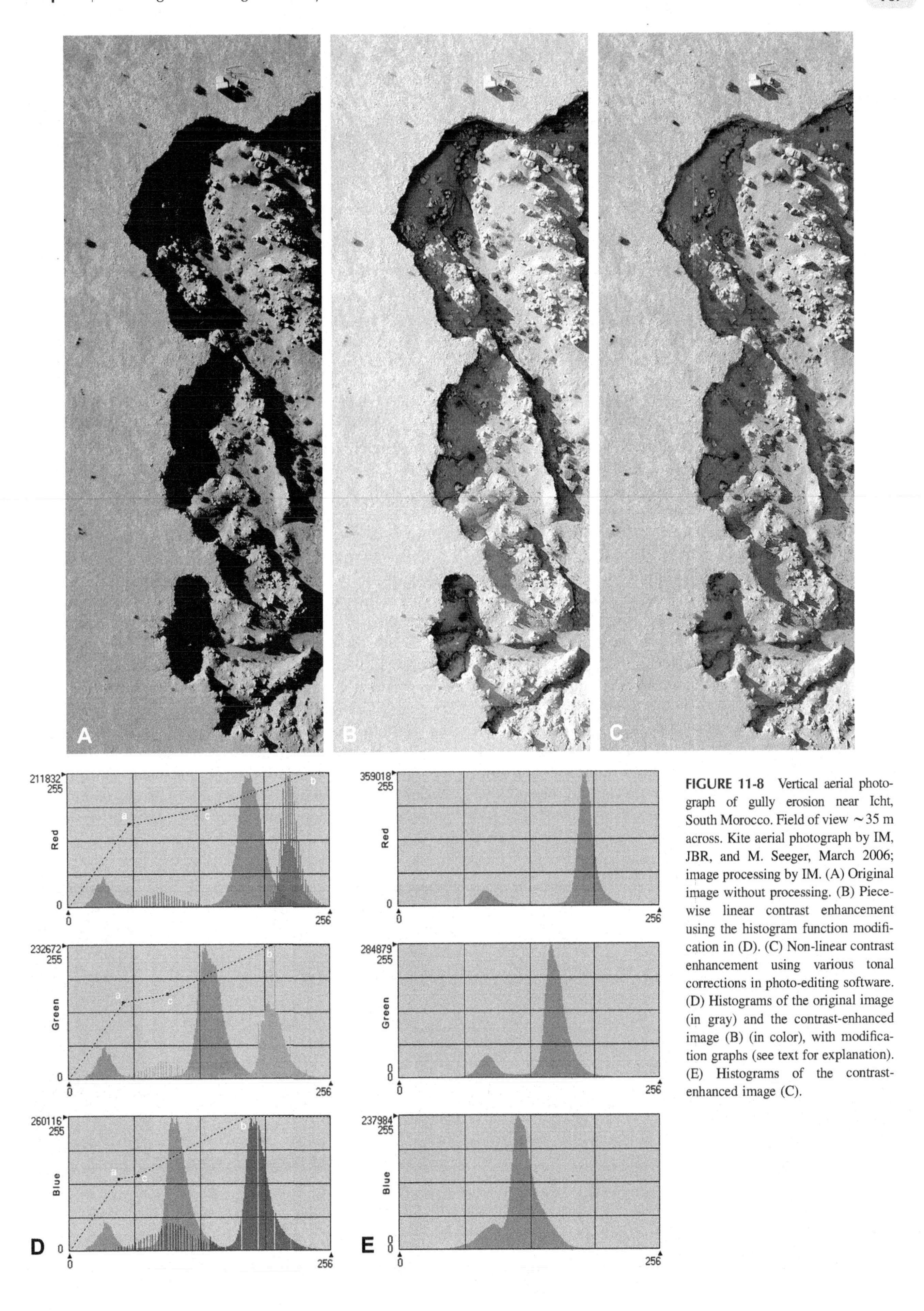

FIGURE 11-8 Vertical aerial photograph of gully erosion near Icht, South Morocco. Field of view ~35 m across. Kite aerial photograph by IM, JBR, and M. Seeger, March 2006; image processing by IM. (A) Original image without processing. (B) Piecewise linear contrast enhancement using the histogram function modification in (D). (C) Non-linear contrast enhancement using various tonal corrections in photo-editing software. (D) Histograms of the original image (in gray) and the contrast-enhanced image (B) (in color), with modification graphs (see text for explanation). (E) Histograms of the contrast-enhanced image (C).

histograms in Figure 11-8D, belonging to the image in Figure 11-8B, are the result of a piecewise contrast enhancement using a graphic histogram modification (see for example Richards and Jia, 2006). Breakpoint *a* increases the brightness values of the shadow peak to a level corresponding to the mean values of the sun peak. To counteract the resulting tonal compression of the sunlit regions, breakpoint *b* raises their highest values to the maximum file value 255. Finally, the curve is flattened a little at the depression between the shadow peak and the sun peak by inserting breakpoint *c*, compressing the original values where few pixel counts exist.

Even less contrast between shadowed and sunlit regions are left in Figure 11-8B (with histograms in Figure 11-8E): The two histogram peaks have moved closer, in the blue channel even melted together. This is the result of a more sophisticated contrast enhancement realized with professional photo-editing software (*Adobe Lightroom*). Here specific tonal ranges were modified separately in succession, addressing the lights, darks, and shadows individually.

The contrast enhancements presented above were carried out on 8-bit images that are already compressed versions of the originally higher dynamic range the sensor had recorded. Stretching parts of the histogram to greater contrasts results in missing brightness values or jumps between color shades (banding) as can be seen from the shadow peaks in Figure 11-8D. For strong adjustments, this effect may even become visible to the naked eye in the enhanced image. It can be avoided by processing the image in RAW format at higher radiometric resolution (12 or 14 bit) and converting it to an 8-bit file afterward. As computer screens and printers (let alone our eyes) cannot make use of the larger image depth anyway, the reduction to 8 bit would not degrade the image quality, but the banding caused by the contrast stretch would be smoothed out to invisibility. Working with RAW imagery offers many advantages for post-processing that are beyond the scope of this book; for more details the user is referred to the wide choice of manuals on digital image processing (e.g. Langford and Bilissi, 2008).

The examples discussed so far are meant to improve color contrasts throughout the image. In some cases, however, it might be useful to enhance or even overenhance only parts of the image in order to maximize the discrimination of tonal ranges associated with certain types of features. This is illustrated with an agricultural scene (Fig. 11-9), where only the image values associated with the fallow field were stretched in the histogram, followed by a strong saturation boost in the IHS color space. As a result, the assumedly homogeneous field now shows clear distinctions of different zones, from blackish-blue, to yellow-brown, to pink, to light cream colored. Such techniques are useful mapping aids for soil survey and may help to reduce the number of bore holes in field sampling by up to 50% (Fengler, 2007).

11.3.3. Image Filtering

Image filtering involves so-called neighborhood operations, in which the value for each pixel in the scene is recalculated based on the values of surrounding pixels. In a simple example, the value for a pixel represents the average of surrounding pixels as defined by a window or convolution mask with the target pixel at the center. Typical square window sizes are 3 × 3, 5 × 5, 7 × 7, and so on (Fig. 11-10). This type of manipulation is called a low-pass or low-frequency filter, and the result blurs the image by smoothing out pixel-to-pixel variations. This approach may utilize simple averages or other statistical functions for calculating the value of each pixel in the image, and windows with other shapes may be employed (Jensen, 2005). In any case, larger windows create more smoothing or blurring of the visual

FIGURE 11-9 Agricultural fields near Nieder-Weisel (Wetterau), Germany. Hot-air blimp photograph by A. Fengler and JBR, May 1998; image processing by IM. (A) Digitized transparency slide, corrected for slight vignetting. (B) Color stretch and saturation boost applied for enhancing fallow land area.

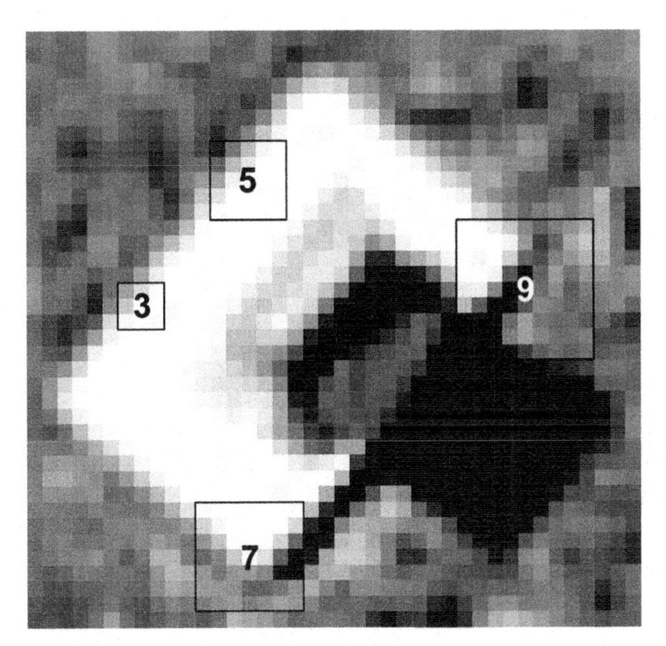

FIGURE 11-10 Highly magnified view of a tombstone from Figure 5-11 displaying individual pixels in the original image. Windows of 3 × 3, 5 × 5, 7 × 7, and 9 × 9 are indicated. Such windows are the basis for low-pass filtering, which blurs or smooths the image.

FIGURE 11-11 Examples of image filtering. (A) Starting image that has been cropped and given brightness enhancement, but without any sharpening or blurring. Helium-blimp airphoto by SWA and JSA over suburban housing district in Manhattan, Kansas, United States, September 2006. All image processing done with *Adobe Photoshop* software. (B) Unsharp mask, amount = 30%, radius = 2 pixels. (C) Gaussian blur, radius = 5 pixels.

image. Low-pass filters may be useful for reducing noise or other undesired high-frequency patterns in order to homogenize areas with similar spectral properties and bring out more clearly the detailed patterns in an image.

The opposite is to enhance small pixel-to-pixel variations, known as high-pass filtering or sharpening. First a low-pass filtered version of the original image is produced, then the low-pass value for each pixel is subtracted from twice the original value (Jensen, 2005). The effect is to magnify the local variations between pixels. A further step is edge enhancement, which is designed to highlight boundaries within the image; many types of edge enhancements exist and are available in image-processing software.

Unsharp mask is a traditional darkroom technique adapted for digital imagery. This filter brings out edge detail and is usually the last operation applied for image processing. For general display, unsharp mask is recommended (Fig. 11-11). However, over or more sharpening should be avoided, as it creates many small artifacts.

11.4. IMAGE TRANSFORMATIONS

Image transformations are operations that re-express the existing information content of an image by generating a new set of image components representing an alternative description of the data (Mather, 2004; Richards and Jia, 2006). Some authors (e.g., Lillesand et al., 2008) group multispectral transformations under image enhancements, because they are designed to increase the visual distinctions between image features and offer a changed look at a scene. Image transformation techniques include spectral ratioing, principal components analysis (PCA), color-space transformations, and Fourier analysis. The latter, which

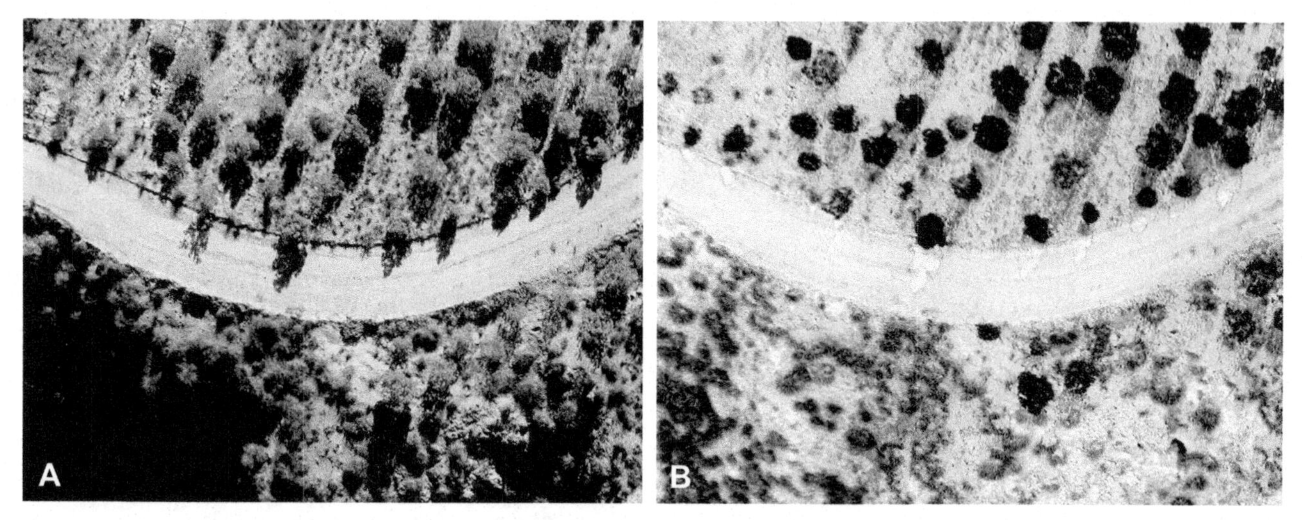

FIGURE 11-12 Open pine forest near Negratín Reservoir, Province of Granada, Spain. (A) Hot-air blimp photograph taken by JBR, V. Butzen, and G. Rock, October 2008. (B) Ratio image of green and blue image band (white: ratio value = 1, black: ratio value > 3). Image processing by IM.

transforms an image from the spatial domain into the frequency domain (Mather, 2004), may be useful for analyzing or eliminating specific patterns from an image (see Fig. 4-34B and Marzolff, 1999); however, it is a complex technique rarely used for SFAP processing and shall not be discussed further here.

11.4.1. Image Ratios and Vegetation Indices

An image ratio is the result of dividing the pixel values of one image band by the corresponding values in another image band. The effect is that spectral variations between the image bands are enhanced while brightness variations are suppressed. This makes image ratios useful for two main purposes: distinguishing more clearly between features with subtle color differences and reducing the shadowing effect in sunlit scenes (usually called "topographic effect" in remote sensing, because relief is the main cause for shadowing in small-scale images).

Owing to their high resolution and low flying height, SFAP images are often strongly subdivided into sunlit and shadowed parts (see Chapter 4). Figure 11-12A shows a vertical view of pine trees in a reforestation area in southern Spain. Stark shadows are cast by the trees onto the bare soil, low shrub, and grass vegetation and also onto neighboring trees. In the bottom part of the image, the steep slopes dropping into a dry valley are also heavily shadowed. Classifying this image, e.g., for analyzing pine canopy and shrub/grass cover, would result in serious misclassifications because the brightness differences by shadowing are much greater than the spectral differences that distinguish the cover types.

In order to suppress the darkening effect, the ratio between the green and blue image bands was calculated after adding a constant value of 25 to the green band (Fig. 11-12B). This offset constant was necessary because of the slightly different shadowing factor due to higher scattering of the blue wavelengths (see also Mather, 2004). In the resulting ratio image, the pine trees clearly stand out against the shadowless background of light soil and grayish shrub and grass cover. The tussocks on the steep slope in the lower left are now even better discernible than their sunlit counterparts. Such ratio images may be useful as an additional image channel in multispectral classification, helping to reduce the spectral confusion between sunlit and shadowed areas (Marzolff, 1999; see also Fig. 11-17).

Even more important is the role of image ratios for enhancing spectral characteristics of minerals or vegetation. Vegetation is a surface cover type with characteristic spectral reflectance curves (see Figs. 4-15 and 4-16). The steep slope between the near-infrared (NIR) and red wavelengths occurs for no other ground-cover types, and describing it with image ratios using these spectral bands has long been an important method for investigating biophysical parameters of vegetation. Vegetation indices are designed to indicate, for example, leaf area index (LAI), green biomass, percentage cover, chlorophyll content, or protein content, while normalizing effects of shadowing and illumination variation. An overview of different vegetation indices is given by Jensen (2007).

The most commonly known vegetation index, the NDVI, calculates as the ratio of the difference between NIR and red divided by their sum. NIR wavelengths can also be captured with small-format cameras (see Chapter 6.5), and a number of SFAP studies have employed this index for research in vegetation conditions and agricultural crop growth parameters (Buerkert et al., 1996; Hunt et al., 2005; Jensen et al., 2007; Berni et al., 2009).

Figure 11-13 shows two images from a time series taken in Kruger National Park for an ongoing research project

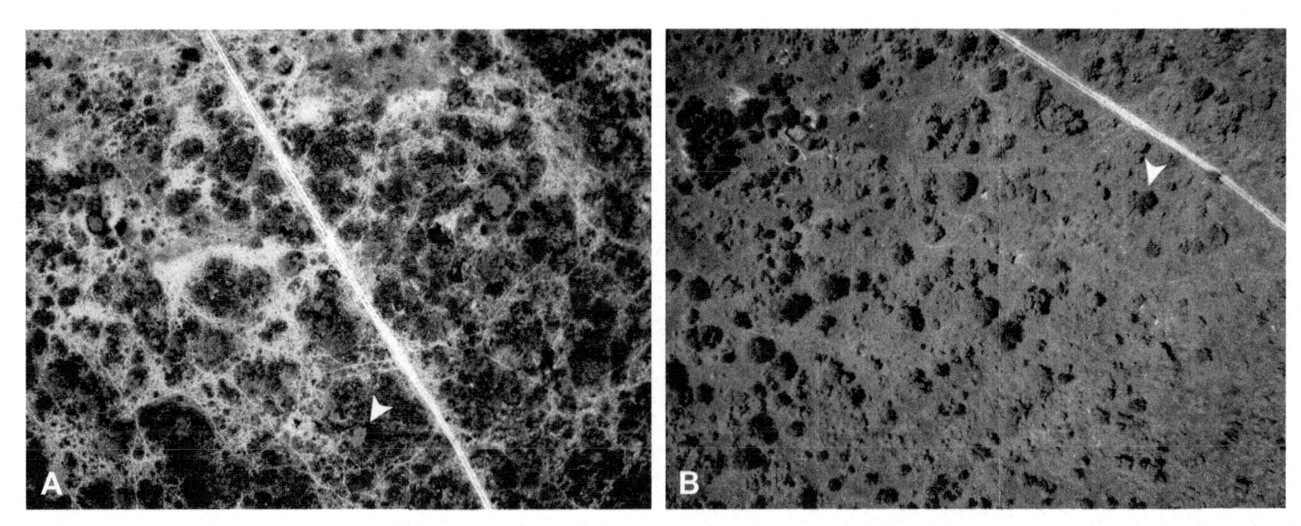

FIGURE 11-13 Savanna landscape in Kruger National Park, South Africa. Color-infrared aerial photography taken from helicopter by M. Delgado, S. Higgins, and H. Combrink. (A) November 2008. (B) March 2009. Early-flushing trees (see arrow for example) clearly stand out in the November image with high infrared reflectance, but appear darker in the March image. The areas covered by the two images do not coincide exactly; use pond in upper left and dirt track for orientation. Images courtesy of M. Delgado.

conducted by the universities of Cape Town (South Africa) and Frankfurt am Main (Germany). Spatial and seasonal patterns of NDVI along a rainfall gradient in savanna landscapes are investigated with small-format color-infrared photography (taken with a *Tetracam* ADC camera) and hyperspectral spectroradiometer measurements. The aim is to improve the understanding of savanna leaf phenology by determining the partial contribution of trees and grasses to the total landscape greenness. Two datasets are derived from the images, a land-cover map distinguishing grasses and early- and late-flushing trees, and an NDVI dataset (Fig. 11-14). NDVI value statistics are then calculated for the three vegetation cover classes. The high temporal and spatial resolution of the images allows, for the first time, for the contributions of grasses and early- and late-flushing trees to NDVI to be distinguished (Combrink et al., 2009).

The Kruger National Park study is a good example for the advantages SFAP can present over conventional remotely sensed images in terms of spatial and temporal resolution. Satellite imagery with medium to low spatial resolution, common in vegetation conditions research,

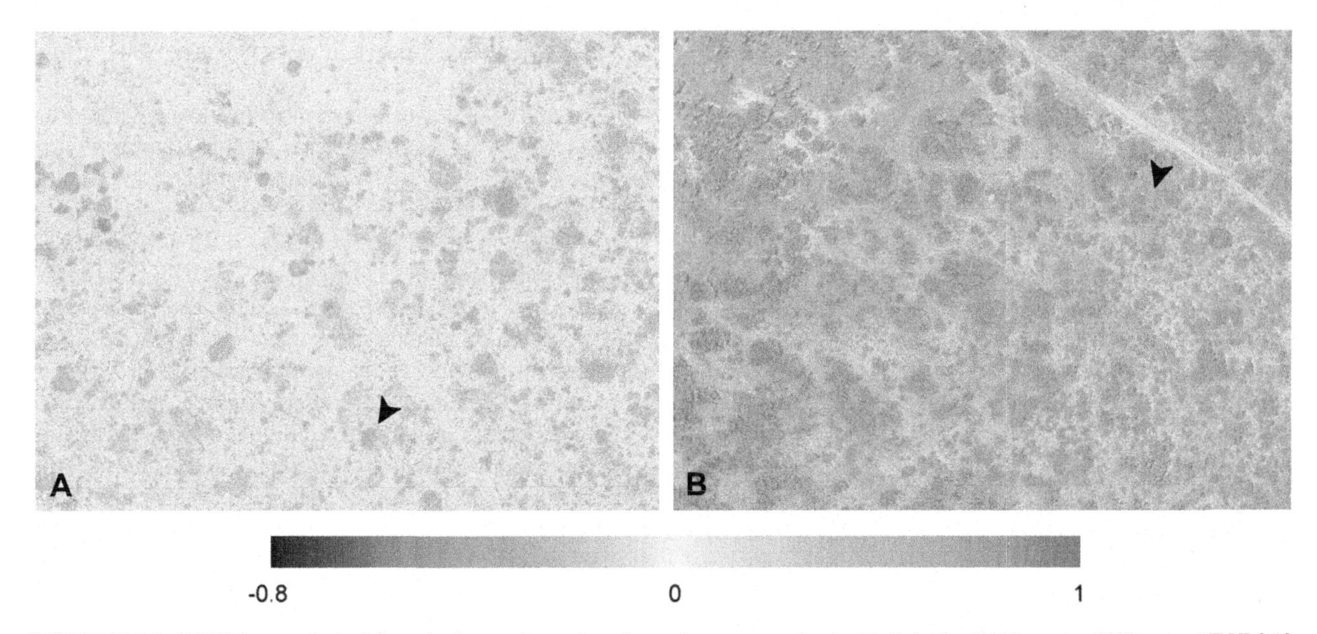

FIGURE 11-14 NDVI datasets derived from the images in previous figure. Image processing by M. Delgado. (A) November 2008, mean NDVI 0.19. (B) March 2009, mean NDVI 0.56. Early-flushing trees (see arrow for example) show the highest NDVI values in the November image, but have been overtaken by the late-flushing trees in March. Images courtesy of M. Delgado.

would not allow distinguishing clearly between the three cover types. High-resolution satellite imagery, on the other hand, is too expensive or too infrequently collected to meet the requirements of a time series adapted to site-specific phenological cycles. For this study, the National Park administration's helicopter was used until recently as the sensor platform, but will be replaced by an unmanned helicopter in the near future because of the logistical and financial efforts associated with the manned platform.

Different digital camera images from the same survey or from varying dates may show radiometric variations due to varying exposure settings, solar illumination, and atmospheric conditions. If ratio values are to be correlated with field measurements or compared over time in a monitoring project, it might therefore be necessary to perform some kind of radiometric calibration. A feasible method is to photograph, at the beginning, end, or during the survey, some unchanging standard objects like colored panels and cross-calibrate the values derived from the image with these reference values. Such procedures have been reported, for example, by Inoue et al. (2000) and Hunt et al. (2005).

11.4.2. Principal Components Analysis and Color-Space Transformations

PCA is often used as a method for data compression in multispectral and hyperspectral imaging with the aim of statistically maximizing the amount of information from the original *n*-dimensional image into a much smaller number of new components (Mather, 2004). SFAP images certainly do not present challenges to the analyst in terms of dimensionality, but PCA still may be interesting for natural-color SFAP, because it transforms the three original image bands (blue, green, and red) into statistically uncorrelated bands. As the correlation of spectral values within the visible range of the spectrum is usually particularly high, this may result in an altogether different, colorful view of the scene. An example is given in Figure 11-15, where the vertical view of the highly patterned, but rather color-monotonous landscape shown in Figure 5-12 was transformed by PCA. Note the striking variations of colors on the allegedly homogeneous fallow field left of the center and the new olive grove in the lower right corner, and the vivid green distinction of the larger trees at the image center and right edge.

Similar effects may be achieved with color-space transformations, e.g., conversion of the image from the RGB to IHS color system (Chapter 5.2.5). As with ratio images, PCA components and IHS image bands may improve classification results for object classes which are otherwise difficult to distinguish in the original RGB bands (Laliberte et al., 2007).

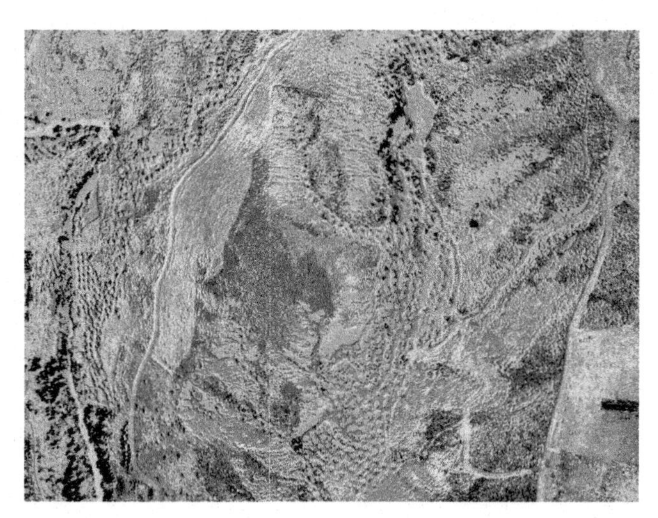

FIGURE 11-15 Principal components image (RGB = PCA1, PCA2, PCA3) of the aerial photograph shown in Figure 5-12. Image processing by IM.

11.5. IMAGE CLASSIFICATION

The pixel values in a digital or scanned analog photograph are the result of spectral reflectance measurements by the sensor or film. Thus, they carry detailed quantitative information about the spectral characteristics of the surfaces or objects depicted, usually in an image depth of 8 or 12 bits per band. No human interpreter is able to appraise the subtle differences that can be recorded with such high radiometric resolution. Also, the small patterns and fine spatial details presented by different ground-cover types often make it impractical, if not impossible, to map their distribution by hand. Image classification techniques aim at the automatic identification of cover types and features using statistical and knowledge-based decision rules (Mather, 2004; Lillesand et al., 2008).

Classical methods of image classification are based on the recognition of spectral patterns by analyzing solely the brightness values of the individual pixels across multispectral bands. However, while spectral reflectance—color—is the most important quality distinguishing individual ground-cover types in an image, it is by no means the only one, and shape, size, pattern, texture, and context also may play important roles for identifying features (see Chapter 10). Elements of spatial pattern recognition, knowledge-based rules, and object-oriented segmentation techniques have gained in importance during the last decade since very high-resolution satellite imagery has become available. Beyond looking at individual pixels only, these techniques analyze groups or patches of pixels in their spatial context.

Well-established methods both for unsupervised and supervised multispectral classification exist. The challenge for the analyst is to describe the object classes that are desired for the final map in terms of well-distinguished spectral classes. In spite of their reputation as uncomplicated, easy-to-use image data, this may be surprisingly difficult with small-format aerial photographs. One of the great assets of SFAP—its high spatial resolution—is also a drawback for multispectral image classifiers. The fine details of color, texture, light, and shadow that are recorded for the objects with *GSDs* usually in the centimeter range often bring about within-class variance that may be higher than the spectral differences between classes. In addition, reflectance values in the visible spectrum are much more highly correlated with each other than in the longer wavelengths recordable with electronic scanners operating in the infrared spectrum. And finally, brightness variations within the image (vignetting effects, multiview-angle effects, shadowing) and between images (exposure differences, viewing-angle differences) are more prominent in SFAP images than for satellite images or metric large-format cameras at high altitudes.

Accordingly, the number of studies employing automatic classification techniques for small-format images is rare compared to those using visual interpretation, manual mapping, and photogrammetric measuring. Some researchers use simple thresholding procedures to classify just two or three classes. Marani et al. (2006), for example, successfully extracted small channel structures from low-altitude helium-balloon photographs for a multi-scale approach to mapping geomorphological patterns in the intertidal zone of the Venice Lagoon. In a study that used SFAP taken from an unmanned aerial vehicle (UAV) for validating low-resolution MODIS satellite observations of Arctic surface albedo variations, Tschudi et al. (2008) also used thresholding for the digital images, as the gray-scale nature of their scenes made more complicated multispectral methods dispensable. They classified three categories—ice cover, melt ponds, and open water—in order to estimate accurate pond coverage in a given region and verify the results of the MODIS image analysis. For another example of this approach involving vegetation, see Chapter 15.

Classification techniques may be used to derive not only qualitative, but also quantitative information from images. Figure 11-16 illustrates the procedure employed for deriving detailed maps of percentage vegetation cover in a study on vegetation development and geomorphodynamics on abandoned fields in northeastern Spain (Marzolff, 1999, 2003; Ries, 2003). Film (35 mm) photographs taken from a hot-air blimp were georectified to 2.5 cm *GSD* (Fig. 11-16A) and classified into 15–20 spectral classes with an unsupervised ISODATA clustering algorithm. These classes were then allocated to percentage vegetation cover based on field observations and close-range reference images (Fig. 11-16B). The aim was to produce a vegetation map with distinct zones that representatively described the cover patterns in a pre-defined

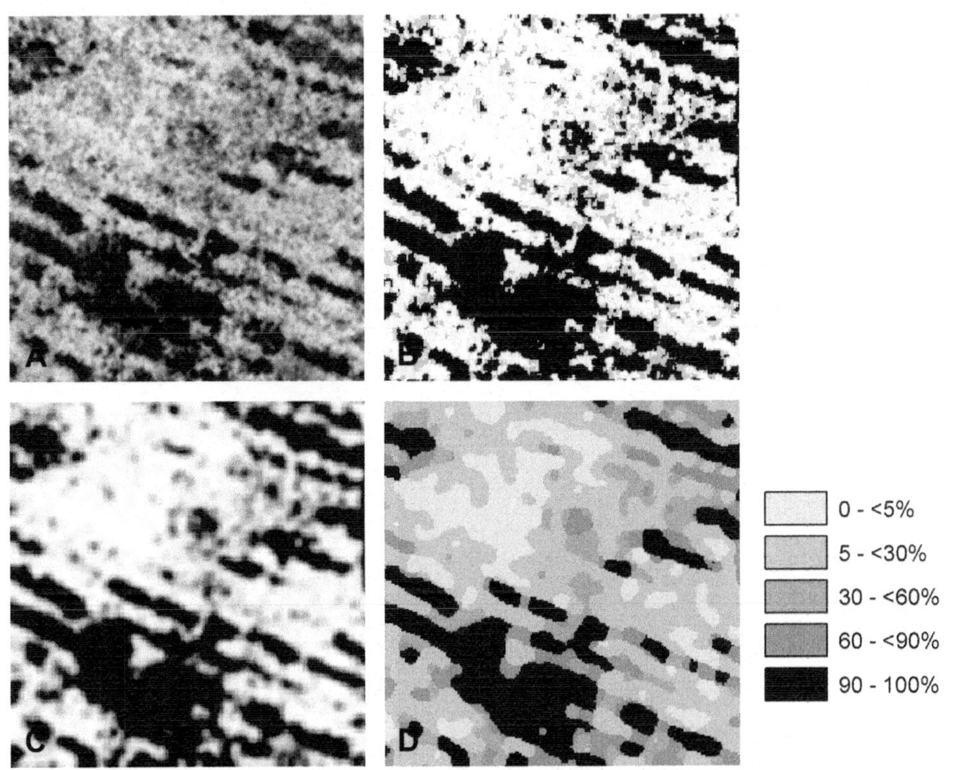

FIGURE 11-16 Procedure for classification of vegetation cover shown with subset of a hot-air blimp photograph taken by IM and JBR near María de Huerva, Province of Zaragoza, Spain, April 1996. Field of view ~6 m across. (A) Georectified image. (B) Classified percentage cover (white = 0%, black = 100%). (C) Mean filtered classification. (D) Final map in five vegetation cover classes. Adapted from Marzolff (1999, figs. 6-4 and 6-5).

classification scheme. Therefore, the classified dataset was filtered with a 5 × 5 mean filter (Fig. 11-16C), recoded into five cover classes, and again filtered with a 7 × 7 majority filter (Fig. 11-16D; see also Marzolff, 1999).

Often, the microstructural details depicted by the high resolution of SFAP together with shadowing effects (cast shadow, internal vegetation canopy shadowing) necessitate more sophisticated classification procedures even for "simple" object classes. Lesschen et al. (2008), for example, found that the supervised maximum likelihood method was the only one that allowed distinguishing bare soil from vegetation patches in vertical SFAP of a semi-arid environment and also classified the shaded areas correctly. With the resulting binary maps of distinct patterns, they calculated spatial metrics for investigating the spatial heterogeneity in vegetation and soil properties after land abandonment in southeastern Spain. The example in Figure 11-17A shows that heavily shadowed images (compare with Fig. 11-12A) may still be difficult to classify correctly. Using an additional ratio band for suppressing shadowing effects significantly improved the results in this case (Fig. 11-17B). Although the ratio-classified image still shows some misclassifications at the edges of cast shadows, shrub and grass cover in shaded areas are classified much better, and the pine tree canopy appears less frayed.

Little work has been done so far with object-oriented classification of SFAP images, although improvements can be expected from these techniques with respect to the difficulties associated with the high image resolution. Promising results with object-oriented segmentation techniques applied to digital camera images were achieved by Rango et al. (2006) and Laliberte et al. (2007) for classification of mixed rangeland vegetation at the Jornada Experimental Range in New Mexico, United States, and by Dunford et al. (2009) for quantifying vegetation units in a Mediterranean riparian forest environment.

11.6. STEREOVIEWING AND PHOTOGRAMMETRIC ANALYSIS

In Chapter 3, the fundamentals and benefits of stereoscopic analysis already have been introduced. However, stereoscopy also might be useful for visual interpretation disregarding quantitative measurements—many aspects of a scene may reveal themselves to the observer more clearly or exclusively in 3D view. Generally, stereoscopic viewing may be achieved either with or without optical aids, including the following approaches (Jensen, 2007).

- Place the images in reading distance (left image to the left and right image to the right side), relax eyes as if looking at infinity until the two images fuse into an unfocussed third 3D image in the middle, and focus back on this stereomodel (try with Figs. 2-9 and 3-14).
- Place the images in reading distance (left image to the right and right image to the left side), let the lines of sight cross until the two images fuse into a third 3D image in the middle. Note: using this method on the stereo-images in this book will cause relief inversion.
- Print one image in red and the other in blue, green, or cyan on top of each other to create anaglyphs, use corresponding anaglyph lenses to view.
- Arrange the images under a lens or mirror stereoscope (see Fig. 2-10).
- Use stereoviewing software and hardware, e.g. electronic shutter lenses.

The first two methods are a matter of practice; some people can do this easily, but most cannot without causing

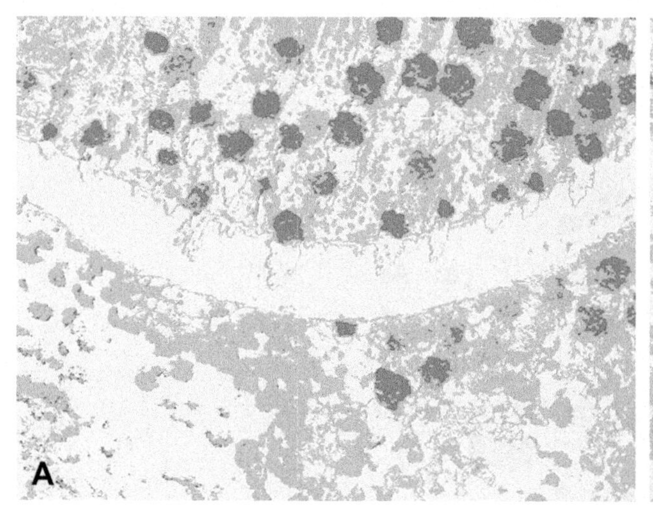

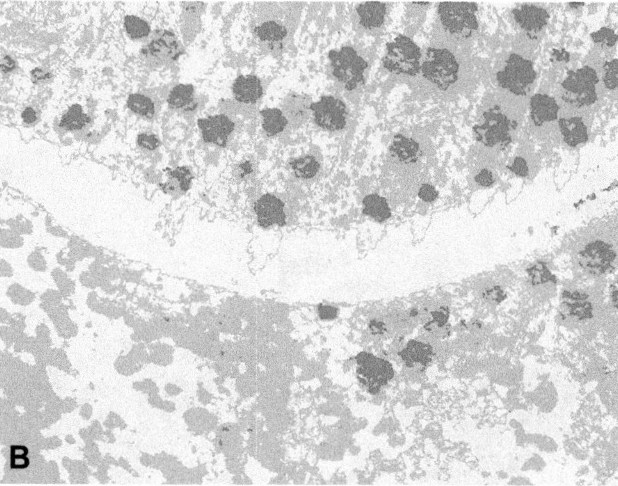

FIGURE 11-17 Supervised maximum likelihood classification of the image shown in Figure 11-12A; image processing by IM. (A) Classification of original RGB image. (B) Classification of composite of green image band and ratio image of green and blue image band shown in Figure 11-12B. Dark green = pine trees; medium green = grasses and shrubs; light green = bare soil.

considerable eye strain. These methods are only useful for an aid-free, quick appraisal of a stereoscene, and zooming in on details is obviously not feasible. Anaglyphs are easy to realize and work well for most people, but they are color-distorting, a real nuisance to the human brain and not recommended for serious work. Stereoscopes and their digital realizations are clearly the best, but also most expensive option.

Irrespective of the viewing method, the images need to be orientated so they reflect their relative position at the moment of exposure. For SFAP images this is usually not as simple as placing them next to each other in parallel, as SFAP surveys rarely result in regularly aligned flightlines with consistent overlaps (see Fig. 3-10). In photogrammetry software, the relative orientation is computed via GCPs and tie points. The manual method for reconstructing the flightline between the two exposures, which needs to be aligned to the eye base of the observer, is illustrated by Figure 11-18. Mark (or for quick casual viewing roughly identify and memorize) the center point (principal point, o) in each image and also its corresponding point (o') in the other. Then lay out both images with the center points at the outside and their corresponding points at the inside so all fall on one straight horizontal line. There are two ways to do this (the whole arrangement may be turned by 180°); in case of sunlit scenes it is best to choose the variant where the shadows are cast toward the lower right rather than the upper left in order not to confuse the brain (see Fig. 5-3). The ideal spacing between the two images depends on the viewing method and the observer's liking.

In the following, three stereoviewing methods of practical use for SFAP are briefly described; the orientation technique presented above works for all of them.

11.6.1. Creating Simple On-Screen Stereoviews

Lack of 3D impression is not something you want to discover after producing a series of A4-sized glossy prints or having spent days on a photogrammetric triangulation project. Before choosing an image pair for printouts, anaglyph creation, or photogrammetric analysis, it is useful to assure oneself of its stereoscopic quality. Even with a decent overlap, which is easy to assess in monoscopic view, the stereo impression might be unsatisfactory owing to unfavorable image tilts or differing scales, both typical situations for SFAP. The latter may be remedied by scaling one image to fit the other prior to printing them, but the former can only be cured by choosing different stereopairs.

Simple freeware image viewers (e.g. *IrfanView*) are useful tools for such checks (Fig. 11-19). Opened side by side, the images can be scaled and rotated following roughly the orientation procedure described above. With a little

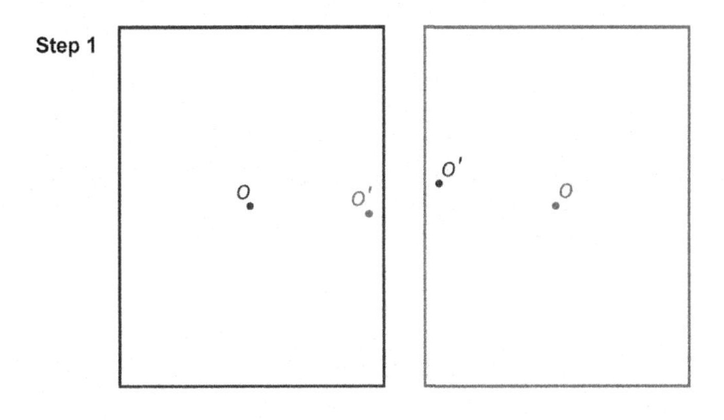

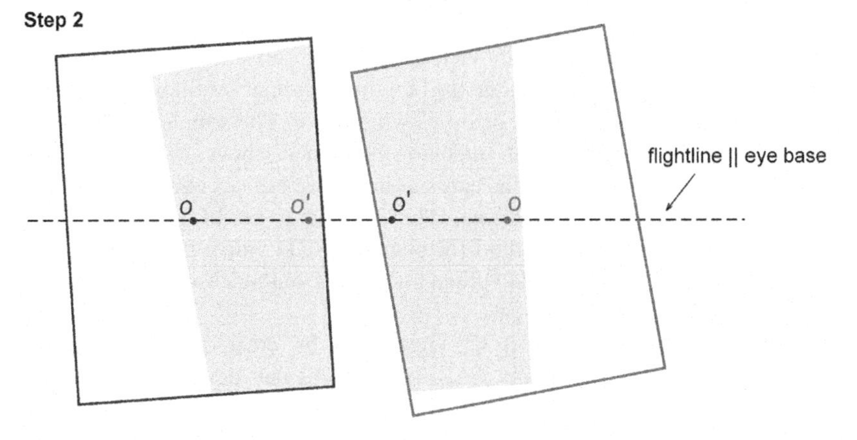

FIGURE 11-18 Relatively orienting two images for stereoscopic viewing. Step 1: identifying the image centers o (by using the image corners) and their corresponding points o' (by using the image contents). Step 2: rotating both images so all points lay on a horizontal line parallel to the eye base; 3D viewing is possible in the stereo-area shaded in gray.

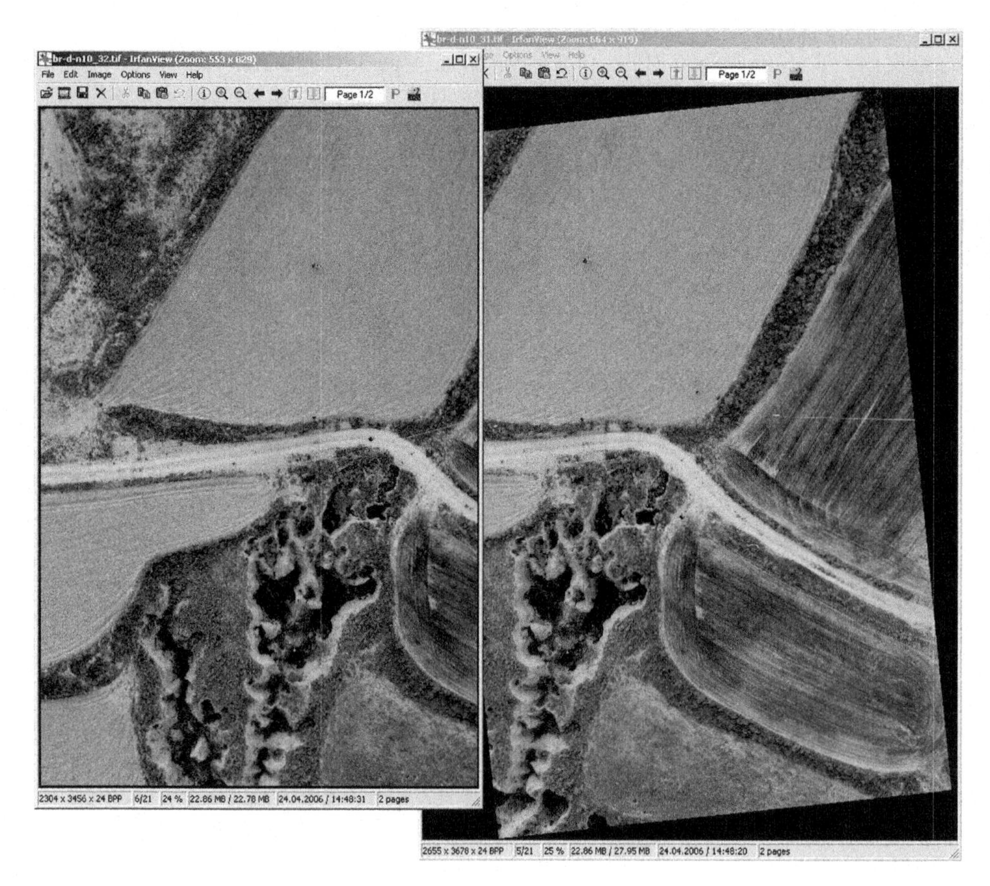

FIGURE 11-19 Two vertical aerial photographs of Barranco Rojo near Botorrita, Province of Zaragoza, Spain, oriented for stereoscopic viewing on a computer screen. The right image has been rotated and slightly enlarged to match up with its stereopair. The parallel-eyes method may be used to view this scene in 3D. Hot-air blimp photographs by IM and JBR, April 2006.

practice, the parallel-eyes method (or, if preferred, the cross-eyed method) is quite helpful for selecting the optimal images from a survey series.

Even more convenient are stereoviewing tools that allow the user to rotate and scale the images for relative orientation and then offer the choice to display them for parallel-eyed, cross-eyed, anaglyph or even shutter-lens viewing. A variety of useful freeware is available on the Internet (e.g. *StereoPhoto Maker*).

11.6.2. Using Printouts Under a Stereoscope

Stereoscopes are optical devices that help to view the two images separately and with relaxed eyes. Lens or pocket stereoscopes permit viewing the full overlap area of prints up to ~9 cm × 13 cm or parts of larger prints. Mirror stereoscopes are designed to accept standard 23 cm × 23-cm metric airphotos and therefore work well also with A4/US letter-sized prints (Fig. 11-20). Most models come with magnifying options up to eight times. They are large enough for allowing the analyst to draw beneath them, so it is possible to trace objects identified in the stereoview with a pen onto one of the images or a transparent overlay. However, remember that this sketch would not have the geometric properties of a map as it contains all distortions present in the base image.

To prepare printouts for use with a stereoscope, checking and if applicable scaling them first as described above is advisable. Especially when intended to be viewed at high magnification, a high printing resolution (600 dpi minimum) is necessary, or image details would be veiled by the printing dots. High-resolution photo-quality printouts or photographic prints are best suited.

11.6.3. Digital Stereoviewing

As with analog stereoviewing, on-screen stereoviewing requires that each of the two images is perceived by one eye only. There are various hardware and software solutions for this that separate the two images either by color, by time, by polarization, or by viewing angle. The simplest solution is the anaglyph method mentioned above; it requires no special graphic cards or hardware besides cheaply available anaglyph glasses. This is also a useful and financially feasible method for displaying 3D views to a larger audience via video projector, and the authors use it frequently in student courses.

Excellent 3D views can be created with dedicated stereographic cards, suitable display devices, and stereo-eyeware. With the active shutter-glasses method, the two images are displayed alternately with a frequency depending on the monitor refresh rate (Fig. 11-21). LCD shutter

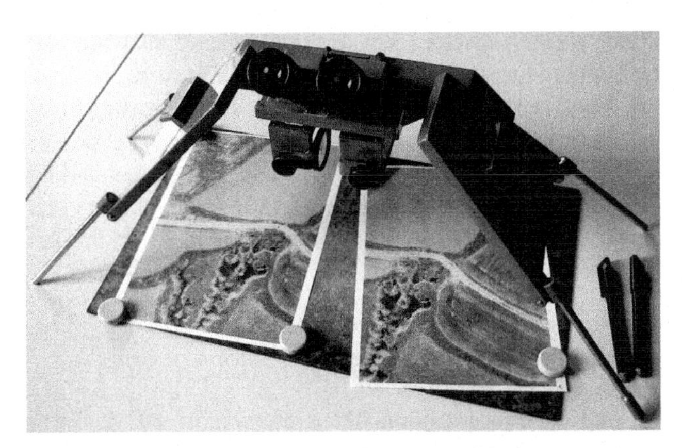

FIGURE 11-20 Two A4 printouts of the images shown in previous figure under a *Zeiss* mirror stereoscope fitted with 8× magnification binoculars.

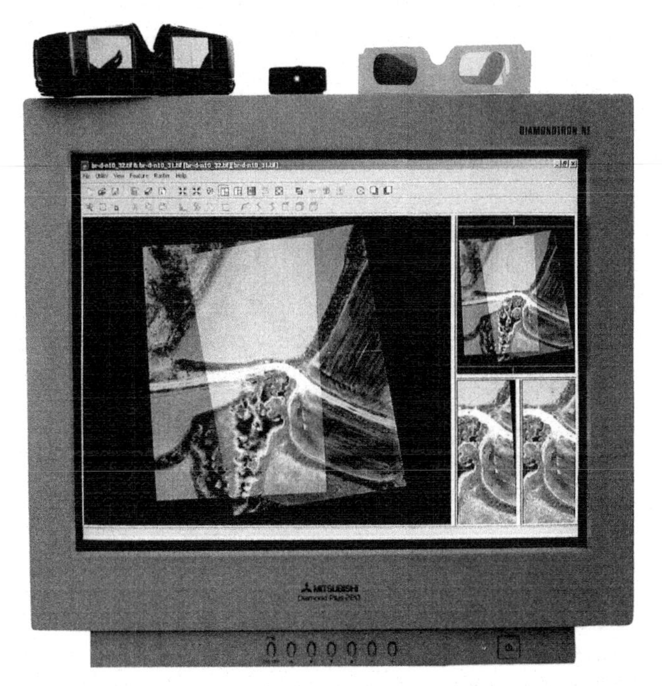

FIGURE 11-21 The stereopair shown in Figure 11-19 displayed in quad-buffered stereo mode on a computer monitor, to be viewed with wireless active shutter glasses activated by an infrared emitter (top left and center). Alternatively, the system shown here can be used for anaglyph viewing (red-cyan glasses top right). Photo by IM.

glasses alternately block one of the images in synchronization with the display. This method relies on high monitor refresh rates (120 Hz minimum are required to avoid flickering) that at the time of writing are still rare with LCD flat-panel monitors and usually only available with the near-extinct large CRT monitors. Another method is the superimposition of the images through polarizing filters and the use of passive polarized glasses. Recent stereo-display systems based on the same technique use two LCD monitors mounted at an angle with a semi-reflecting beamsplitter mirror between them, a dual-head graphics card, and polarized glasses for viewing.

Finally, autostereoscopic display devices have for the last decade or so offered the possibility of aid-free 3D viewing. A stereogram is composed of vertically interleaved lines from the two stereo-images, and an equally slitted barrier or lenticular plate is placed in front of this view. When viewing the screen from the right distance, the barrier or the lenses on the lenticular plate block the view of the left image's lines to the observer's right eye and vice versa. Thus, a true stereoscopic image using the human 3D perception ability is created. Shan et al. (2006) have found that autostereoscopic technology can be employed for photogrammetric stereo-measurements but yield 16–25% lower precision than the more commonly used shutter-glasses methods.

11.6.4. Stereoscopic Measuring and Mapping

Costs for the high-end technologies for stereoviewing briefly introduced above are usually in the four-figure range, and simple 3D viewing and interpreting of relatively oriented stereopairs probably do not merit investing in them. However, they are widely used in digital photogrammetry and 3D visualization of virtual GIS models.

Stereomapping tools work with absolutely oriented stereopairs—the interior and exterior orientations of each image need to be known. They may be determined either by camera calibration and direct measurements during image acquisitions (hardly applicable for SFAP) or as a result of photogrammetric triangulation. The techniques for manual measuring and mapping from stereomodels have already been discussed in Chapter 3.4.2. Using appropriate stereoviewing technology, the operator can collect measurements and features in 3D view, usually storing them in common GIS vector file formats. Stereoscopic mapping as opposed to mapping from 2D imagery has three major advantages.

- Result is distortion-free because the floating-mark positions are automatically converted to object coordinates.
- Terrain height information (spot heights, contour lines) may be mapped.
- Identifiability and delineability of three-dimensional objects are much improved.

Figure 11-22 is an example for a large-scale topographic map of the gully also shown in Figure 11-1. It was digitized with *IMAGINE Stereo Analyst* using a photogrammetric block file created with *Leica Photogrammetry Suite* and prepared for cartographic visualization with *ESRI ArcGIS*. This map could not have been compiled in the same detail from the already presented 2D imagery because features like the drainage lines, the gully edges, or the heaps of soil material broken off the gully walls are difficult or impossible to identify and delineate precisely without depth information. The distribution of collapsed gully wall material and vegetation cover within the gully provides indications of recent and past gully development, helping to

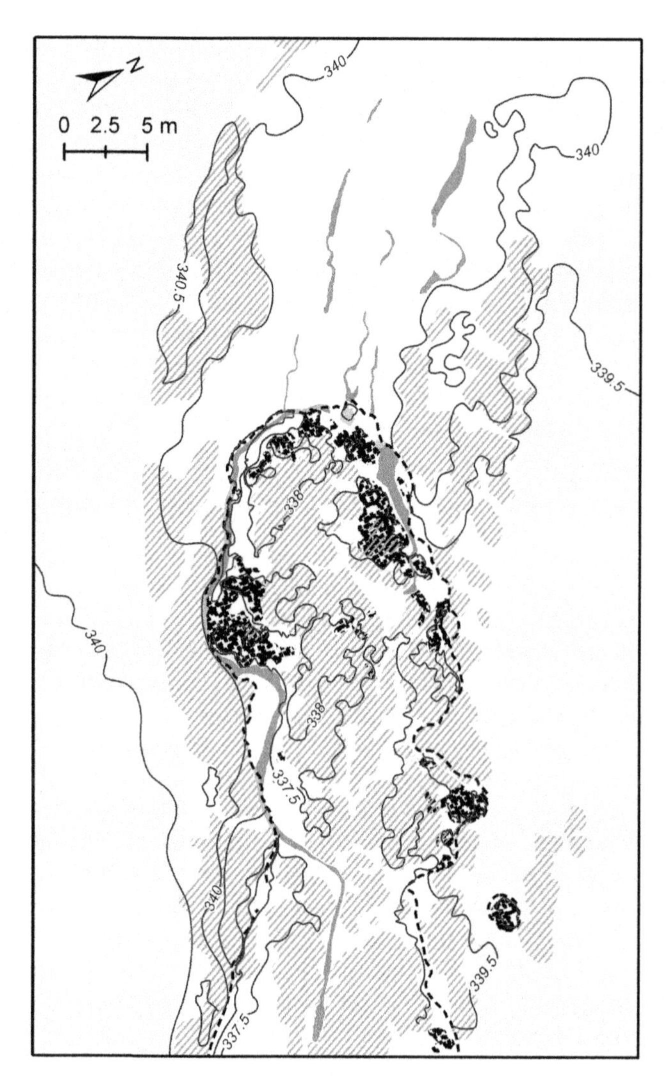

FIGURE 11-22 Topographic map of gully Bardenas 1, Bardenas Reales, Province of Navarra, Spain (compare Fig. 11-1). Based on kite aerial photography taken by JBR, M. Seeger, and S. Plegnière, March 2007; photogrammetric analysis and cartography by IM.

look back in time from this first image of an ongoing monitoring project (see also Giménez et al., 2009).

One problem associated with stereoviewing SFAP image blocks is that the variations in scale and orientation are often higher than usual with traditional aerial photography—even kite aerial images intended to be "vertical" may easily be tilted 5–10° from nadir, and differences in flying height of a few meters between consecutive images also are not uncommon. Depending on the stereoviewing software used, this may result in the whole stereomodel appearing tilted and distorted, which is difficult to view. Choosing stereopairs with strong tilt or >5% scale difference should therefore be avoided. Quick-check techniques like the one shown with Figure 11-19 are helpful for looking through a large image series.

11.6.5. DEM Generation

An introduction to automatic extraction of DEMs from digital stereomodels was given in Chapter 3.4.3. Generally, creating DEMs from SFAP is associated with the same procedures and difficulties as creating them from traditional photography, and various geoscientists, including the authors, have extracted detailed elevation models from SFAP images (e.g., Hapke and Richmond, 2000; Henry et al., 2002; Marzolff et al., 2003; Scheritz et al., 2008; Marzolff and Poesen, 2009; Smith et al., 2009). However, there are some specific characteristics of SFAP as compared to metric aerial photographs with practical consequences for photogrammetric processing that are briefly summarized in the following.

- Small-format digital photographs are usually taken in color with a Bayer pattern sensor (see Fig. 6-4). When using image matching algorithms for DEM extraction that work on monospectral data (single image bands), the green band will be likely to give better results as only 50% of the pixel values are interpolated by demosaicking algorithms (unlike 75% in the red and blue bands).
- The DEM extraction is controlled by a set of strategy parameters defining search and correlation window sizes, correlation coefficient limits, etc. Because the default values for these parameters are designed for the standard airphoto case, they might not be ideal for the SFAP case. Slight adaptions of the values might improve point density and quality (Marzolff and Poesen, 2009; Smith et al., 2009). For example, the comparatively higher remaining y parallax resulting from lower triangulation accuracy may be taken into account by increasing the search size in y direction that limits the distance of the corresponding point search to the epipolar line. Higher image noise and resolution may result in low point density, which can be counteracted carefully by increasing the correlation window size and decreasing the correlation coefficient limit. An increased search size in x direction may also be necessary if high terrain differences, causing larger x parallaxes, are present in the scene (see also below).
- Differences in image scale, which are common for SFAP series, may hamper not only visual stereoscopic analysis

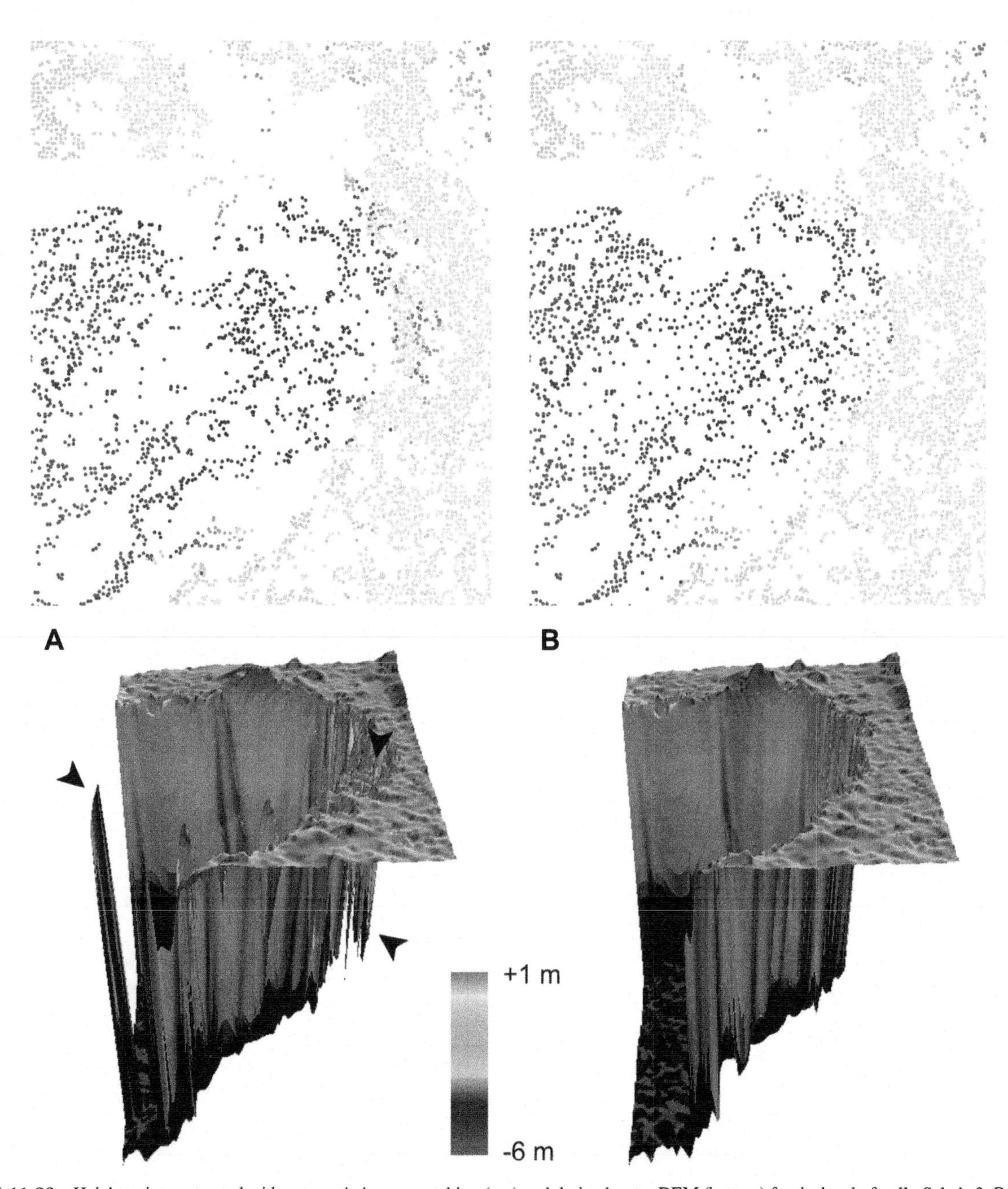

FIGURE 11-23 Height points extracted with automatic image matching (top) and derived raster DEM (bottom) for the head of gully Salada 3, Province of Murcia, Spain. Derived from hot-air blimp aerial photography taken by IM, JBR, and M. Seeger, April 1998; photogrammetric processing by IM. (A) Uncorrected data showing numerous errors due to steep walls, overhangs, and vegetation (see arrows for examples). (B) Manually edited height points and derived DEM.

but also automatic DEM extraction—the image matching procedure may fail if the *GSDs* differ too much. Again, this may be avoided only by rejecting images that do not meet these requirements.

- Because SFAP images capture basically the same objects as medium to small-scale traditional aerials, but with much smaller *GSD* and from considerably lesser distance, some error-prone situations are actually amplified (see Fig. 3-16). The user should be aware that the number of erratic points in DEM extraction may increase with higher terrain complexity in terms of elevation variability and vegetation cover (Fig. 11-23). It is therefore highly advisable to provide for the possibility of visual 3D assessment of the results as well as terrain editing when choosing software for small-format aerial photogrammetry. Although encapsulated black-box workflows for direct DEM generation from stereopairs are fast and tempting, they tend to conceal the deficiencies in the raw dataset of extracted height points.

TABLE 11-1 Summary of imaging software usable for SFAP and supported image-processing tasks.

SFAP image processing	Simple image-viewing software	Camera software	Professional image-editing software	GIS software	Remote sensing software	Digital photogrammetry software
Import of camera-specific image formats	○	●	○	—	—	—
Camera-specific image enhancements *	—	●	○	—	—	—
Radiometric enhancements	●	●	●	○	●	●
Image scaling / resolution change	●	●	●	●	●	○
Image rectification by GCPs	—	—	—	○	●	○
Image ortho-rectification	—	—	—	—	○+	●
Combination with GIS vector data	—	—	—	●	○	○
Map layout	—	—	—	●	●	○
Image classification	—	—	○	○+	●	—
Vector object extraction	—	—	○	○+	○+	○+
Stereoscopic viewing	—	—	—	○+	○+	●
3D analysis / DEM extraction	—	—	—	—	—	●
Software examples	Irfan View	All digital camera brands	Adobe Photoshop, Adobe Lightroom, NIH image, GIMP	ESRI ArcGIS, GRASS, Quantum GIS, Intergraph GeoMedia	ERDAS Imagine, IDRISI, ENVI, GRASS, PCI Geomatica	Leica Photogrammetry Suite, SocetSet, VirtuoZo, LISA, PhotoModeler

— Not available; ○ Sometimes, indirectly or partly available; ○+ Usually available as an add-on module; ● Available

White balance, optional lens-distortion correction, chromatic aberration, vignetting.

11.7. SOFTWARE FOR SFAP ANALYSIS

So which software is most suitable? This depends to a large degree on the image-processing tasks to be carried out. A huge range of computer programs exists from quite simple to highly sophisticated, and from free shareware to five-figure expenses. With respect to SFAP applications, image-processing software falls in two important main categories: (a) those understanding geocoordinate systems, and (b) those that do not. Another possible distinction is between programs for image visualization and optical enhancement only, and programs offering additional techniques for extracting secondary information, both qualitative and quantitative, about the objects and areas pictured.

Table 11-1 is an attempt at summarizing and comparing different software types with respect to typical image-processing tasks. Exemplary software packages named here are those most often used by or known to the authors, but do not imply any general preferences, recommendations, or quality judgements. In addition, dedicated freeware or shareware packages exist for nearly all individual tasks listed that are specialized on certain image-processing procedures (e.g., *PCI Geomatica FreeView* for viewing numerous geospatial file formats, *PTLens* for lens-dependent corrections, *StereoMaker* for stereoviewing).

11.8. SUMMARY

Image-processing techniques are mathematical or statistical algorithms that change the visual appearance or geometric properties of an image or transform it into another type of dataset. Many established image-processing techniques exist in the remote sensing sciences, and they may be applied to small-format aerial photography (SFAP) just as to satellite imagery or conventional large-format airphotos.

Geometric correction and georeferencing are a prerequisite for most measurement and monitoring applications, and planning and preparing ground control prior to the survey is essential in these cases. Precise geocorrecting of SFAP taken over varied terrain may not be a trivial task, possibly necessitating photogrammetric techniques for digital elevation model (DEM) extraction. Image mosaics, controlled or uncontrolled, are commonly created to cover larger areas with high-resolution photographs taken from low heights.

Some degree of image enhancement is applied to most SFAP in order to improve its visual appearance or correct lens-dependent aberrations. Histogram adjustments are particularly useful to emphasize the color variations present in a scene or highlight selected parts for discriminating feature characteristics otherwise unnoticed.

New datasets that show an image in another light may be derived by image transformations, e.g., ratioing, principal components analysis (PCA), or color-space transformations. One of the best-known methods is the computation of a vegetation index from color-infrared images, which enhances the characteristic spectral reflectance of plants and may be correlated with biophysical parameters such as leaf area index, biomass, or chlorophyll content.

Converting an image into a raster map identifying different ground-cover types or other categories is accomplished with image-processing techniques. The small *GSDs*, high within-class variances, and low range of spectral bands that are typical for SFAP present, however, a challenge to traditional classification algorithms. Classifying beyond simple thresholding tasks for two or three categories might require individual adaptation of standard classification methods. A promising approach for the future is given by object-oriented segmentation techniques developed for high-resolution imagery.

Stereoscopic images offer additional possibilities of image analysis, both visual and automatic. Various methods for stereoviewing from quite basic to high-tech solutions exist. Digital photogrammetry stations allow stereoscopic measuring and mapping and automatic height-point extraction for DEM creation. SFAP may, thus, be the base for extremely detailed high-resolution representations and virtual models of 3D terrains and features.

Which software to use for image processing of SFAP depends to a large degree on the intended analysis. Sophisticated tasks may require expensive specialized remote sensing or photogrammetry software, but various freeware and shareware packages also exist that may be employed successfully for advanced processing.

Chapter 12

Glacial Geomorphology

12.1. INTRODUCTION

Modern glaciers and ice sheets cover approximately 10% of the world's land area. Of this, most glacier ice is found in Antarctica and Greenland with all other areas accounting for only about 5% of the total. During the Ice Age (Pleistocene Epoch) of the last one million years, glaciers and ice sheets expanded dramatically and repeatedly over large portions of northern Eurasian and North American lowlands and in mountains and high plateaus around the world. At times, the volume of glacier ice during the Pleistocene was at least triple that of today (Hughes et al., 1981). Global sea level declined by at least 120 m, which allowed ice sheets to spread over broad continental shelves, particularly north of Eurasia (Svendsen et al., 1999; Polyak et al., 2000).

Geomorphology is the study of the Earth's surficial landforms both on land and on the seafloor. This study is both descriptive and quantitative; it deals with morphology, processes, and origins of landforms. Glacier ice is a powerful agent that created many distinctive landforms that are well preserved nowadays in regions of former ice expansion. Glaciers modify the landscape in three fundamental ways by erosion, deposition, or deformation (Fig. 12-1). A given site may be subjected to each or all of these processes during the advance and retreat of a glacier, and repeated glaciation may overprint newer landforms on older ones. In addition, glacial meltwater is also an effective geomorphic agent that may erode or deposit conspicuous landforms in connection with glaciation. The results are complex landform assemblages that represent multiple glaciations during the Pleistocene.

Aerial photography has long been utilized to illustrate, describe, interpret, and map the diverse types of landforms created by glaciation (e.g. Gravenor et al., 1960). Traditionally this approach is based on medium-scale, panchromatic, vertical airphotos taken from heights of several thousand meters (Fig. 12-2). In recent years, satellite imagery has been utilized increasingly for glacial geomorphology (Williams, 1986). Low-height, oblique airphotos also have proven effective for recognizing and displaying various types of landforms (e.g. Dickenson, 2009), including distinctive glacial landforms such as eskers and drumlins (Prest, 1983). The advantage of oblique views is the ability to visualize the three-dimensional expression of individual landforms within the surrounding terrain (Fig. 12-3).

The following small-format aerial photography (SFAP) was acquired with kites and a small helium blimp at formerly glaciated sites across the United States and several countries in northern Europe. All represent landforms created during the last major glaciation in the late Pleistocene, some 10,000 to 25,000 years ago. Because of their young age, the landforms are fresh and well expressed in the landscape. They are grouped according to the main geomorphic process involved in their formation.

12.2. GLACIAL EROSION

Glacial valleys and fjords are among the most spectacular examples of combined erosion by glacier ice and meltwater. Such valleys may be 100s to >1000 m deep and extend for 10s to >100 km in length. They are typically found in mountains or rugged upland areas that were invaded by ice sheets or subjected to local valley glaciation and are

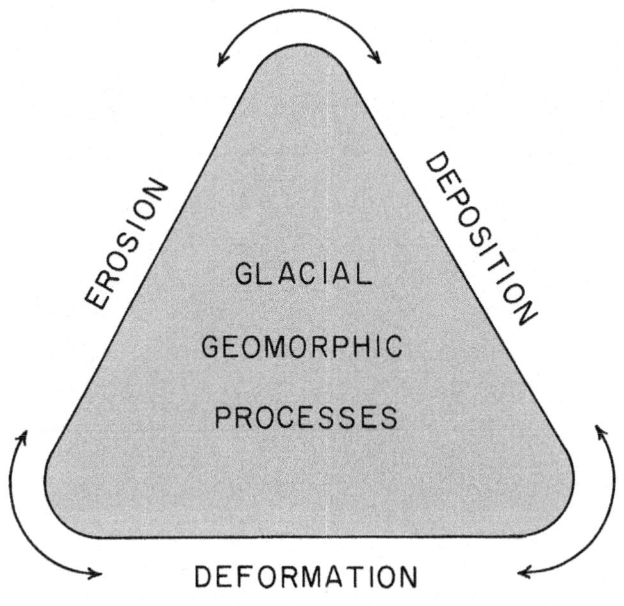

FIGURE 12-1 Triad of effects created by glaciation, on which modern glacial geomorphology is based. Taken from Aber and Ber (2007, fig. 1-5).

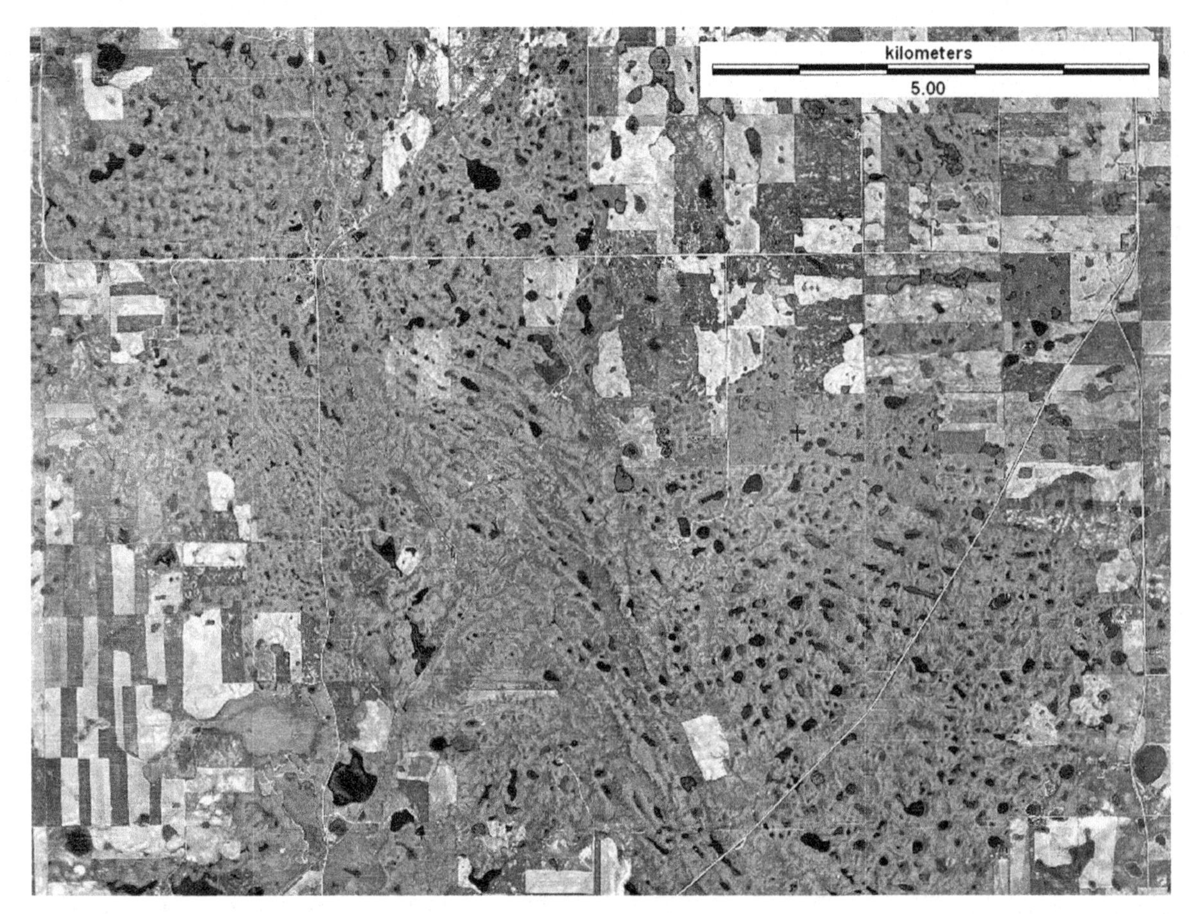

FIGURE 12-2 Conventional, panchromatic airphoto of the Crestwynd vicinity, southern Saskatchewan, Canada. Large ice-shoved ridges cross the scene from NW to SE. Photograph A21639-7 (1970); original photo scale = 1:80,000. Reprocessed from the collection of the National Air Photo Library, Natural Resources Canada. Taken from Aber and Ber (2007, fig. 2-20).

FIGURE 12-3 Palisades State Park in eastern South Dakota, United States. Glacial meltwater eroded a spillway channel across a bedrock ridge composed of hard quartzite. The vertical wall of the spillway can be seen on the right side of this oblique view toward the southwest. Kite aerial photo by JSA, July 1998.

especially common where montane glaciers descended into the sea or large lakes. Famous examples include Hardangerfjord, Norway; Königssee, Germany (Fig. 12-4); and Lake Okanagan, British Columbia, Canada.

The Finger Lakes occupy a series of long, straight valleys that penetrate the Appalachian Plateau south of the Lake Ontario lowland in west-central New York, United States (Fig. 12-5). The Finger Lakes are often described as inland fjords because of their deeply eroded bedrock valleys and thick sediment infill. The valleys are generally wide and shallow toward the north with thin sediment fill, and the troughs become narrow, deep gorges to the south with thick sediment fill. The Finger Lakes troughs were eroded by strong ice-stream flow coming from the north enhanced by high-pressure subglacial meltwater drainage (Mullins and Hinchey, 1989).

Among the individual Finger Lakes, Keuka Lake is the most unusual because of its branched shape (Fig. 12-6). SFAP was conducted with a small helium blimp at Branchport at the northern end of the west branch of the lake. Most of the surrounding land is heavily forested or agricultural (Fig. 12-7), which limited ground access for SFAP, so an open field at a school was utilized as the spot to

FIGURE 12-4 Königssee, a lake in a deep ice-carved valley on the northern side of the Alps, Berchtesgaden National Park, southern Germany. Ground photo by JSA, July 2007.

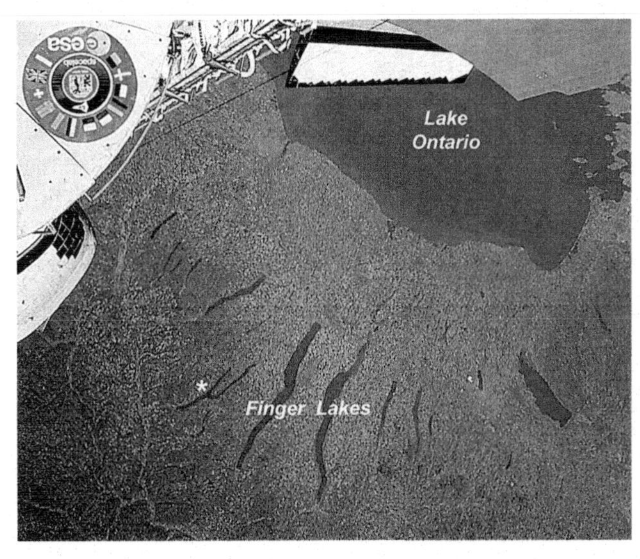

FIGURE 12-5 Astronaut photograph taken from the space shuttle over the Finger Lakes district of western New York. Asterisk indicates Keuka Lake. STS 51B-33-028, April 1985. Hasselblad, 70-mm film, near-vertical view. Courtesy K. Lulla, NASA Johnson Space Center.

launch the blimp. Oblique photographs were acquired with a primary focus on the valley of Sugar Creek to the north and West Branch Keuka Lake to the south (Fig. 12-8). These views emphasize the long, straight nature of the valley bounded by steep bluffs incised into the upland plateau.

The Tatra Mountains in southernmost Poland and northern Slovakia are part of the Carpathian Mountain system of east-central Europe. The Tatras experienced significant tectonic uplift within the past few million years, and highest peaks exceed 2500 m elevation. The Tatras supported numerous alpine glaciers during the Pleistocene, and these glaciers left behind a classic assemblage of

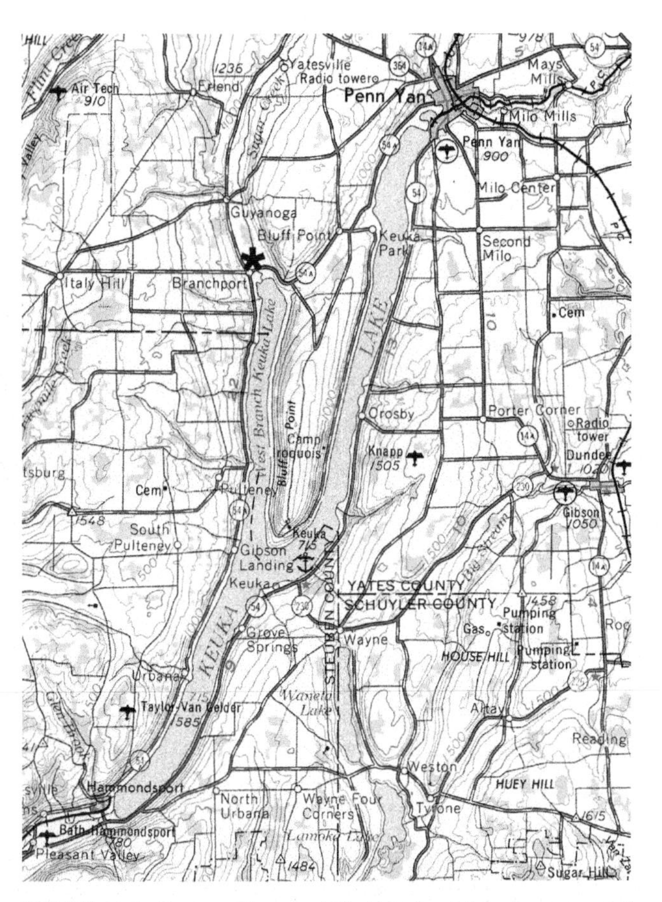

FIGURE 12-6 Topographic map of Keuka Lake vicinity in west-central New York. Length of Keuka Lake from Penn Yan to Hammondsport is ~32 km; elevations in feet; contour interval = 50 feet (~15 m). Asterisk indicates SFAP ground site. Adapted from Elmira NK 18-4, New York, 1:250,000 (1973), U.S. Geological Survey.

ice-carved valleys, moraines, and outwash deposits. The combination of tectonic uplift and glaciation led to rapid erosion, and extensive alluvial fans were deposited from streams and glacial meltwater along the southern flank of the range.

Kite aerial photography (KAP) was conducted on both sides of the Tatra Mountains in order to document and understand better the geomorphic features connected with glaciation (Aber et al., 2008). On the southern flank, broad alluvial fans form a conspicuous apron that slopes southward from the Tatra range into lowlands (Fig. 12-9). Much of the surficial sediment was derived from deep glacial erosion of valleys within the mountains. Among the largest of these valleys is Vel'ká Studená dolina, which is some 6 km long and more than 1000 m deep (Fig. 12-10).

The appearance of the Tatra Mountains on the Polish side differs considerably. The mountain front ends abruptly along a linear trend adjacent to a relatively low flanking valley without prominent alluvial fans to mark the transition (Fig. 12-11). Within the Polish Tatras, glaciated valleys are abundant (Fig. 12-12), but generally are not so long nor so deep as on the Slovak side. Kite aerial

FIGURE 12-7 Panoramic ground view of Keuka Lake looking toward the northeast from the western side. The lake is about 1.2 km wide to right (south) and branches into two major arms toward left. Note heavily forested terrain. Adapted from Aber and Ber (2007, fig. 9-11).

FIGURE 12-8 Oblique views of glaciated valley at Branchport, New York, United States. (A) View over West Branch Keuka Lake looking south toward the sun. (B) View northward along the valley of Sugar Creek (visible in lower right corner). The upland plateau in right background stands ~90 m above the valley floor in foreground. Helium-blimp aerial photos by JSA and SWA, August 2005.

FIGURE 12-9 Wide-angle view of the southern Tatra Mountains looking toward the northwest from near Stará Lesná, Slovakia. The broad terraces in the foreground and left background represent alluvial fans built of gravelly sediment washed out of the Tatra Mountains, in part by glacial meltwater. Kite aerial photo by JSA and SWA, July 2007.

photographs emphasize the geomorphic differences between the northern and southern sides of the Tatra Mountains, a situation not readily obvious from conventional maps or satellite images.

In the general scheme of alpine glaciation in the northern hemisphere, glaciers on northern sides tend to be larger and, thus, have more geomorphic impact than glaciers on southern sides of mountain ranges. But this is not the case in the Tatra Mountains, as demonstrated by SFAP, where recent tectonic uplift and fault movements have played important roles for glaciation and landform development.

12.3. GLACIAL DEPOSITION

Glacial deposits underlie many notable landforms, of which drumlins and eskers are among the most distinctive. Drumlins are streamlined hills ideally having the shape of a teardrop or inverted spoon. They occur in fields containing dozens or hundreds to thousands of individual drumlins. They are arranged en echelon in broad belts or arcs behind conspicuous ice-margin positions, and the pattern of drumlins is thought to indicate ice flow direction. Drumlins have complicated origins involving deposition, erosion,

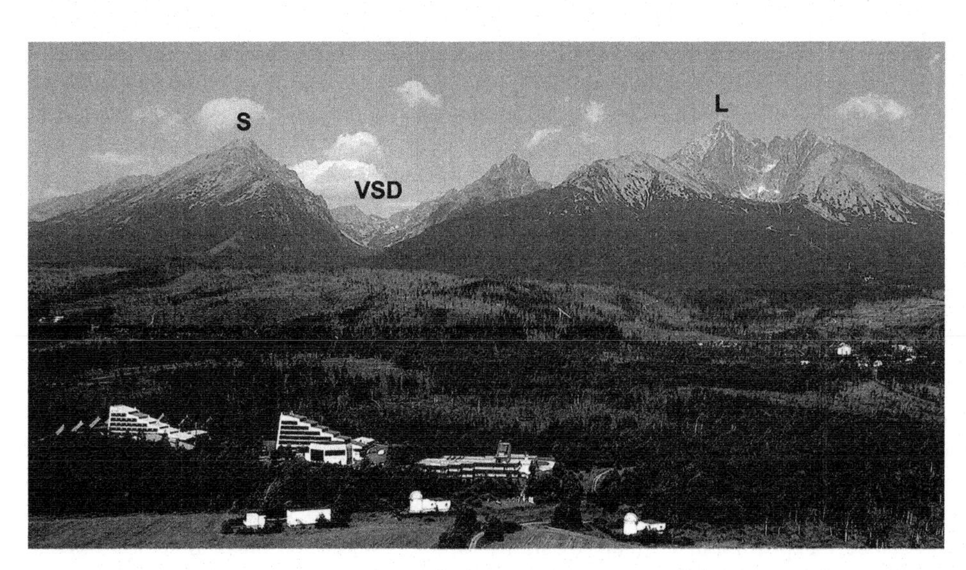

FIGURE 12-10 Close-up view of the Tatra Mountain front near Stará Lesná, Slovakia. High peaks include Slavkovsky (S) at 2452 m and Lomnicky (L) at 2634 m, separated by a major glacial valley, Vel'ká Studená dolina (VSD). The glacial valley extends directly into the mountain range, as seen in this view. Kite aerial photo by JSA and SWA, July 2007.

FIGURE 12-11 Overview of the Polish Tatra Mountains looking toward the southwest from near Toporowa Cyrhla; the city of Zakopane is visible in right background. Notice the linear boundary between the mountain front and the valley in the foreground. Kite aerial photo by SWA and JSA, July 2007.

FIGURE 12-12 View toward the southeast from Kopa Królowa Wielka in the Polish Tatras. Żółta Turnia (ZT) peak is 2087 m elevation, and to its right is the head portion of Gasienicowa Dolina (GD), an ice-carved valley. On far right, the foot trail to Hala Gasienicowa can be seen. Kite aerial photo by SWA and JSA, August 2007.

FIGURE 12-13 Lake Saadjärv drumlin field in eastern Estonia. Lake Saadjärv (left) occupies an elongated trough, and a long, smooth drumlin extends northwestward into the distance on right. Another lake, Soitsjärv, can be seen at extreme upper right. Kite aerial photo by JSA and SWA, September 2000.

deformation, and meltwater action beneath ice sheets (Menzies and Rose, 1987). Drumlins are common in many formerly glaciated regions, including eastern Estonia (Fig. 12-13).

Eskers are long, fairly narrow ridges of sand and gravel. They may be straight or sinuous, continuous or beaded, single or multiple, sharp- or flat-crested. They vary from a few meters to 10s of m high, and may be <1 to 100s of km in length. Eskers are deposited from various types of meltwater streams under the ice or at the margin of retreating glaciers (Banerjee and McDonald, 1975).

The island of Vormsi and adjacent seafloor in northwestern Estonia are especially well known for eskers (Aber, Kalm, and Lewicki, 2001). These eskers were deposited in subglacial tunnels during the final phase of late Pleistocene ice-sheet glaciation of the region. One esker in particular can be traced from the Austergrunne peninsula on the northern side of the island, southward to the peninsula at Rumpo, and across the shallow seafloor to the islets of Rukkirahu and Kuivarahu (Fig. 12-14). Although slightly modified by post-glacial sea action, the morphology of the esker is still quite distinct (Fig. 12-15).

12.4. GLACIAL DEFORMATION

The combined pressure of glacier loading and forward movement of ice deformed soft sedimentary substrata in many locations, which resulted in conspicuous ice-shoved

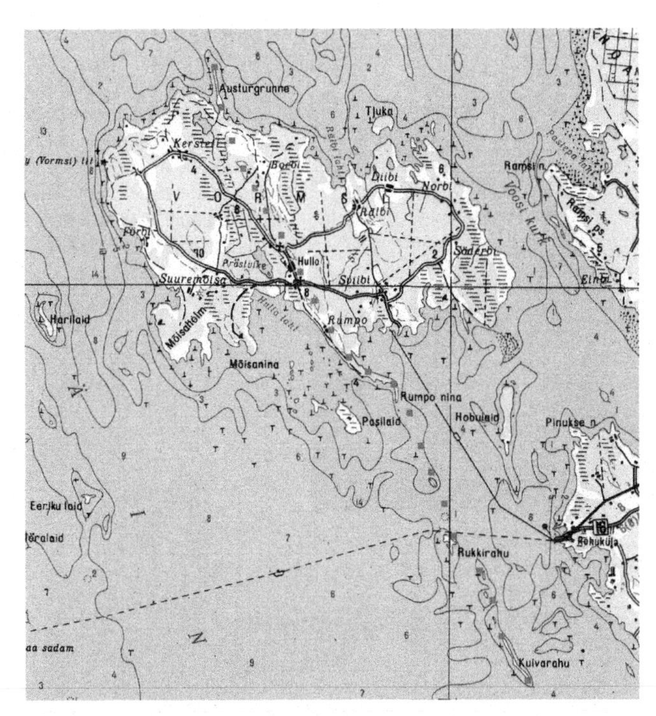

FIGURE 12-14 Topographic map of Vormsi and surroundings showing the trace of a long esker system (red dots) that extends at least 26 km from Austergrunne to Kuivarahu. Adapted from Tallinn Eesti topograafiline kaart, nr. 1, 1:200,000 (1992), Estonia.

FIGURE 12-15 High-oblique views looking over the esker at Rumpo on the island of Vormsi, Estonia. (A) View toward southeast. The road and pine forest follow the crest of the esker, which forms a peninsula in the shallow sea. (B) View northward across the island. The villages, agricultural fields, and road occupy the crest of the esker. Kite aerial photos by JSA and SWA, August 2000.

hills that may rise 100–200 m above surrounding terrain. In many cases, a depression marks the source of materials that were pushed into adjacent ridges. The combination of ice-scooped basin and ice-shoved hill is a basic morphologic form called a hill-hole pair (Aber and Ber, 2007). A good example of a hill-hole pair is Devils Lake Mountain in northeastern North Dakota. The ice-shoved hill is a slightly arcuate, continuous, single ridge that parallels a narrow source basin on its northwestern side (Fig. 12-16). The ice-shoved ridge is approximately 4 km long, 1 km wide, and stands >50 m above the source basin (Fig. 12-17).

Denmark possesses many well-developed and long-studied glacial deformations and ice-shoved hills of various types (e.g. Pedersen, 2000, 2005). Denmark is also a country famous for its wind power, which is most suitable for KAP. This method was employed by the authors for documenting ice-shoved hills in the Limfjord district of northwestern Denmark. The Limfjord is an inland estuary that was excavated in part by glacial thrusting of soft bedrock into adjacent ice-shoved hills. Feggeklit is a small ice-shoved hill that is connected by a narrow peninsula to the larger island of Mors (Fig. 12-18). Dislocated, folded, and faulted bedrock is exposed in a cliff along the eastern side of Feggeklit (Fig. 12-19). These strata were thrust up from the Limfjord basin by ice movement from the north (Pedersen, 1996).

Another larger example is Flade Klit, which consists of multiple, parallel, ice-shoved ridges on the northern side of Mors (Fig. 12-20). Flade Klit is approximately 3.5 km long

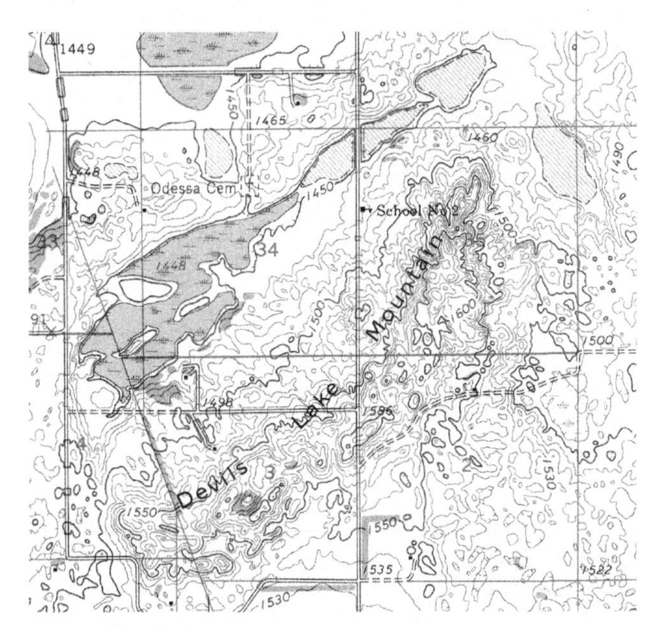

FIGURE 12-16 Topographic map of Devils Lake Mountain vicinity, northeastern North Dakota, United States. Elevations in feet; contour interval = 10 feet (~3 m); each grid square represents one square mile (~2.6 km^2). Taken from Hamar quadrangle, North Dakota, 15-minute series, 1:62,500 (1962), US Geological Survey.

FIGURE 12-17 Devils Lake Mountain seen from the northwestern side with the source depression in the foreground. Superwide-angle image; nearly all of the ice-shoved hill and source basin (lake) are visible. Helium-blimp aerial photo; adapted from Aber and Ber (2007, fig. 4-3).

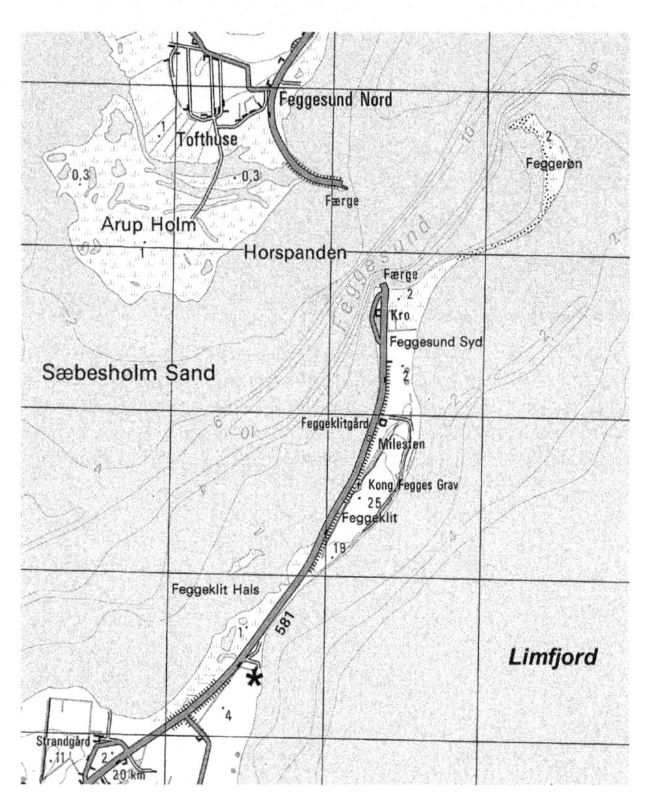

FIGURE 12-18 Topographic map of Feggeklit vicinity, northwestern Denmark. The hilltop is ~30 m above the floor of the Limfjord. Location for kite aerial photography indicated by asterisk (*). Each grid square is 1 km^2; adapted from 1116 I Thisted, Danmark, 1:50,000 (1983), Geodætisk Institut, Denmark.

FIGURE 12-19 Northeastward view over the flat-topped hill at Feggeklit, northwestern Denmark. Cliff on the eastern side exposes deformed bedrock thrust up from the Limfjord in the background. Kite aerial photo by JSA and SWA, September 2005.

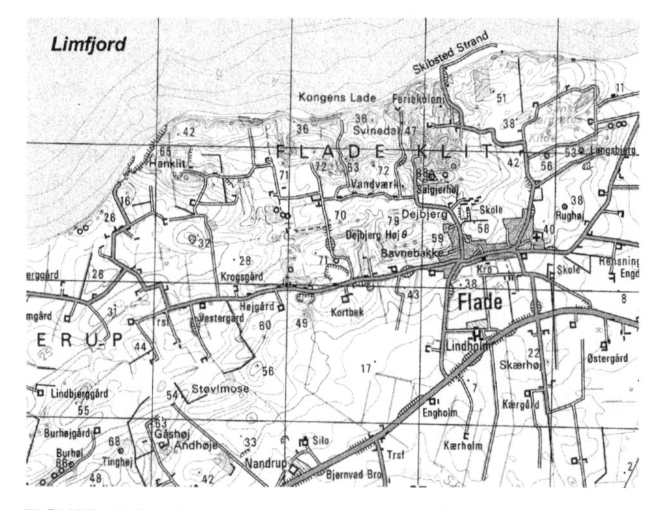

FIGURE 12-20 Topographic map of Flade Klit vicinity, northwestern Denmark. Hanklit is a cliff exposure of deformed bedrock at the western end of Flade Klit. Each grid square is 1 km^2; adapted from 1116 I Thisted, Danmark, 1:50,000 (1983), Geodætisk Institut, Denmark.

(E–W) and 1.5 km wide. The ridges comprise a conspicuous hill with a gentle crescentic shape, concave toward the north. Maximum elevation within Flade Klit reaches 88 m at Salgerhøj, which is ~100 m above the floor of the Limfjord estuary (Fig. 12-21). The 50-m-high cliff exposure at Hanklit reveals three folded masses of dislocated bedrock that were thrust out of the Limfjord basin from the north (Klint and Pedersen, 1995).

FIGURE 12-21 View toward the northeast over ice-shoved ridges of Flade Klit. Hanklit is a cliff exposure of dislocated bedrock on the western end of Flade Klit; Salgerhøj is the highest point (88 m). Kite aerial photo; adapted from Aber and Ber (2007, fig. 5-24).

12.5. SUMMARY

Small-format aerial photography provides low-height, large-scale imagery that complements conventional aerial photography, satellite imagery, and ground observations for recognizing, mapping, and interpreting glacial geomorphology. Oblique views are particularly useful for depicting individual landforms created by glacial erosion, deposition, and deformation. Such views help to visualize complex landform assemblages that are typical of many formerly glaciated regions. In like manner, SFAP may be applied to other types of geomorphic situations, ranging from coral reefs to permafrost landforms.

Chapter 13

Gully Erosion Monitoring

13.1. INTRODUCTION

Gullies are permanent erosional forms that develop when water concentrates in narrow runoff paths and channels and cuts into the soil to depths that cannot be smoothed over by tillage any more. They occur all over the world mostly in semi-arid and arid landscapes where high morphological activity and dynamics can be observed. Semi-arid climate conditions and precipitation regimes encourage soil erosion processes through low vegetation cover and recurrent heavy rainfall events. Torrential rains with irregular spatio-temporal distribution result in high runoff rates, as the crusted and dry soil surface inhibits infiltration. In addition, widespread land-use changes of traditional agriculture toward both more extensive use, such as abandoned agricultural fields used for sheep pasture, and less sustainable use, such as almond plantations, prepare the ground for soil crusting, reduced soil infiltration capacity, and increased runoff, which together aggravate the risk of linear erosion downslope and cause considerable offsite impairment such as reservoir siltation (Poesen et al., 2003).

In this context, gullies link hillslopes and channels, functioning as sediment sources, stores, and conveyors. From a review of gully erosion studies in semi-arid and arid regions, Poesen et al. (2002) concluded that gullies contribute an average of 50–80% of overall sediment production in dryland environments.

Gullying involves a wide range of subprocesses related to water erosion and mass movements, such as headcut retreat, piping, fluting, tension crack development, and mass wasting (Fig. 13-1), and it is the complex interaction of these subprocesses on different temporal and spatial scales that complicates reliable forecasting by gully erosion models (Poesen et al., 2003).

The evaluation of gully development rates under different climatic and land-use conditions provides important data for modelling gully erosion and predicting impacts of environmental change on a major soil erosion process. Numerous authors have investigated (non-ephemeral) gully development in different environments (e.g., Burkard and Kostaschuk, 1997; Oostwoud Wijdenes and Bryan, 2001; Vandekerckhove et al., 2001; Archibold et al., 2003; Betts et al. 2003; Martínez-Casasnovas et al., 2004; Avni, 2005; Wu, Zheng, et al., 2008), but still both methodological problems and a lack of comparability across study areas exist.

Poesen et al. (2002, 2003) therefore stressed the need for more detailed and more precise monitoring and modelling of gullies. Lane et al. (1998) pointed out that the monitoring of the changes in form may provide a more successful basis for understanding landform dynamics than monitoring the process driving those dynamics, particularly when spatially distributed information on process rates can be acquired. In this context, remote sensing is an obvious choice for monitoring gully erosion, as it allows the rapid and spatially

FIGURE 13-1 Gully erosion in southeastern Spain. (A) Active headcut of small gully ~2.5 m wide and <1 m deep. (B) Large gully with remains of piping hole and fluted walls typical for higher dispersible sub-horizons. (C) Large gully filled with debris of collapsed sidewall. Ground photos taken by IM in 2006.

continuous coverage of a site. However, the measurement precision and repeat rate attainable with conventional remotely sensed images are not able to correspond with the process magnitudes and dynamics that are required for recording and investigating the short-term spatial and temporal variability of gully retreat (Ries and Marzolff, 2003). Therefore, most gully monitoring studies using remotely sensed imagery resort to standard large-format aerial photographs with medium to small image scales, looking at medium- to long-term intervals (e.g., Burkard and Kostaschuk, 1997; Vandaele et al., 1997; Nachtergaele and Poesen, 1999; Vandekerckhove et al., 2003).

Bridging the resolution gap between terrestrial and conventional aerial photography, small-format aerial photography (SFAP) has proven to be an excellent tool for gully erosion monitoring in several research studies conducted by the authors (Marzolff, 1999; Marzolff et al., 2003; Ries and Marzolff, 2003; Marzolff and Ries, 2007; Marzolff and Poesen, 2009). The following sections summarize some of the work done with hot-air blimps and kites across semi-arid landscapes in southern Europe and West Africa, illustrating the benefits of SFAP not only for quantifying but also for understanding gully erosion processes by selected examples.

13.2. STUDY SITES AND SURVEY

The study areas are situated in Spain, Morocco, and Burkina Faso along a transect running from semi-arid to sub-humid Mediterranean and sub-tropical desert climate to tropical wet-dry regions (Fig. 13-2). Thus, they cover climatic regions where precipitation regimes are the most favorable for gully erosion. The gully environments range from agricultural areas and rangeland to deserts. Beginning between 1995 and 2008, 23 gullies of varying types, sizes, and ages have been monitored in intervals of usually one or two years.

Depending on wind conditions, a hot-air blimp (see Chapter 8.3.3) or a large rokkaku-type kite (see Chapter 8.4.1) is employed as a platform for analog (up to 2002) or digital single-lens reflex (SLR) cameras, allowing for photographic surveys from flying heights up to 350 m. The survey is usually conducted at various heights and scales in order to collect detailed photographs as well as overviews with stereoscopic overlap. At all gully sites, ground control points (GCPs) measured with a total station in a local coordinate system are permanently installed with steel pipes and marked with signals for the photographic survey (see Chapter 9.4). From the resulting stereoscopic images with extremely high ground resolutions of 0.5–10 cm, various techniques of analysis for measuring gully development and loss of soil material have been carried out with image-processing systems, geographic information systems (GIS), and digital photogrammetry stations.

13.3. GULLY MAPPING AND CHANGE ANALYSIS

During the early years of study, 2D measurements of headcut retreat and gully growth were accomplished by rectifying and—if required—mosaicking the photographs using GCPs installed in the surroundings of the gully. Thus, detailed mapping of the gully edges was possible despite the remaining relief displacement in the gully interior, which was not at the focus of the investigation and could not have been geocorrected accurately with polynomial rectification (see Chapter 11.2.2). In later project stages, 3D mapping in a stereoviewing environment and digital elevation model (DEM) extraction using digital photogrammetry software made it possible to map the 3D gully forms even more precisely (see Fig. 11-22) and to create 3D models for measuring gully volume (see Fig. 11-23). Change analysis from the multitemporal datasets for quantifying linear retreat and increase of gully area and volume is performed with GIS software.

Three exemplary maps of gully development illuminate the variability in headcut retreat behavior and gully evolution. Gully Gorom was monitored with SFAP over a period of three seasons in a project terminated in 2002; its change map in Figure 13-3 is complemented with smaller-scale satellite image data from 2007. Note the much coarser outlines of the last gully stage; high-resolution satellite data were a viable supplement in this case only because of the large size of the gully and the fact that it is the fastest growing in the whole dataset. The gully edges retreat along the full length of the gully, reshaping the gully form during each rainy season. The surrounding glacis area has little relief that might lead to preferential linear flow paths, and it is only in the upslope direction of the shallow main drainage line that the gully clearly grows faster.

Along the gully edges, mass wasting is the main reason for gully growth, and large clods of soil come to rest beneath the gully rim before being washed away by further erosion. Piping (subsurface erosion), often promoted by desiccation cracks at the headcut vicinity, is also involved in the retreat process. Despite the great width of the gully, incision is only shallow and the heights of the sidewalls are only between 40 and 100 cm. The maximum depth of the central drainage channel is ~1.2 m. Within the part of the mapped gully, areal growth is quite regularly around 800 m^2 per year, but maximum linear headcut retreat varies depending on the spatial development of the gully's shape.

At the Barranco Rojo, situated at the verge of the Ebro Basin in northeastern Spain, the situation is quite different (Fig. 13-4). Although the gully is of a similar elongated dendritic shape with many secondary headcuts along the sidewalls, active headcut retreat occurs only locally and gully growth is much less. Piping has played an important role in the past and present gully development, as can be

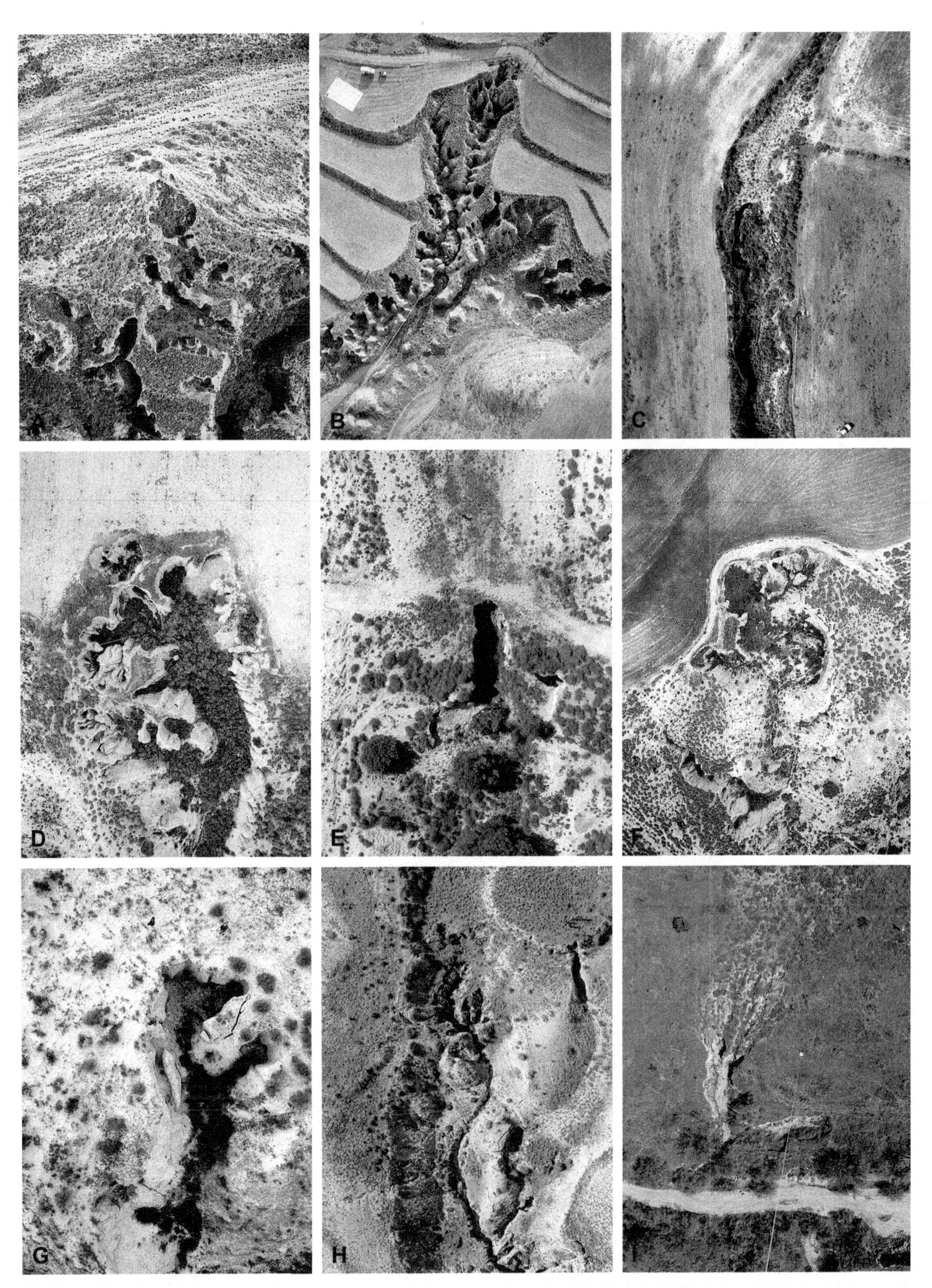

FIGURE 13-2 Selected gullies monitored by SFAP in Spain (A–K), Morocco (L–N), and Burkina Faso (O–P). Kite and hot-air blimp aerial photographs by IM, JBR, and collaborators, taken between 1996 and 2008. (A) Barranco de las Lenas. (B) Barranco Rojo. (C) Bardenas 1. (D) Salada 1. (E) Salada 3. (F) Salada 4. (G) Luchena 1. (H) Freila. (I) Casablanca.

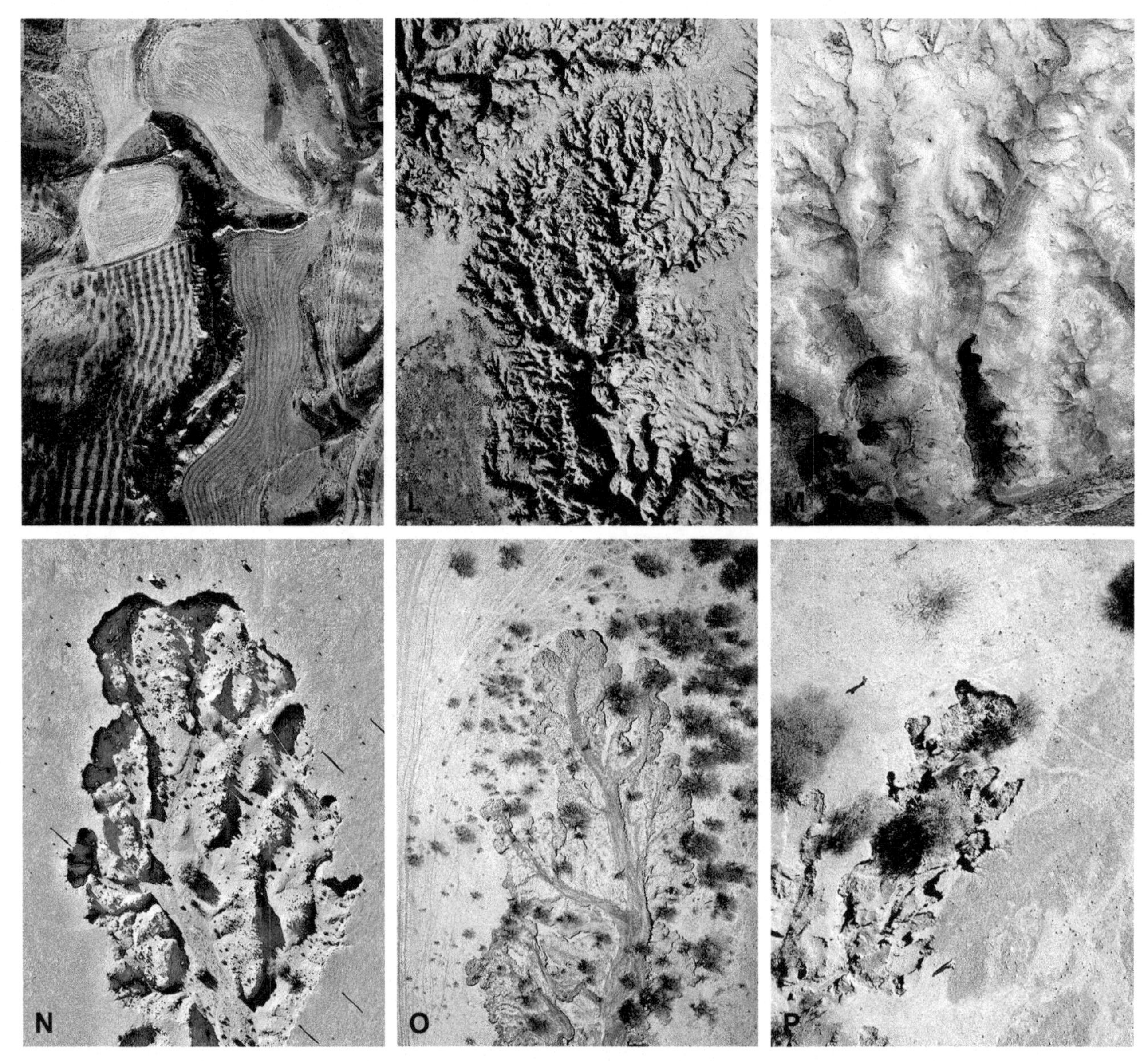

FIGURE 13-2—Cont'd, (K) Belerda 1. (L) Talaa. (M) Foum el Hassane. (N) Icht. (O) Gorom-Gorom. (P) Oursi.

seen from the location of changed gully areas and in the underlying airphoto from many remains of collapsed pipes in the gully interior. Within the gully, a continuous surface drainage line is not visible; its interior is divided into numerous sub-catchments drained by piping holes.

At the Barranco Rojo, the formation of the various individual headcuts and branches is obviously an alternating process rather than a continuous one as at Gully Gorom. This may be due to the processes involved and the spatio-temporal variability of the gully surroundings. Although areal change measured from the aerial photographs amounts to only a fraction of the change observed at Gully Gorom, volumetric soil loss may be comparatively larger, owing to the greater depth, up to 6 m in the part mapped, and abundant subsurface erosion features of Barranco Rojo. With piping as a major process of gully development, areal change would be evident only when the surface has caved in, and the gully can appear inactive at the surface in spite of active subsurface erosion processes beneath.

The greater variability of the topography and land use around the Barranco Rojo also may be potentially important for the different behavior of the two gullies depicted in Figures 13-3 and 13-4. Agricultural terraces, which are also subject to piping processes, are separated by short steep slopes, and a dirt track dissects the gully catchment just upslope from the main headcut. Land use varies both spatially and temporally between cereal fields, clean summer-fallow, and older, shrub-covered fallow land. Rainfall simulations and infiltration experiments have shown a great range of runoff coefficients, infiltration and erosion rates on these surfaces (Seeger et al., 2009). Thus, runoff coefficients and flow paths in the headcut vicinity are

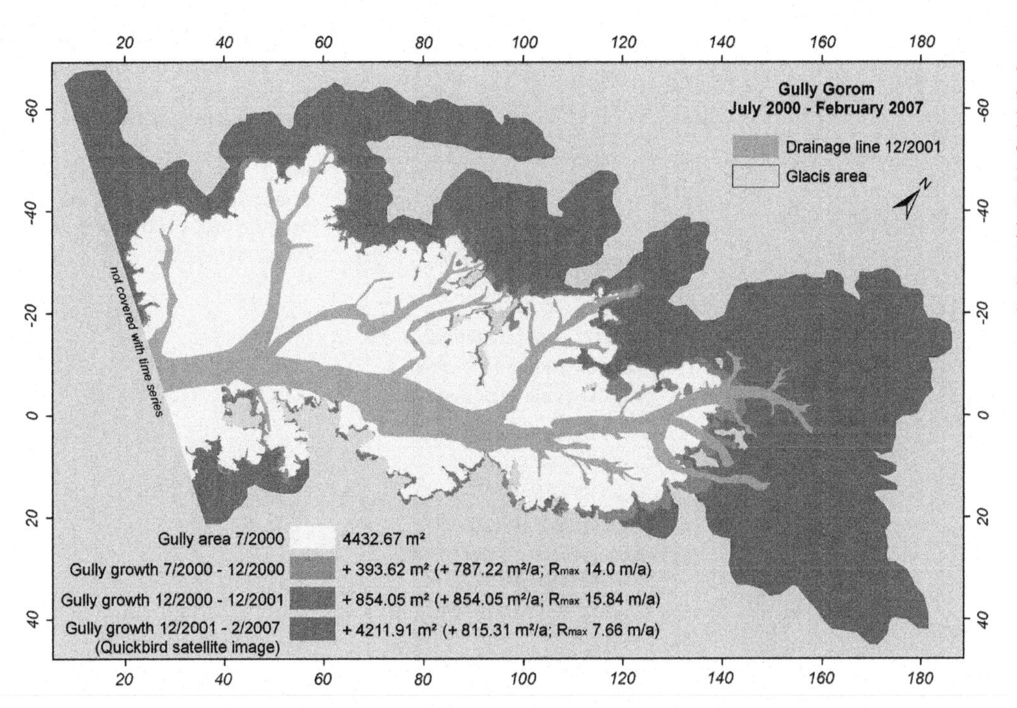

FIGURE 13-3 Two-dimensional change analysis of Gully Gorom-Gorom, Province of Oudalan, Burkina Faso. Gully growth quantified between rainy seasons 2000 and 2001 (SFAP) and 2007 (*Quickbird*). Based on kite aerial photography by IM, JBR, and K.-D. Albert; image processing and cartography by K.-D. Albert and IM.

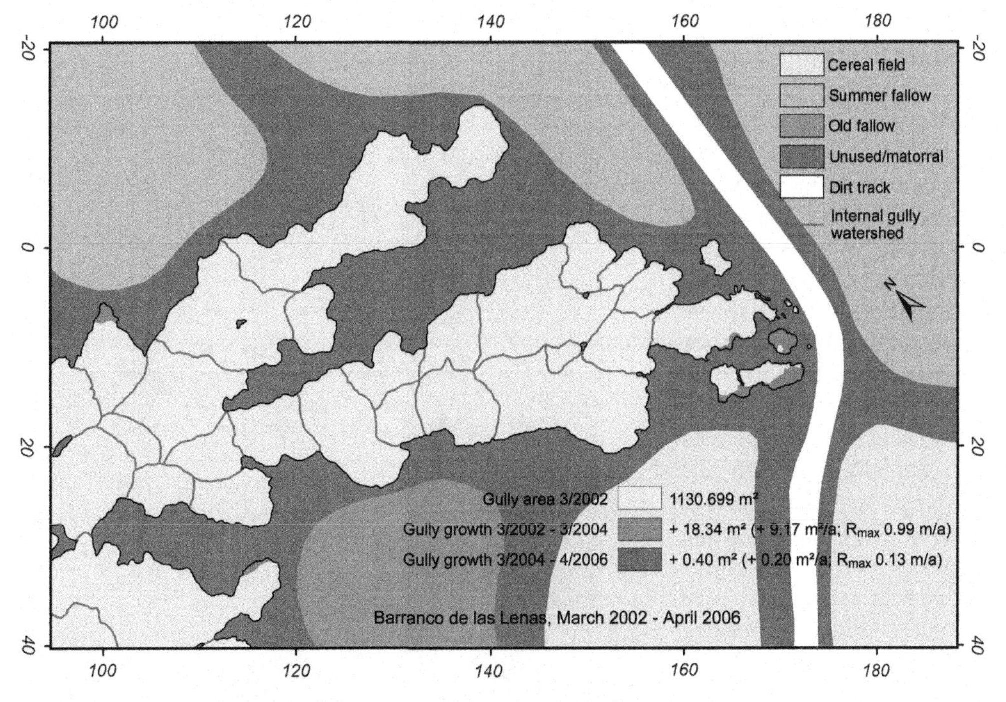

FIGURE 13-4 Two-dimensional change analysis of Barranco Rojo, Province of Zaragoza, Spain. Gully growth quantified between March 2002 and April 2006 based on hot-air blimp photographs by IM, JBR, and M. Seeger; image processing and cartography by IM. Adapted from Marzolff and Ries (2007, fig. 4).

constantly changing over the years, contributing to the spatial variability of gully development.

The assumption that gully growth is strongly related to the characteristics of topography, substratum, and land use in the near vicinity is also confirmed by the example of Gully Salada 1 (Fig. 13-5), where human interaction plays an additional role. Until 2004, this large gully cutting into the Quaternary deposits of the Guadalentín Basin in southeastern Spain was the fastest growing of the 14 Spanish gullies being monitored. The immediately adjoining almond plantation, which was established only a few years before monitoring began, is kept weed-free by regular plowing. Soil crusting results in very high runoff rates and in the formation of ephemeral gullies between the almond trees. To bar the gully from retreating farther into the plantation, the farmer had plowed up an earthen dam around the gully margin—a measure with limited success. Runoff from the plantation collected at the earthen dam

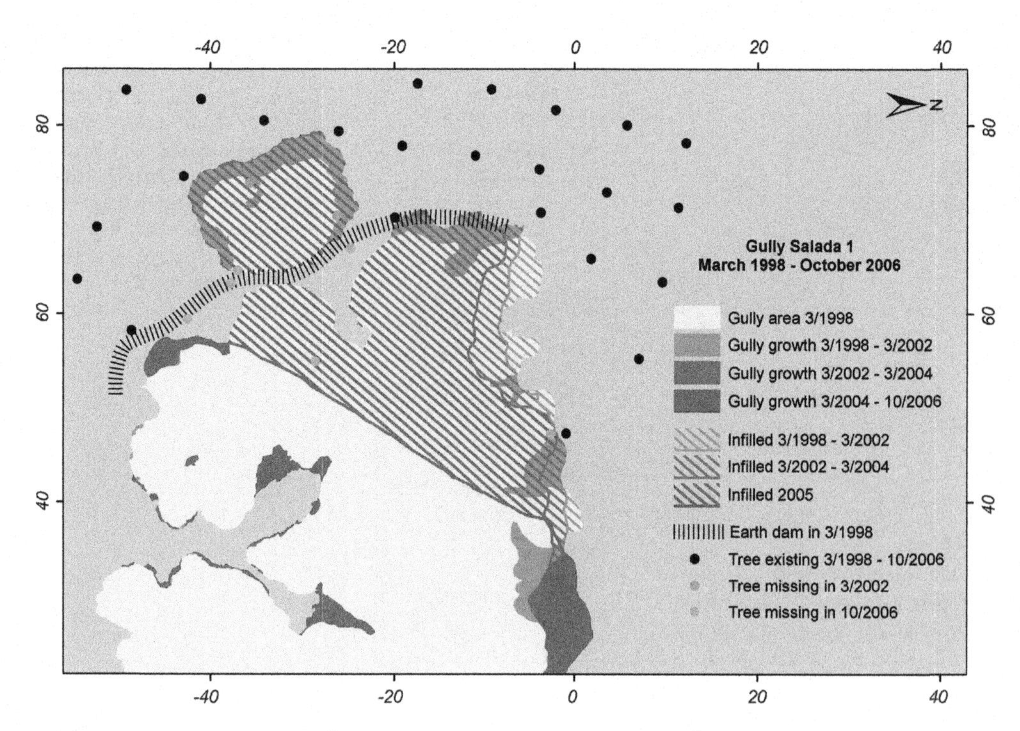

FIGURE 13-5 Two-dimensional change analysis of Gully Salada 1, Province of Murcia, Spain. Gully growth quantified between March 1998 and October 2006 based on hot-air blimp and kite photographs by IM, JBR, and M. Seeger; image processing and cartography by IM.

subsequently resulted in subsurface erosion processes, creating a growing piping hole, which drained beneath a bridge remaining of the former dam (see also Vandekerckhove et al., 2003).

The piping hole increased rapidly in the following years, while the main gully was used as a rubbish dump, and building rubble was shoved repeatedly over its northern edge by tractors. In 2005, the upper part of the gully including the piping hole was completely filled with rubble and soil material up to the level of the former surface (sloping into the remaining gully beyond the mapped infill boundary (Fig. 13-6A). This re-established the former border of the field; the missing almond trees were replanted between 2006 and 2008. By 2006, the infilling already had begun to subside and show large settlement cracks (Fig. 13-6B), preparing the ground for resumed piping processes in the future. Ground observations in September 2009 revealed a recent dam break with a large and deep erosion rill cutting from the almond field into the infilling where the former piping hole had been.

Using a hybrid method combining automatic height-point extraction with manual 3D editing and digitizing (Marzolff and Poesen, 2009), high-resolution DEMs (5–10 cm pixel size) were created for selected gullies. Figure 13-7 shows the example of a surface model of Gully Bardenas 1 with the corresponding orthophoto (see Figs. 11-1 and 11-22).

Cut-and-fill operations in a GIS environment enable determining the total soil loss at the gully site as well as volumetric changes between monitoring dates; 3D modelling from SFAP is clearly superior here to traditional

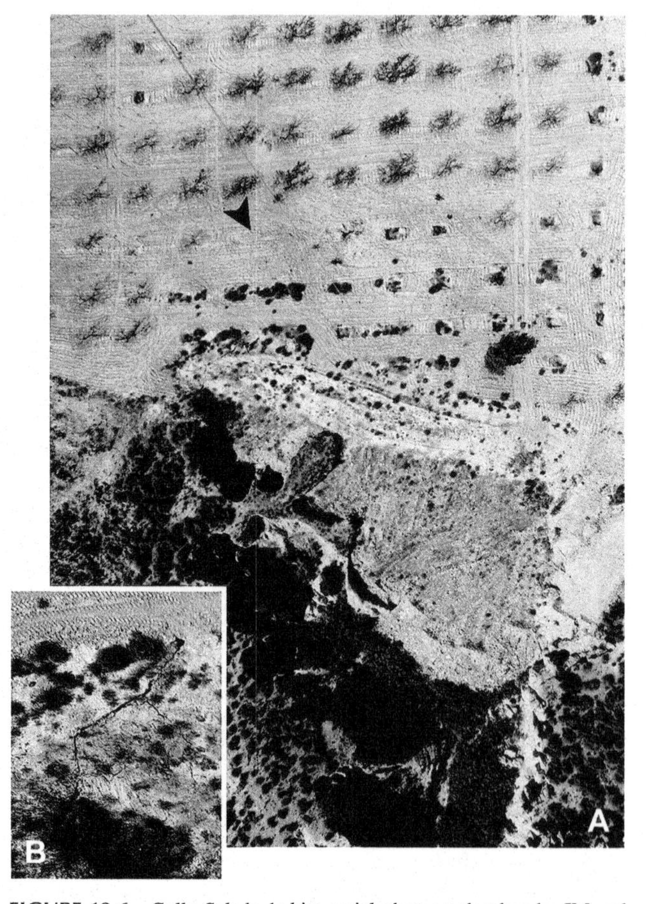

FIGURE 13-6 Gully Salada 1; kite aerial photograph taken by IM and JBR in October 2006. (A) Overview of gully after refilling of the upper part (compare with Figs. 13-2D and 13-5). Arrow indicates former position of piping hole. (B) Detail of large settlement crack near former gully edge.

FIGURE 13-7 Orthophoto draped over photogrammetrically derived DEM of Gully Bardenas 1, Province of Navarra, Spain. Based on kite aerial photographs taken by JBR, M. Seeger, and S. Plegnière in March 2007; photogrammetric processing by IM. Until February 2009, this gully increased its 35 m length shown here by 2.5 m, eroding just over 20 m^3 on an area of 16.6 m^2.

terrestrial measurement methods. Modelling the complete form rather than taking sample measurements of gully extent and depth allows the stratification of volumetric change in erosion and deposition aspects, yielding results for the gully's own sediment balance. The example of Gully Salada 3—a typical bank gully of the simplest, single-headcut U-profile form (Fig. 13-8)—illustrates the complexity of the patterns often simplified as "headcut retreat."

Between 1998 and 2002, ~4.4 m^2 area, corresponding to 13 m^3 soil material, was lost to backward erosion at the headcut. In the same period, the surface height within the gully increased as 45% of the material eroded at the headcut was deposited on the gully bottom. Most of this came to lie close to the headcut, but some was washed farther downslope at the gully bottom. Limited erosion only took place on the gully floor, and its longitudinal profile, which can be estimated from the 3D models shown in Figure 13-8A and B, obviously experienced an increase of gradient not due to downslope erosion, but due to upslope deposition.

Figure 13-8C also shows some of the difficulties associated with creating surface models for highly variable terrain (see Chapter 3.4.3). Along the sidewalls, which contributed to the gully growth with another 0.7 m^2, the volume change visible in the change map must be considered somewhat inaccurate (note the improbable gain at the sidewall to the lower right of the calculation area limit). Here, the camera's perspective eye could not obtain an unobstructed view of the narrow and deep corridor carved by the gully, causing less inclined slopes in the 2002 model.

More details about the gully development rates and their relation with local topography, substratum, and runoff and infiltration behavior in the gully headcut surroundings are given by Geißler (2007), Marzolff and Ries (2007), Ries and Marzolff (2007), and Seeger et al. (2009). In summary, the development rates of the gullies vary between individual study areas along the transect, and particularly the Spanish gullies retreat more slowly than expected. Maximum linear retreat per year (R_{max}/a) varies by a factor of 100, ranging from below 0.1 m/a to nearly 10 m/a. Within the same study area, the variability is much lower with a factor of only 10. When ranking the study regions according to soil erosion parameters assessed by experimental measurements, a clear association exists between the resulting order, both for maximum and minimum values of runoff coefficients and material output with the maximum retreat rates observed at the gully headcuts.

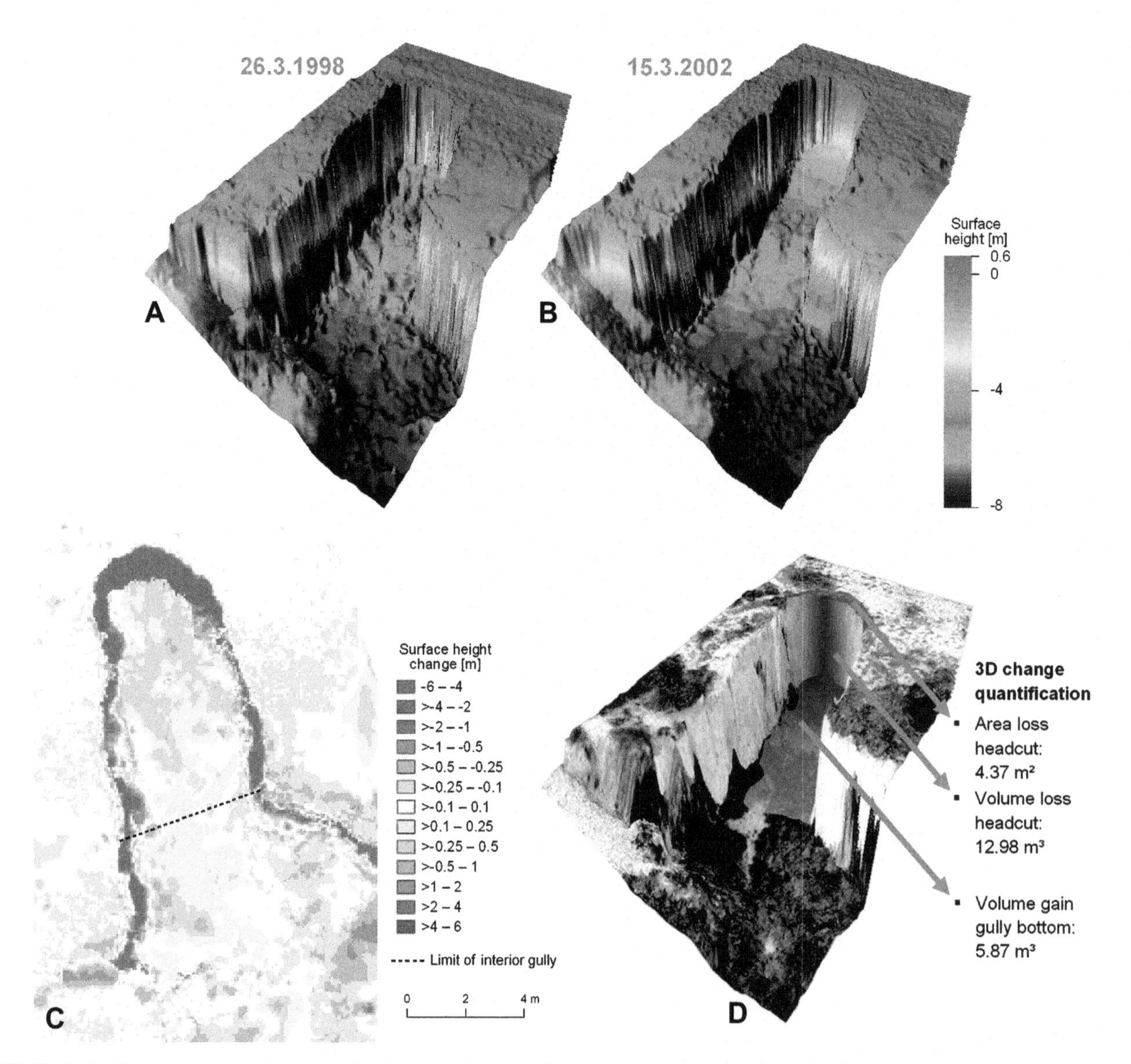

FIGURE 13-8 Three-dimensional change quantification of Gully Salada 3, Province of Murcia, Spain. Based on hot-air blimp photographs taken by IM, JBR, and M. Seeger. All heights are given relative to local datum. (A) DEM in March 1998. (B) DEM in March 2002. (C) Surface difference map. (D) Orthophoto draped over 1998 DEM with summary of change measurements within the interior gully area. Adapted from Marzolff and Poesen (2009, fig. 9).

13.4. SUMMARY

Gullies are permanent erosional forms that develop in many parts of the world, particularly in arid and semi-arid environments. Gullies function as sediment sources, stores, and conveyors that link hillslopes to downstream channels. Human land use, and especially changes in land use, may accelerate gully expansion by head cutting, sidewall collapse, piping, floor erosion, and other processes, which lead to widespread land degradation and potential damage to human structures and activities.

The results achieved with small-format aerial photography for monitoring gully erosion continue to demonstrate that SFAP can be considered an advantageous alternative to field methods or conventional aerial photography. Change quantification based on the detailed maps and DEMs provides additional information on the differences in headcut retreat behavior which cannot be described by simple linear measures, and the spatially continuous survey of the entire form offers the possibility of distinguishing different zones and processes of activity both at the gully rim and within the gully interior. In addition, more than any other measurement method, photographic capture of a transient situation allows for retrospective interpretation of the spatial processes leading to the actual gully form, and new parameters of interest may be derived even years after the survey, owing to the spatial continuity and sample density of the SFAP inventory of a site.

Chapter 14

Wetland Environments

The use of aerial photography has resulted in an essential advance in the understanding of mire structure.

(Aaviksoo et al. 1997, p. 5)

14.1. OVERVIEW

Wetlands include myriad environmental types: bayou, bog, fen, mangrove, marsh, moor, muskeg, pan, playa, sabkha, swamp, tundra, and several more terms in other languages. Many of these names are united under the general term mire nowadays, but that still does not specify the basic characteristics of wetlands. A definition for what is a wetland often depends on who is asking the question and what development or study is proposed for a particular wetland site. The fact that wetlands may dry out from time to time complicates the attempt to describe wetlands in a simple fashion, and some wetlands may be dry more often than they are wet (Tiner, 1997).

Many definitions for wetlands have been proposed and utilized over the years. Among the most widely accepted is that of Cowardin et al. (1979), which was adopted by the U.S. Fish and Wildlife Service.

> *Land where an excess of water is the dominant factor determining the nature of soil development and the types of animals and plant communities living at the soil surface. It spans a continuum of environments where terrestrial and aquatic systems intergrade.*

This definition comprises three aspects—water, soil, and living organisms, which are accepted by wetland scientists as the basis for recognizing and describing wetland environments (Schot, 1999; Charman, 2002).

- *Water*: Ground water (water table or zone of saturation) is at the surface or within the soil root zone during all or part of the growing season (Fig. 14-1A).
- *Soils*: Hydric soils are characterized by frequent, prolonged saturation and low oxygen content, which lead to anaerobic chemical environments where reduced iron is present (Fig. 14-1B).
- *Vegetation*: Specialized plants are adapted for growing in standing water or saturated soils, such as moss, sedges, reeds, cattail and horsetail, rice, mangroves, cypress, and cranberries (Fig. 14-1C).

This triad is the modern approach for wetland definition under many circumstances that include greatly different environments. Notice that water quality is not specified—salinity varies from fresh, to brackish, to marine, to hypersaline; acidity may span the entire scale of naturally occurring pH values. Emergent vegetation ranges from heavily forested swamps to nearly bare playas and mudflats. Wetlands are present in all climatic and topographic settings around the world, covering substantial portions of the land and shore areas of the Earth. Existing wetlands comprise an estimated 7–10 million km^2 or 5–8% of the land surface of the globe (Mitsch and Gosselink, 2007). However, as much as half of the world's wetlands have been lost to human development during the past few millennia.

For various environmental and economic reasons, much scientific research is directed toward wetlands, and many different techniques have been utilized to collect, compile, analyze, and synthesize data. In recent years, traditional ground mapping methods have been supplemented with the use of geographic information systems and remote sensing techniques for wetland research (Jensen et al., 1993; Juvonen et al., 1997; Ahvenniemi et al., 1998; Barrette et al., 2000).

Tiner (1997) noted the difficulty of wetland airphoto interpretation because of highly variable water, topographic, and vegetation conditions that may apply. He considered color-infrared imagery as best for recognizing vegetated wetlands, such as marshes, swamps, and bogs, and he emphasized the importance of photographic scale for setting spatial limits on wetland mapping. In this regard, small-format aerial photography (SFAP) has distinct advantages for certain types of wetland investigations (Aber and Aber, 2001).

- High-resolution (2–5 cm), large-scale imagery is suitable for detailed mapping and analysis.
- Equipment is light in weight, small in volume, and easily transported by foot, vehicle, or small boat under difficult field conditions—peat, mud, and water.

FIGURE 14-1 Primary aspects of wetlands. (A) Water pools of various sizes and shapes occupy much of the surface of Nigula bog in southwestern Estonia. A narrow, wooden footpath can be seen across bottom of view, September 2001. (B) Salt crust forms as soil moisture evaporates from mudflat at Dry Lake in western Kansas, May 2008. (C) Curly dock (*Rumex crispus* L.) forms a distinct, rust-colored band along the shore of Lake McKinney in southwestern Kansas, October 2006. Kite aerial photographs by SWA and JSA.

- Minimal impact on sensitive habitat, vegetation, and soil. Silent operation of kites and blimps does not disturb wildlife (Fig. 14-2, see also Fig. 1-3).
- Repeated photography during the growing season and year-to-year documents changing environmental conditions.
- Visible and infrared imagery in vertical and oblique vantages in all orientations relative to the sun position, shadows, and ground targets (Aber et al., 2002).
- Special lighting effects, such as sun glint, may aid in recognition of small water bodies (Amsbury et al., 1994).
- Lowest cost by far relative to other manned or unmanned types of remote sensing to achieve comparable high spatial, spectral, and temporal resolutions.

The following examples are drawn from raised bogs of Estonia in north-central Europe and prairie marshes and playas of the central Great Plains in Kansas, United States. SFAP for the former was undertaken exclusively with kite aerial photography (KAP), and the latter combines kites and a helium blimp for lifting camera rigs. Climatic, hydrologic, and geographic settings for these two regions are substantially different, as demonstrated by their wetland conditions.

14.2. RAISED BOGS, ESTONIA

Estonia is a small country rich in wetlands. Located at the far eastern end of the Baltic Sea, more than one-fifth of the territory is covered by swamps, marshes, fens, and bogs (Orru et al., 1993), and more than 40 wetland sites are protected as national parks, nature preserves, or mire conservation areas. Masing (1997) called bogs *monuments of nature*, because they preserve in their peat deposits a record of past climates and environments. Indeed much of what we know about Scandinavian climate and environment of the past 10,000 years has been gleaned from intensive investigations of bogs.

On the ground, a raised bog presents a rather austere view. It appears to be a haphazard collection of peaty hummocks, muddy hollows, and water-filled pools. As seen from above, however, mires display complex and intricate patterns of vegetation and water bodies, which have resulted

FIGURE 14-2 Superwide-angle view looking northward over Rachael Carson National Wildlife Refuge surrounded by suburban housing. Nesting birds in the salt marsh are particularly sensitive to nearby activity on the ground or overhead, which they may interpret as predators. Special permission was necessary to conduct SFAP using a tethered helium blimp that had to be flown at least 90 m and no more than 150 m above the ground. Silent blimp operation at this height minimized possible disturbance of birds on their nests. Helium-blimp aerial photo by SWA, JSA, and V. Valentine, August 2009; Maine, United States.

largely from organic evolution within the bog environment. For the bird's-eye view, various techniques of remote sensing have been applied for mire research in Estonia. These techniques range from satellite imagery (Aaviksoo, 1995; Aaviksoo et al., 2000) and conventional airphotos (Aaviksoo, 1988) to high-resolution, near-surface aerial photography (Aaviksoo et al., 1997).

In order to understand better the detailed spatial geometry of vegetation and water within bogs, KAP was employed at two mire complexes, Endla and Nigula, both designated as Ramsar wetlands of international importance (Fig. 14-3; Aber and Aber, 2001; Aber et al., 2002).

14.2.1. Endla Nature Reserve

The present reserve was created in 1985 as an expansion of the previous smaller Endla-Oostriku mire reserve. It is located immediately south of the Pandivere Upland in east-central Estonia (Fig. 14-4). The Endla mire complex grew up in the depression of former Great Endla Lake (Allikvee and Masing, 1988). Several remnants of this lake still survive, notably Endla Lake and Sinijärv (Blue Lake). The Endla mire system covers ~25,000 ha and contains several bogs separated by narrow rivers, and significant springs rise in the western part of the complex. The lakes, bogs, and springs are important sources of recharge for the Põltsamaa River. Among the bogs, Männikjärve has been investigated intensively since the early 1900s. A small meteorological station is located in the bog, and an elevated, wooden walkway allows visitors to travel across the bog without disturbing the surface (Aaviksoo et al., 1997).

Oblique views across the bog display overall patterns of hummock ridges, dwarf pines, hollows, and water-filled pools (Fig. 14-5). In close-up oblique and vertical views, it

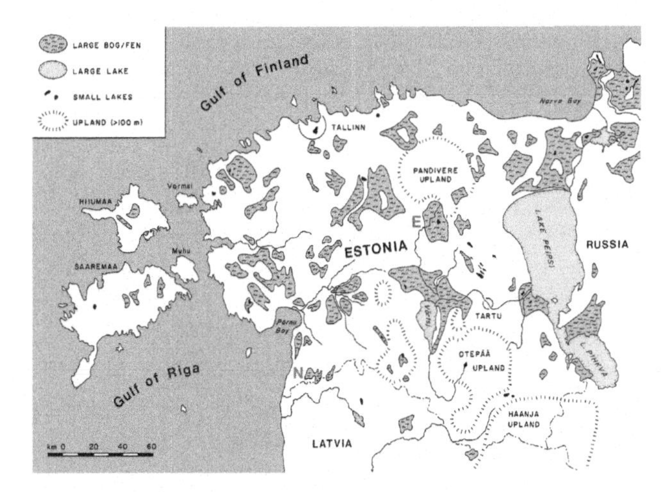

FIGURE 14-3 Generalized distribution of large mire complexes in Estonia and adjacent territories. Endla mire complex (E) and Nigula bog (N). Adapted from Aber and Aber (2001, fig. 5); based mainly on Orru et al. (1993).

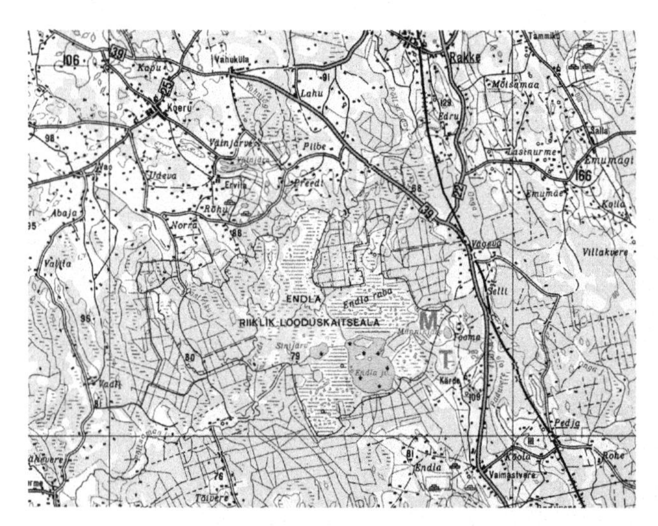

FIGURE 14-4 Topographic map of Endla vicinity, Estonia. Kite aerial photography was conducted at Männikjärve (M) and Teosaare (T) bogs. Map derived from Narva, Eesti Topograafiline Kaart, sheet 2, scale 1:200,000 (1992).

FIGURE 14-6 Vertical shot of Teosaare bog illustrating the complex distribution of pools, hummocks, moss, and dwarf pine trees. White marker in lower center is 1 m^2. Kite aerial photo by SWA and JSA, September 2001.

FIGURE 14-5 Oblique westward view of Männikjärve bog showing public observation tower and raised boardwalk. The center of the bog contains water-filled pools in hollows between elongated hummocks that support dwarf pine trees. Kite aerial photo by SWA and JSA, September 2001.

FIGURE 14-7 Close-up vertical picture of pools and hummocks in the central portion of Männikjärve bog. A, *Sphagnum cuspidatum* floating in water; B, *S. cuspidatum* at pool shore (barely emergent); C, *Sphagnum rubellum* above water table; D, Scots pine trees on hummocks along with dwarf shrubs—*Empetrum nigrum*, *Chamaedaphne calyculata*, *Andromeda polifolia*, and *Calluna vulgaris*. The boardwalk is ~60 cm wide. Taken from Aber et al. (2002, fig. 2)

is possible to identify individual small trees, moss hummocks, faint trails, small potholes, and other structures (Fig. 14-6). Varieties of peat moss are distinct in their autumn coloration—bright red, reddish orange, and greenish yellow (Fig. 14-7). Color-infrared images emphasize active moss around the margins of pools (Fig. 14-8).

14.2.2. Nigula Nature Reserve

Nigula is a typical plateau-like bog covering ~2340 ha with sparse trees and numerous small pools in southwestern Estonia (Fig. 14-9). So-called mineral islands rise within the bog and support deciduous trees on nutrient-rich soil. Mire formation began as a result of infilling and overgrowing of an ancient lake following retreat of the last ice sheet from the region (Loopmann et al., 1988). The bog has been the subject of numerous investigations of its characteristics (e.g., Ilomets, 1982; Koff, 1997; Karofeld, 1998). A narrow footpath of wooden boards laid directly on the moss circles through the bog, although part of the path is closed to the public, and an observation tower is open for public use.

Oblique views depict the relationships of various components of the bog—pools, hummocks, vegetation

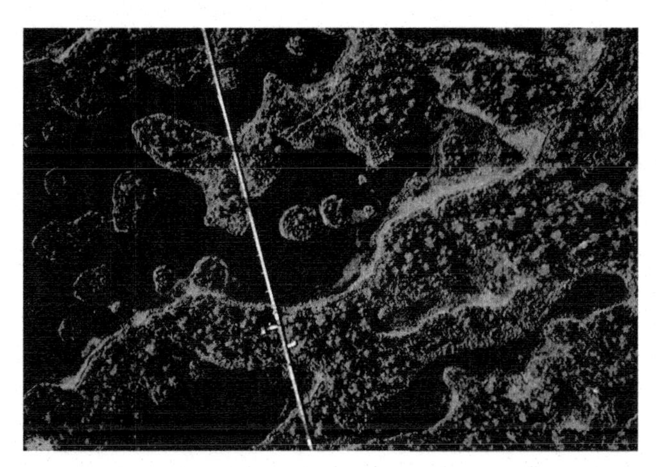

FIGURE 14-8 Color-infrared view of Männikjärve bog. Active moss photosynthesis is concentrated at the margins of pools, as shown by bright pink-red in this false-color image. Compare with previous figure; taken from Aber et al. (2002, fig. 8).

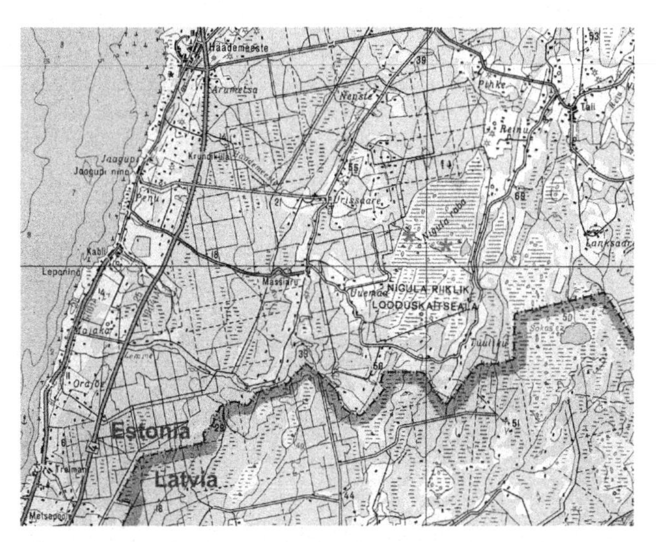

FIGURE 14-9 Topographic map of Nigula vicinity, Estonia. Kite aerial photography was conducted at two marked (*) sites within the bog. Map derived from Pärnu, Eesti Topograafiline Kaart, sheet 3, scale 1:200,000 (1992).

FIGURE 14-10 Oblique overview of Nigula bog looking toward the west. Narrow wooden footpath (~40 cm wide) is visible in lower left portion, and a mineral island appears in the right background. Kite aerial photo by SWA and JSA, September 2001.

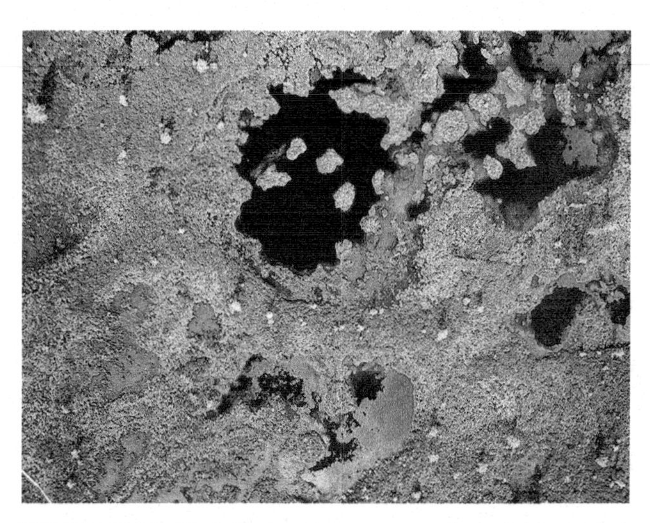

FIGURE 14-11 Vertical view in eastern portion of Nigula bog. *Sphagnum cuspidatum* forms a silvery green "mat" floating in parts of some pools. The reddish-brown zones include *Sphagnum magellanicum* and *S. rubellum*. Heather covers much of the mottled green surface along with a few small pines (note shadows). A portion of the footpath is visible in lower left corner. Taken from Aber et al. (2002, fig. 3).

zones, and mineral islands (Fig. 14-10), and vertical views reveal intricate spatial patterns (Fig. 14-11). Distinct vegetation zones are developed around the mineral island; these zones reflect variations in soil moisture and nutrients (Fig. 14-12).

14.2.3. Discussion

Wooden walkways provide good access to some portions of the study bogs, but wading through soft peat and knee-deep water is necessary to reach many areas. In this regard, the high portability of equipment is critical for successful KAP in the bog environment. The combination of late summer weather and early autumn vegetation color makes for excellent results in September. However, widespread burning of agricultural waste renders October photographs hazy (see Fig. 4-28), and low sun angle (at 58°N latitude) creates excessive shadowing after fall equinox.

High-resolution SFAP may be applied for microstructural investigations and analysis of mires at scales of 1:100–1:1000 (Masing, 1998). The civilian imagery interpretability rating scale has 10 rating levels (0–9; see Table 10-1 and Chapter 10.2). The resolution of vertical digital kite aerial photographs (2–5 cm) provides for interpretability ratings of 7–8 (Leachtenauer et al., 1997). This ground resolution is an order of

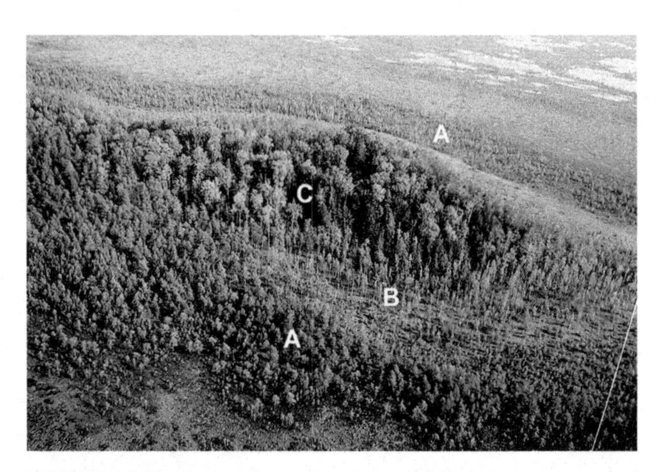

FIGURE 14-12 Low-oblique view over Salupeaksi, a tree-covered mineral island (a drumlin), in the middle of Nigula bog. Notice the marked forest zones of the island. A, pine; B, birch (partly bare); C, ash, elm, maple, and other deciduous hardwoods, some of which are displaying fall colors. Kite line crosses the lower right corner of photograph. Taken from Aber et al. (2002, fig. 4).

magnitude more detailed than conventional airphotos. The Estonian examples demonstrate remarkably intricate and complex spatial patterns and depict abundant open water (pools) developed within bogs at the microstructural level (see Fig. 5-21). On this basis, the areas shown by vertical SFAP (~1 ha) could be used as training sites for improved interpretation and classification of land cover depicted in conventional airphotos and satellite images.

A direct comparison of color-visible and color-infrared images favors the visible portion of the spectrum for revealing overall variations and details for all types of land cover in bogs. The natural colors and relative ease of interpretation are advantages for color-visible imagery in both vertical and oblique views. Photosynthetically active green plants strongly absorb red (0.6–0.7 μm) light and strongly reflect near-infrared (NIR, 0.7–1.0 μm) energy (Colwell, 1974; Tucker, 1979), which is shown vividly in color-infrared photographs. Moss (*Sphagnum* sp.), however, has a considerably lower NIR reflectivity, in general, compared to trees and grass. The seasonal peak of NIR reflectivity for moss occurs in late summer, whereas most deciduous trees and grass have their peaks in late spring and early summer (Peterson and Aunap, 1998). Thus, interpretation of bog vegetation takes some care in terms of spectral characteristics and seasonal conditions.

Mires are normally rather dark features in most conventional airphotos and satellite images. The reason for this is apparent from examination of the color-infrared image (Fig. 14-8). The zone of active photosynthesis is distributed in narrow strips or clumps, no more than 1–2 m wide, at the margins of pools and hollows. Recognition of these patterns and photosynthetic activity requires submeter spatial resolution. In lower-resolution imagery, however, strong near-infrared reflections from such narrow moss zones are blended with weak reflections from adjacent water, mud, and hummocks to create an average low value for each pixel in the image. These results suggest that while plant activity is low overall in late summer, bogs contain narrow zones within and around pools that support a high level of photosynthesis. This finding may have significant implications for accumulation of peat biomass, growth of bog microtopography, methane flux, and related environmental factors.

14.3. PRAIRIE MARSHES AND PLAYAS, KANSAS

Kansas is located in the Great Plains in the geographic center of the coterminous United States. The predominant natural land cover is prairie grass, which is mostly replaced today by agricultural cropland and rangeland. The state experiences a strongly continental climatic regime; the eastern portion enjoys a general water surplus, whereas the western part lies in the rain shadow of the southern Rocky Mountains and is relatively dry. Temperature undergoes a large seasonal range from below −20 °C to above 40 °C, and droughts and floods are recurring events.

FIGURE 14-13 (A) Quivira National Wildlife Refuge, a Ramsar site. Salty marshes and pools in sand hills terrain of Stafford County, central Kansas, May 2008. (B) Playa amid wheat fields in a rare wet mode following heavy winter snow and spring rain. Scott County, west-central Kansas, May 2007. Kite aerial photographs by SWA and JSA.

Given the highly variable conditions imposed by continental climate, Kansas wetland environments are in a constant state of flux dictated primarily by short- and long-term changes in available water. In spite of this stressful situation, many types of large and small wetlands are found throughout the state (Fig. 14-13). The primary advantage of SFAP in this situation is repeated photography to document ephemeral events and year-to-year variations in these ever-changing wetlands. Two sites are subjects for long-term SFAP observations, Cheyenne Bottoms and Dry Lake (Fig. 14-14).

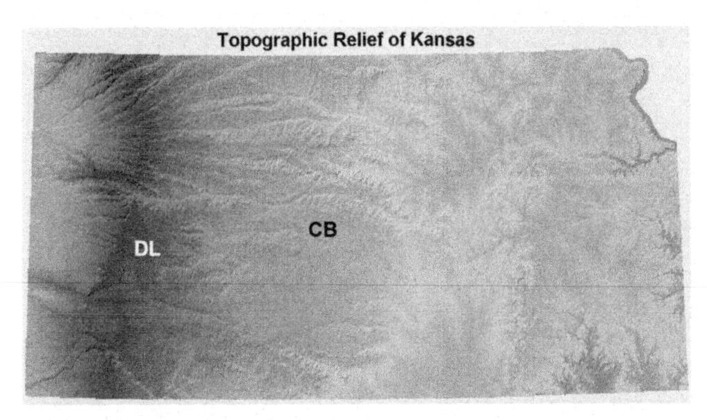

FIGURE 14-14 Generalized elevation map of Kansas showing locations of two wetland study sites, Cheyenne Bottoms (CB) and Dry Lake (DL). Elevation slopes from more than 4000 feet (1220 m) along the western edge (purple) to less than 700 feet (215 m) on the southeastern border (blue). State boundaries correspond approximately to latitudes 37° and 40°N and longitudes 94.6° and 102°W. Image derived from United States 30-second digital elevation dataset.

14.3.1. Cheyenne Bottoms

Cheyenne Bottoms is the premier wetland of Kansas. Located in the center of the state, it is considered to be among the most important wetland sites for migration of shorebirds and waterfowl in North America (Zimmerman, 1990). The bottoms occupy a large, oval-shaped depression (~166 km^2) that is the terminal point for an enclosed drainage basin (Fig. 14-15). Cheyenne Bottoms is managed in part by the Kansas Department of Wildlife and Parks and partly by The Nature Conservancy (TNC) as well as by other private landowners. Beginning in the early 1990s, TNC started to acquire land in the upstream portion of Cheyenne Bottoms, north and west of the state wildlife area (Fig. 14-16). The management goal of TNC is to protect habitat for shorebirds and waterfowl through reclamation of natural marshes, wet meadows, mudflats, and adjacent grassland. In pursuing this goal TNC has undertaken substantial alterations of the previous agricultural land use in the areas it owns and manages.

An important concern of recent years was the expansion of cattail (*Typha* sp.) thickets during the 1990s, which threatened to overspread open marshes and mudflats, thus rendering the habitat less suitable for many migratory birds. Cattail may be controlled in several ways. However, TNC lacked heavy equipment for mechanical removal, had no means to artificially regulate water levels in the marshes, and rejected herbicides in this sensitive environment. Thus, TNC adopted a patient strategy to exploit recurring drought episodes, when dead cattail thatch might be removed, as the primary means to control cattail

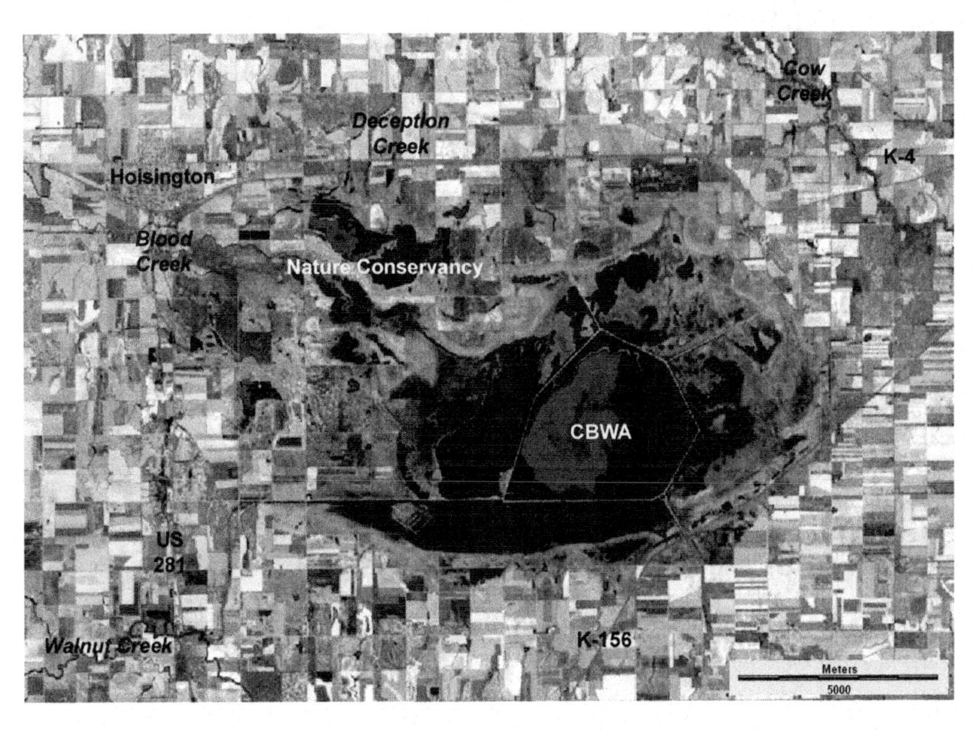

FIGURE 14-15 Satellite image of Cheyenne Bottoms and surroundings in central Kansas, United States. CBWA = Cheyenne Bottoms Wildlife Area managed by the state of Kansas. Landsat false-color composite based on Thematic Mapper bands 2, 5, and 7 color coded as blue, green, and red; 10 July 1989.

FIGURE 14-16 High-resolution, natural-color Ikonos satellite image of Nature Conservancy land in the northwestern portion of Cheyenne Bottoms, Kansas, United States. TNC marshes are subjects of long-term SFAP observations. A mixture of moist and dry mudflats, active and dormant vegetation, and residual water pools is visible in this typical mid-summer, drought scene, July 2003.

FIGURE 14-17 Healthy cattail thickets (dark green) cover much of TNC marsh study site prior to drought. View toward west, May 2002. Taken from Aber et al. (2006, fig. 4).

infestation of its wetlands. SFAP was utilized to monitor TNC marshes over a period of several years (Aber et al., 2006) and to provide improved ground truth for better interpretation of satellite images and conventional airphotos.

The spring of 2002 marked the end of the favorable period for cattail expansion, and thickets filled much of TNC marshes (Fig. 14-17). Later in 2002, a two-year drought began, and much of the emergent wetland vegetation, including most cattails, had died by the early summer of 2003 (Fig. 14-18). By the spring of 2004, the marshes were largely dry and most emergent wetland vegetation had died.

FIGURE 14-18 Partially dead cattail thickets, June 2003. (A) Normal color shows that small patches and narrow zones of cattails survived (green), particularly around the margins of thickets. Taken from Aber et al. (2006, fig. 6). (B) Color-infrared close-up view highlights active zone of emergent vegetation (red).

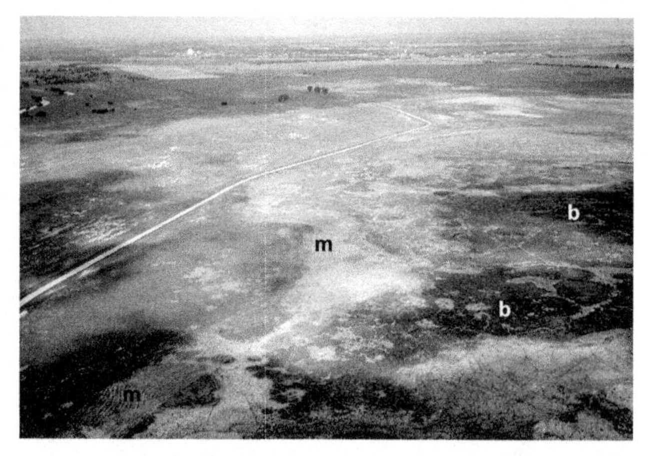

FIGURE 14-19 Dry conditions and dead cattails (reddish brown) in TNC marsh study site. Areas treated by mowing (m) and burning (b). View toward west, May 2004; compare with Figure 14-17. Taken from Aber et al. (2006, fig. 7).

At this time, TNC conducted an experiment in removal of the dead cattail thatch. Both mowing and burning were attempted with limited success (Fig. 14-19). Soon after, heavy rains and runoff flooded the marshes (Fig. 14-20), and by spring of 2005 a healthy mixture of emergent wetland vegetation had become established, including great bulrush (*Scirpus validus*), blunt spike rush (*Eleocharis obtusa*), and small stands of cattail (Fig. 14-21). At this point in time, it appeared that TNC strategy was partly successful for restoring open-water marsh and mudflat habitats attractive for migrating shorebirds and waterfowl.

FIGURE 14-20 Revegetation of former dry mudflat by blunt spike rush (*Eleocharis obtusa*), which appears dark green at scene center. Muddy water from recent flooding fills the marsh in the foreground, and cattle are grazing on wet meadow in the background, July 2004. Taken from Aber et al. (2006, fig. 8).

Later in 2005, another dry period began. This drought continued and became severe by the autumn of 2006, and TNC marshes were completely dry. Based on previous limited success, TNC staff launched a more ambitious experiment to simulate the impact of heavy buffalo (*Bison bison*) grazing on the marsh complex. Dry mudflats were disked (plowed) up and dead vegetation thatch was mowed down (Fig. 14-22).

The spring of 2007 brought heavy rain, the marshes rapidly filled to overflowing (Fig. 14-23), and rains continued well into the summer leading to flooding of historic magnitude. Cheyenne Bottoms was inundated and became a huge, shallow lake for several months. By 2008, the bottoms once again became accessible via roads, although water levels remained high (Fig. 14-24).

As water receded and emergent wetland vegetation grew up, it became apparent that marsh treatments (disking and mowing) in the fall of 2006 had been successful for limiting cattail infestation. TNC staff found that periodic SFAP was the best means available to them for evaluating the impacts of their management techniques. However, continued maintenance will be necessary in future years, as cattails are remarkably hardy plants.

FIGURE 14-21 Mosaic of emergent wetland vegetation types with extensive open water in Nature Conservancy marsh. Superwide-angle view, May 2005. Kite flyers are visible in lower right corner. Taken from Aber et al. (2006, fig. 9).

FIGURE 14-22 Severe drought conditions culminated with completely dry marshes in October of 2006. Main image: A, vegetation thatch is being mowed, and B, mudflats are disked (plowed). View northward; kite aerial photo by SWA and JSA. Inset image: detail of the tractor mowing the thatch.

FIGURE 14-23 Waxing flood conditions at TNC marshes in the spring of 2007 following disking and mowing of previous autumn. Compare with picture above. View northward; kite aerial photo by SWA and JSA, May 2007.

FIGURE 14-24 Waning flood conditions at TNC marshes in 2008 following several months of inundation. (A) June, and (B) September; compare with pictures above. Views northward; kite aerial photos by JSA and L. Buster.

14.3.2. Dry Lake

Dry Lake is an ephemeral water body at the terminal point of an enclosed drainage basin on the High Plains of west-central Kansas (Fig. 14-25). This region is semi-arid, typically receiving less than 50 cm of precipitation per year. At least 2000 playa basins are found in the High Plains of western Kansas (Steiert and Meinzer, 1995). These shallow depressions are dry more often than not, but may contain water following heavy rains or snow melts. Many opinions have been advanced concerning the origin of playa

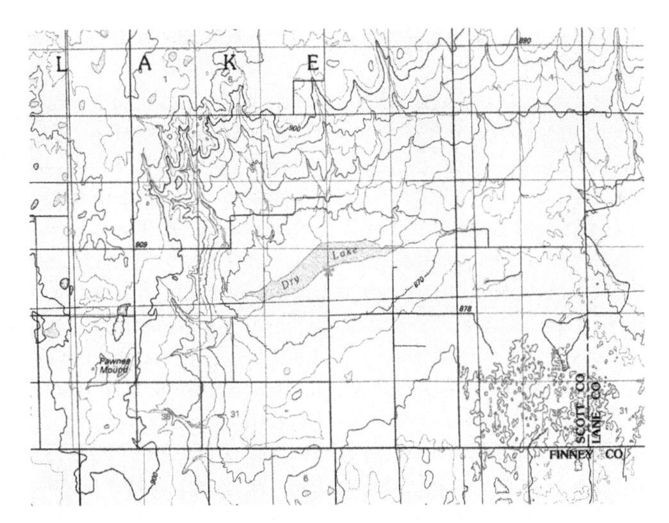

FIGURE 14-25 Topographic map of Dry Lake vicinity in Scott County, west-central Kansas. Elevations in meters; contour interval = 5 m. Each grid square represents one square mile (~2.6 km^2). Location for KAP is given by green asterisk. Map derived from Scott City, Kansas, 1:100,000 scale, U.S. Geological Survey (1985).

depressions, ranging from subsurface solution to wind erosion or buffalo wallows.

Dry Lake is ~4 km long and about half a kilometer wide, when full, which is a rare occurrence. In contrast to Cheyenne Bottoms, which has attracted a great deal of scientific attention, little is known about the geological circumstances of Dry Lake or its wetland environment.

Preliminary KAP at Dry Lake suggests extremely variable environmental conditions from year to year. In 2007, for example, heavy winter snow and spring rain combined to produce significant runoff that filled the lake (Fig. 14-26A). One year later, in contrast, most of the surface consisted of saturated mudflats with a salt crust around the margin and wet surface toward the middle of the basin (Fig. 14-26B). Only a shallow puddle of water remained on the former lake floor in the spring of 2008, and strandlines of receding water are clearly visible along the margins of the lake basin. By the spring of 2009, however, the lake once again held considerable water (Fig. 14-26C), although not as much as in 2007.

14.3.3. Discussion

Wetlands in the central and western portions of Kansas clearly are subject to large seasonal and interannual changes in their characteristics owing to climatic variations. According to the Nature Conservancy land steward, SFAP is the most effective type of remote sensing for management applications at Cheyenne Bottoms, superior to conventional aerial photography or satellite imagery. This assessment is

FIGURE 14-26 Dry Lake looking toward the northeast from slightly different vantage points. (A) Exceptionally wet year, May 2007. Lake full of muddy water. (B) Dry year, May 2008. A broad salt crust surrounds a wet mudflat in the center of basin. (C) Wet year, May 2009. Lake with water and a narrow mud/salt flat around the edge. Kite aerial photos by JSA and SWA.

based on high spatial resolution and frequent temporal coverage that depict details of land management activities and their consequences for habitat conditions. Rapid response, relative ease of use, and low cost make SFAP an attractive tool for monitoring dynamic environments at Cheyenne Bottoms, Dry Lake, and other wetland sites in the Great Plains region.

14.4. SUMMARY

Wetlands cover substantial portions of the world's land and shore regions and represent significant components of the Earth's environment. Thus, wetlands attract considerable public interest and scientific attention, for which small-format aerial photography (SFAP) is well suited to provide low-cost imagery with high spatial, temporal, and spectral resolutions. Equipment is lightweight, compact, and easily transported into wetland environments, and silent operation of some platforms—kites, blimps, balloons—does not disturb sensitive wildlife. SFAP is a means to overview restricted wetland preserves that may be closed to the public (Fig. 14-27).

SFAP can be taken in all possible orientations and in all directions relative to the ground target and sun position. This gives the capability to acquire images quite different from conventional airphotos, and so increases the potential for recognizing particular ground-cover conditions. The combination of high spatial resolution and frequent repeated photography provides a means for highly focused investigations of specific sites in various types of wetland environments.

Masing (1998) envisioned a multilevel approach in mire research and mapping that ranges in scale from 1:10 (most detailed) to 1:10,000,000 (most generalized). Conventional airphotos and satellite imagery span the scale range 1:1000–1:100,000 and smaller. SFAP fills the scale range 1:100-1:1000 and, thus, bridges the gap between ground surveys and traditional remote sensing. This level of scale and resolution is best suited for permanent wetland study plots and control sites. SFAP represents one level of data acquisition in a multistage approach that includes ground observations, conventional airphotos, and satellite images.

FIGURE 14-27 Panoramic view over Laudholm beach and the mouth of the Little River at the Wells National Estuarine Research Reserve on the Atlantic coast of southeastern Maine, United States. Aside from the beach, a path through the forest (lower left), and a small observation platform (*) this wetland environment is closed to the public. Two helium-blimp aerial photographs stitched together for this superwide-angle view; JSA, SWA, and V. Valentine, August 2009.

Chapter 15

Biocontrol of Salt Cedar

15.1. SALT CEDAR PROBLEM

Salt cedar is a shrub or small tree that is native to Asia, the southern Mediterranean, and northern Africa. It comprises several species within the genus *Tamarix*, commonly called tamarisks, which were introduced in the United States beginning in 1823 (DeLoach, 2004). During the Dust Bowl of the 1930s, salt cedar was planted widely for windbreaks and to control stream erosion in the Great Plains and western United States. Since then, however, salt cedar has spread rapidly becoming an invasive plant throughout arid and semi-arid portions of the western United States (Fig. 15-1) and northern Mexico, where it is responsible for a major ecological disaster (Deloach et al., 2007).

FIGURE 15-1 Salt cedar (*Tamarix*) growing beside an irrigation canal near Fallon, Nevada. Salt cedar is a large bush or small tree, up to 4 m tall, with attractive pink flowers, often used as an ornamental garden tree. Salt cedar has become an invasive plant that grows in dense thickets along streams, rivers, and wetlands in the western United States. Taken from Aber et al. (2005, fig. 1).

Salt cedars form dense thickets along waterways; these thickets crowd out native vegetation, degrade wildlife habitat, harm some 50 endangered or threatened native species, cause increases in wildfires and soil salinity, and impede recreational use of parks and natural areas. In addition to its disruption of native ecosystems, salt cedar is also a phreatophyte that consumes large amounts of ground water—a valuable resource in the western United States (Zavalata, 2000).

Most conventional methods for controlling salt cedar infestations are generally ineffective. Once established, *Tamarix* is extremely tenacious and difficult to eradicate through mechanical (cutting) or chemical (herbicide) means. Because of this growing problem, the U.S. Bureau of Reclamation (USBR) is participating with several other governmental agencies in research on the biocontrol of salt cedar at one of several approved field research sites in the Saltcedar Biological Control Consortium (DeLoach and Gould, 1998).

A leaf-eating beetle, *Diorhabda elongata*, imported from Eurasia, is a natural enemy of salt cedar (Fig. 15-2). In their larval and adult stages, *D. elongata* eat the leaves and outer layers of stem tissue of salt cedar and may defoliate the plant to such an extent that it eventually dies (Fig. 15-3). The carefully managed international biocontrol program has proceeded in several steps (DeLoach et al., 2007).

- Ten years of testing *D. elongata* under quarantine conditions in Texas and California, starting in 1986.
- Release of *D. elongata* imported from China into secure field cages at authorized research sites in 1997.
- Field release into the open within circumscribed areas at 10 approved sites in six western states in 2001.
- Additional field releases of *D. elongata* imported from Crete, Greece in two southwestern states in 2003.
- Release of Greek *D. elongata* at study sites in northeastern Mexico along the Rio Grande in 2007.

The intent eventually is to distribute *D. elongata* widely to the public for biocontrol of salt cedar, once thorough research and review of its effectiveness and safety are demonstrated. Remote sensing has been employed as part of

FIGURE 15-2 *Diorhabda elongata deserticola*, imported from northern China, on salt cedar. Third instar larva (~8 mm long). Taken from Aber et al. (2005, fig. 2).

FIGURE 15-4 Overview of U.S. Bureau of Reclamation salt cedar biocontrol study site near Pueblo, Colorado, United States. The study field (*) is situated on the floodplain of the Arkansas River between a fish hatchery and the river (behind). Kite aerial photo by JSA and SWA, May 2003.

FIGURE 15-3 Salt cedar largely defoliated by beetles at Pueblo, Colorado study site. This ground picture was taken near the end of the growing season in October 2003. Salt cedar bush stands about 4 m high. Taken from Aber et al. (2005, fig. 3).

this interdisciplinary research, primarily for identifying stands of salt cedar and evaluating the effectiveness of the biocontrol efforts (e.g., Everitt et al., 2006; Pu et al., 2008). A variety of methods has been applied including standard color photography, videography, hyperspectral imaging, and vegetation indices—data and imagery acquired from conventional manned aircraft at medium altitudes.

15.2. USBR STUDY SITE

Field release of Chinese *D. elongata* took place at the USBR study site near Pueblo, Colorado in 2001 (Fig. 15-4; see also Fig. 6-2). Ongoing research at this site has focused on the biology and behavior of the beetle and the effects of the release on wildlife, salt cedar, and non-target vegetation (Eberts et al., 2003). Biocontrol insect data include information on identification, reproduction, and movements of the beetles, which undergo two life cycles during the annual growing season. Birds, butterflies, and bats are also included in the wildlife monitoring program. Vegetation study primarily has involved collection of baseline data and includes mapping of nearby *Frankenia jamesii*, which is the only local native species related to *Tamarix*.

In order to acquire low-height, high-resolution imagery, kite aerial photography (KAP) was conducted twice during the 2003 growing season in May and August (Aber et al., 2005). Owing to the relatively high altitude (~1400 m) and thin atmosphere, large rigid delta and rokkaku kites were employed to provide sufficient lift (see Chapter 9.3.4).

All manner of photographs was taken, including vertical and oblique views, color-visible film and digital images, and color-infrared slides. Photographs were acquired from heights of 50–150 m above the study site. The bulk of images was collected in vertical position with a small digital camera operated from a radio-controlled rig. A 4-m-long survey arrow was oriented on the ground near the center of the study site to provide accurate scale and north direction for image processing.

The photographs revealed little evidence of the effects of *D. elongata* in late May, as supported by ground observations. By early August, however, beetle impact had become quite noticeable. Preliminary visual assessment of August images indicated that color-visible digital photographs provided the best display of defoliated salt cedar, which appears in a distinctive reddish-brown color (Fig. 15-5).

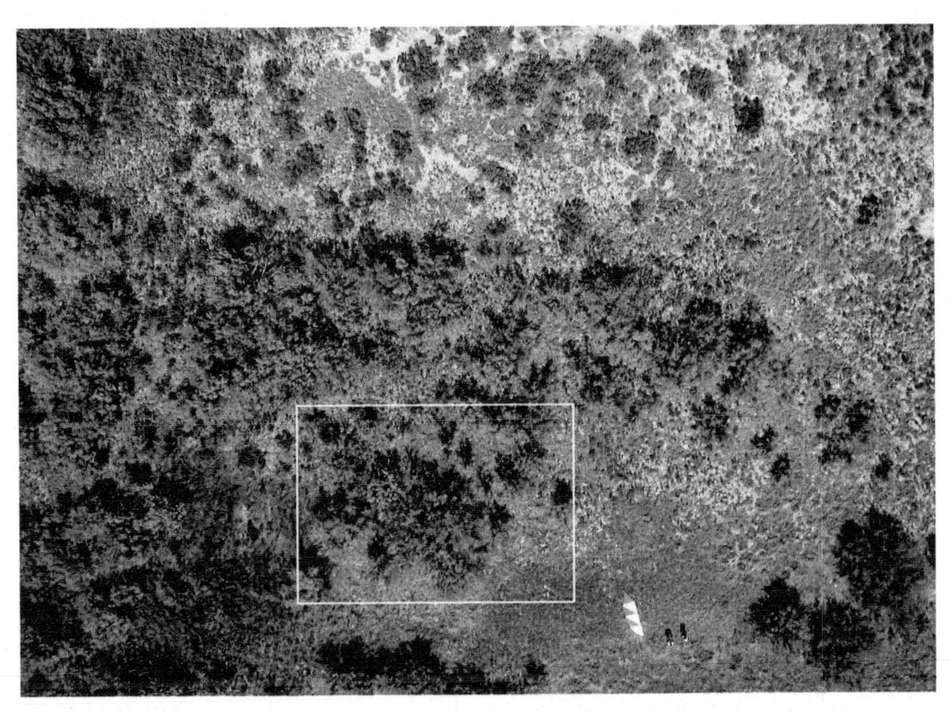

FIGURE 15-5 Near-vertical kite aerial photograph over central portion of study site, August 2003. Ground cover is a patchwork of salt cedar thickets, salt sacaton grass, rabbitbrush and assorted weeds, and bare sandy soil. Camera operators are standing next to the north arrow, which is 4 m long. The white rectangle indicates approximate position of the close-up photograph in next figure. Taken from Aber et al. (2005, fig. 4).

A close-up vertical image of a salt cedar thicket was selected for further quantitative analysis (Fig. 15-6). This image has 1.5 cm ground sample distance (*GSD*), a scale that clearly depicts microstructural (1:100 to 1:1000) details of vegetation (Masing, 1998). The selected photograph was processed using *Adobe Photoshop* and *Idrisi* image-analysis software in the following steps.

- Extract red, green, and blue color bands from original digital image.
- Separate (assign) the reddish-brown colors associated with defoliated salt cedar based on visual examination of the pixel values and setting thresholds for the reddish-brown color class.

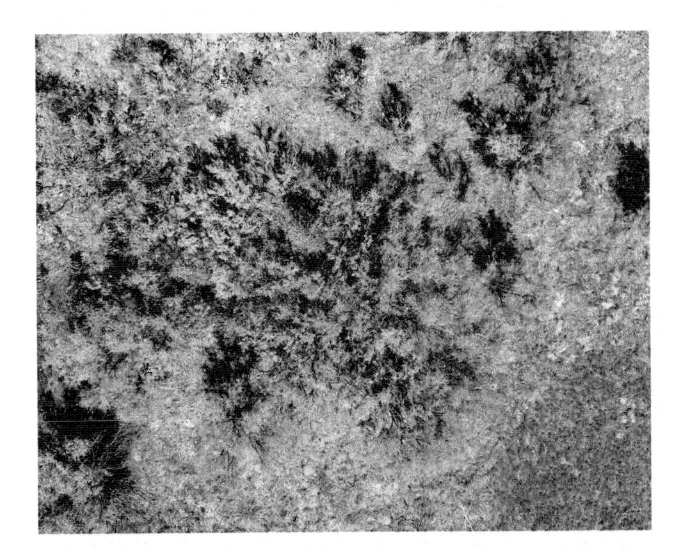

FIGURE 15-6 Close-up vertical kite aerial photograph of salt cedar thicket. Defoliated salt cedar has a distinctive reddish-brown color. Taken from Aber et al. (2005, fig. 5).

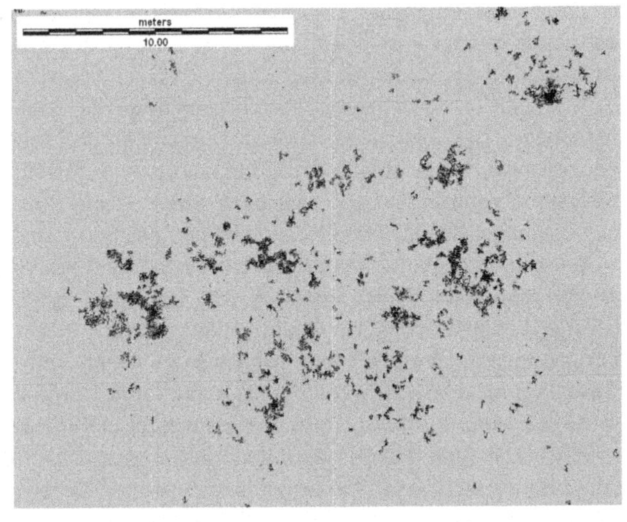

FIGURE 15-7 Classified image derived from photograph shown in previous figure. Zones of defoliated salt cedar are indicated in black, which accounts for approximately 3% of the total scene. Taken from Aber et al. (2005, fig. 6).

- Group contiguous like pixels and determine area (m^2) for each group.
- Eliminate (reclass) those groups less than 1 dm^2 in area.
- Determine area for remaining groups and create final display.

The final classified image reveals the amount of defoliation (Fig. 15-7). Each black group in the image represents the distinctive reddish-brown color associated with dead leaves of defoliated salt cedar. A minimum size threshold (1 dm^2) was set to eliminate numerous tiny groups located in prairie portions of the scene and not associated with salt cedar. On this basis, defoliated salt cedar covers approximately 3% of the image area.

15.3. ANALYSIS OF KAP RESULTS

Digital kite aerial photographs clearly revealed the impact of *D. elongata* on salt cedar at the USBR Pueblo study site. In August, defoliated salt cedar exhibited a distinctive reddish-brown color that was visually obvious and could be separated digitally for quantitative analysis. However, some caveats must be taken into account. The visible color associated with defoliated salt cedar also appeared in the open prairie vegetation around salt cedar thickets. Presumably this color indicated dead or senescent vegetation. For the most part, this color occurred as individual pixels or quite small pixel groups in the prairie zone and could be removed by setting a minimum size threshold (1 dm^2). Most of the suspect pixels were removed from the prairie portion in this manner. However, some small pixel groups may have been eliminated from the salt cedar portions also.

Salt cedar grows to a typical height of 3–4 m. *D. elongata* consume some leaves, but they kill more foliage than they actually consume. Their method of feeding disrupts water transport to non-consumed foliage and causes leaves to die and turn reddish brown. Adult beetles move about the tree, laying eggs, and larvae move also as they grow. Their feeding appears to be driven somewhat randomly following oviposition patterns. A salt cedar bush under attack, thus, may contain a 3D patchwork of healthy and dead portions.

As seen from above, healthy branches near the top may obscure defoliated lower branches. On the other hand, defoliated upper branches could allow healthy lower portions to show through and be visible from above. Thus, defoliation must be nearly continuous in the vertical profile of the salt cedar bush in order to appear clearly in vertical aerial photographs. Furthermore, dead leaves that fall off would not be included in the scene. Based on these factors, quantitative analysis of vertical kite aerial photographs was considered to provide a minimum areal estimate for the amount of defoliation in salt cedar (Aber et al., 2005). The actual volume of salt cedar defoliation could be substantially greater, which might be indicated to better advantage in oblique photographs.

The amount of impacted foliage displayed in vertical images in early August amounted to only a few percent of the study area. This low result reflects the relatively early stage of defoliation exhibited at this phase in the growing season. Within two weeks of taking airphotos, ground checking revealed substantially more defoliation, and by the end of the growing season many salt cedars had no live foliage (Fig. 15-3). KAP could be used to monitor this progression at frequent intervals during the latter part of the growing season.

Salt cedar trees defoliated by beetle attack are not dead and could recover and green up the following year. However, repeated beetle attacks over several (at least 3) years would eventually kill salt cedar. Thus, multi-year monitoring of beetle release sites would be necessary to document the long-term impact of this biocontrol effort. Such monitoring on the ground is time-consuming and labor intensive. Furthermore ground observations are limited by dense thickets of salt cedar that are difficult to penetrate for direct study. KAP offers a bird's-eye view that allows rapid acquisition of large-scale imagery over dense thickets covering several hectares.

The initial cost of equipment and software would be recouped by repeated use at multiple study sites during a period of years. For example, Pitt and Glover (1993) did a detailed cost-benefit comparison of traditional ground survey versus small-format aerial photography (SFAP) from a tethered blimp for evaluating forestry research plots in eastern Canada. They found that low-height aerial photography was not only less expensive, but also produced more objective results and provided a permanent record of seasonal vegetation conditions.

Based on this preliminary success, KAP was adopted by the USBR as a means to document biocontrol study sites. KAP also was utilized by scientists conducting similar biocontrol research at New Mexico State University. KAP proved effective for acquiring the type of low-height, high-resolution imagery necessary to monitor the effects of insect defoliation at the level of individual salt cedar trees. Relatively low cost of equipment and operation combined with flexible scheduling, convenient transportation to distant sites, and ease of use in the field make KAP a technique that could be employed widely for similar biological investigations elsewhere.

15.4. SUMMARY

Salt cedar (*Tamarix* sp.) is an invasive plant in the western United States and northern Mexico that is causing serious environmental harm for native plants and animals as well as human land use. Traditional mechanical and chemical means for control are largely ineffective. As part of a consortium of governmental agencies, the U.S. Bureau of Reclamation is field testing a Chinese beetle for biocontrol of salt cedar. Kite aerial photography was employed to obtain large-scale, high-resolution imagery of the study site in order to document the extent of beetle defoliation of salt cedar during the growing season. *GSD* of 1–2 cm was achieved in vertical airphotos, which allowed identification of defoliated salt cedar patches as small as 1 dm^2.

Quantitative analysis of vertical kite aerial photographs provided a minimum areal estimate for the amount of defoliation in salt cedar. Consideration of the 3D geometry of salt-cedar defoliation and its appearance in 2D vertical imagery demonstrates how complex such a seemingly simple problem can be; any such SFAP analysis should be taken with care. Based on this preliminary success, KAP was adopted by the USBR and other agencies for monitoring and documenting biocontrol study sites across the western United States.

Chapter 16

Vegetation and Erosion

16.1. INTRODUCTION

In his introduction to the vegetation and erosion conference session at the European Geosciences Union's General Assembly 2009, John Thornes (2008), known internationally for his research in soil erosion and desertification processes particularly in semi-arid areas (Thornes, 1990), asserted that vegetation cover long has been accepted as a key factor for controlling overland flow, runoff, and soil erosion. Since the 1940s, in fact, empirical, experimental, and modelling studies have confirmed the general relationship of vegetation cover and the intensity and extent of geomorphodynamic processes and have investigated the roles of changing land use, various agricultural crops, and grazing behavior.

Based on the work by Elwell and Stocking (1974), numerous studies recording and evaluating land degradation have implicitly assumed a "clear" relationship between vegetation and erosion (Ries, 2000). Their interdependence, which was established mainly for cropland and grassland, needs, however, to be seen more differentiated in the case of Mediterranean fallow land, abandoned fields, and shrubland (e.g., García-Ruiz et al., 1996; Molinillo et al., 1997; Ries, 2002). In these environments, the small-scale variability and heterogeneity as well as the substantial change of vegetation cover associated with land-use change and vegetation succession (Fig. 16-1) give rise to a high complexity of geomorphological processes.

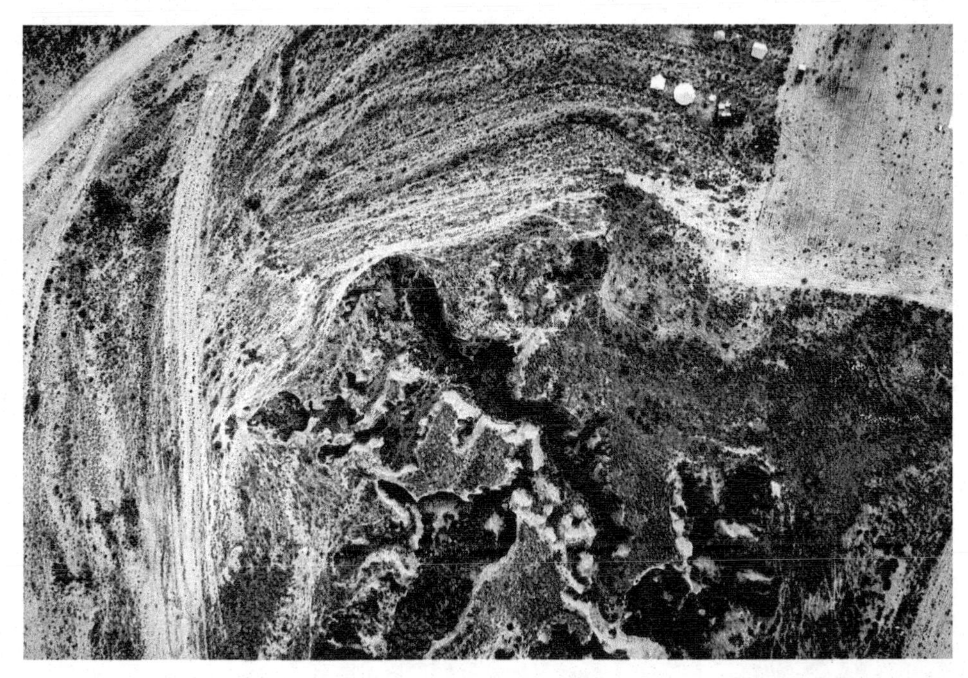

FIGURE 16-1 Abandoned fields and gully erosion near María de Huerva, Province of Zaragoza, Spain. The terraced fields in the upper part of the image were abandoned in the 1930s, while the former cereal field in the upper right corner has lain fallow for only six years. Vegetation cover is patchy to extremely low on both areas following a period of several years with precipitation below average. Drought, grazing, soil sealing, and crusting (and to a trivial extent geographers clearing their tent pitches) are among the factors that keep vegetation cover sparse. The large gully, which drains into the Val de las Lenas, is among those monitored for many years in another study by the authors (see Chapter 13). Hot-air blimp photograph by IM and JBR, April 1996.

16.2. MONITORING VEGETATION AND EROSION TEST SITES

It is this high spatial and temporal variability of vegetation, runoff, and erosion patterns which render small-format aerial photography (SFAP) an especially suitable tool for documenting and monitoring them. Hot-air blimp and kite aerial photography (KAP), among other methods, have been employed by JBR, IM, and their groups since 1995 for investigating geomorphological processes and their relationships to vegetation development on areas under extensified land use, mostly abandoned fields and set-aside land in Spain. In particular, six test areas, 24 m × 36 m in size, and their surroundings were intensively monitored in 6-month to 12-month intervals for the EPRODESERT project (Ries et al., 1998; Marzolff, 1999; Ries, 2000), yielding several hundred images at scales between 1:200 and 1:10,000 (areal coverage approximately 35 m^2 to 10 ha).

The image series taken at the test site María de Huerva 1 (MDH1, Fig. 16-2) documents the development of

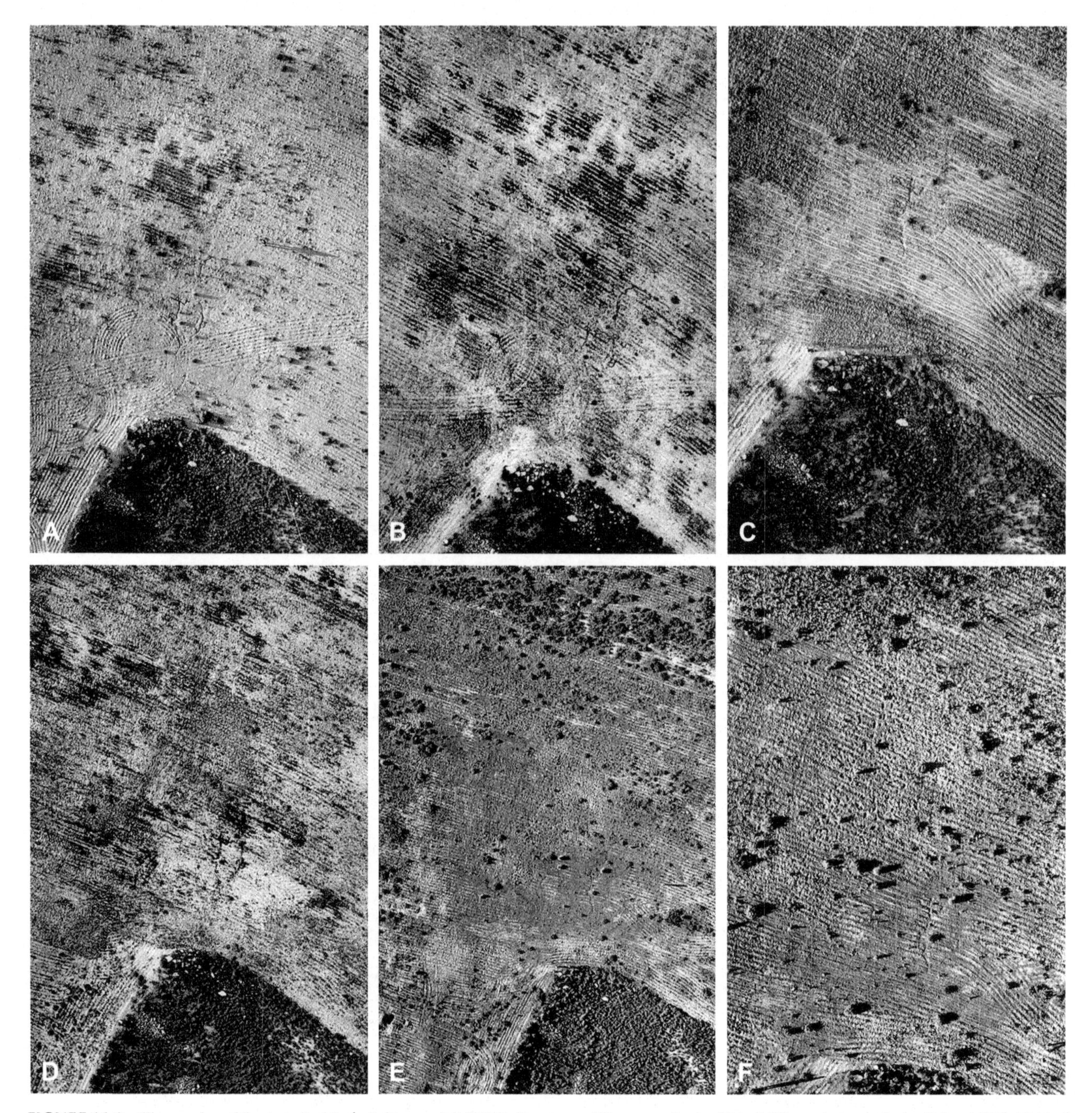

FIGURE 16-2 Time series of the test site María de Huerva 1 (MDH1), Province of Zaragoza, Spain. Hot-air blimp photographs by IM and JBR. Field of view ~40 m across. (A) October 1995. (B) April 1996. (C) August 1996. (D) April 1997. (E) April 1998. (F) September 1998.

a former cereal field in the semi-arid Inner Ebro Basin. At the time of the first image (Fig. 16-2A), this field had been set aside as fallow land for five years under the European Union's subsidized set-aside program. The extremely dry period in the preceding years—158 mm precipitation in 1995—kept vegetation cover as low as 7%. Soil sealing and crusting led to sheet and rill erosion and surface runoff coefficients up to 81% (Ries and Hirt, 2008). Clearly visible are the ridges and furrows of the last tillage operation that are dissected by a large rill system in the image center (see also Fig. 10-11).

The following images document the development of the site for the next three years: six months later in April 1996 after a wet winter (Fig. 16-2B); another five months later in August 1996, following a weed-control tillage dictated by the EU set-aside program (Fig. 16-2C); in spring 1997

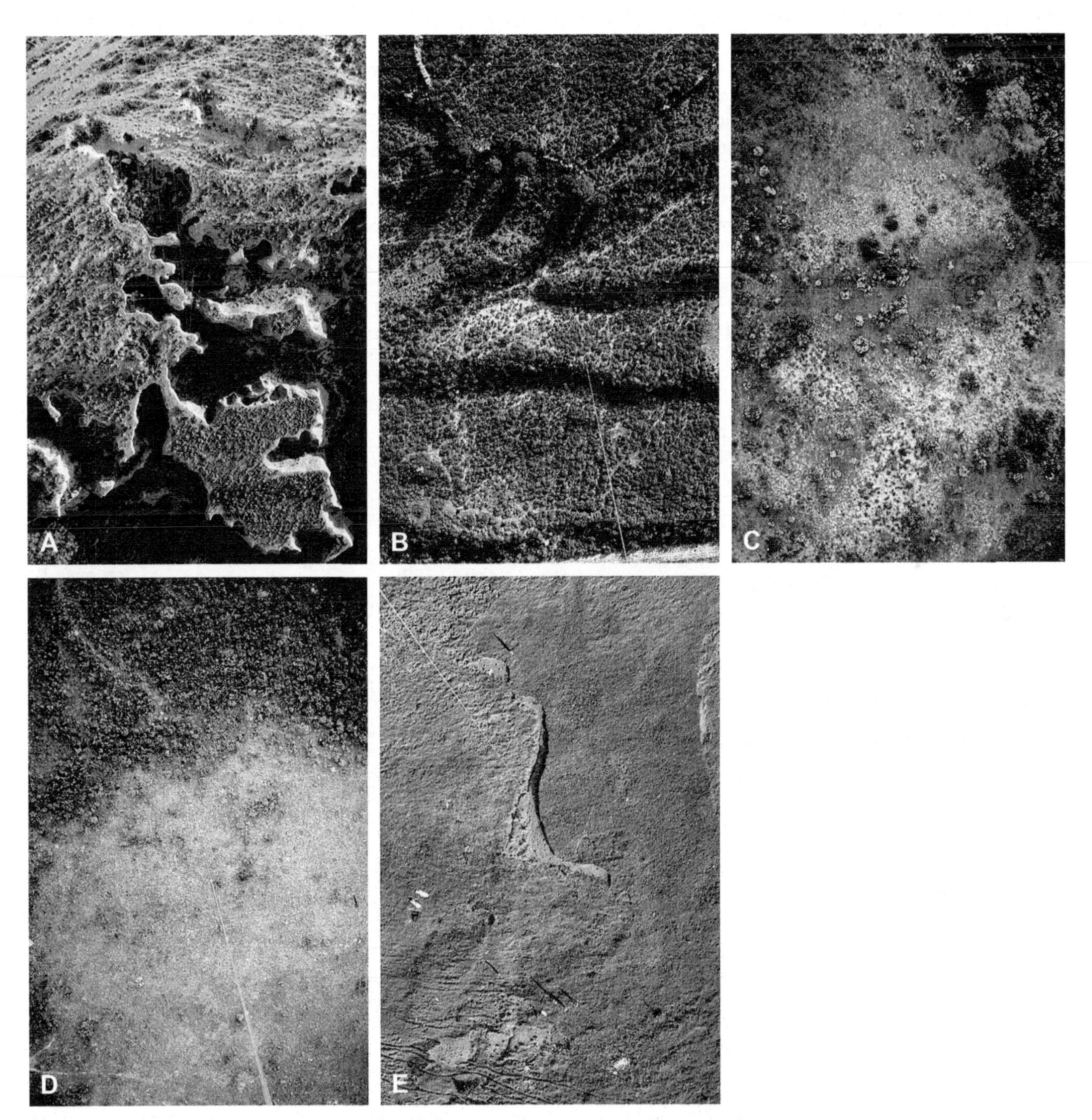

FIGURE 16-3 Further test sites monitored by the EPRODESERT project in northeastern Spain. Hot-air blimp photographs by IM and JBR. Field of view ~40–60 m across. (A) María de Huerva 2, Inner Ebro Basin. (B) Sabayés, Pre-Pyrenees. (C) Bentué de Rasal, Pre-Pyrenees. (D) Arnás, Central Pyrenees. (E) Aísa, High Pyrenees.

(Fig. 16-2D); one year later in spring 1998 (Fig. 16-2E); and in late summer 1998 (Fig. 16-2F). Note the high spatial concurrence of the images and scale that can be achieved with the hot-air blimp platform in repeated surveys. Similar time series were taken at five further test sites (Fig. 16-3).

Using image-processing and geographic information system (GIS) software for rectification and image analysis, a process-geomorphological information system was developed based on digital georeferenced test area maps with 2.5-cm resolution (Fig. 16-4; Marzolff, 1999, 2003). Visual photo interpretation combined with on-screen digitizing, digital image classification (see Fig. 11-16), and hybrid visual/digital classification methods enabled the detailed mapping of geomorphological processes, density and patterns of vegetation cover, as well as plant life forms. Change maps for vegetation and erosion were created by intersecting operations. Texture and Fourier analysis were employed to delineate automatically micromorphological structures caused by plowing (see Fig. 4-34). After calibration of the cameras, digital elevation models (DEMs) could be generated from stereoscopic images by photogrammetric analysis. The DEMs with 25-cm resolution were used for computing slope and curvature maps as well as for simulation of potential flow paths over the test areas. Examples for the resulting maps are shown for one monitoring period of the test site MDH1 in Figure 16-5.

Ten months after the soil crusts had been broken by tillage and encouraged by a wet winter, annual herbs and grasses started to recolonize the fallow land, leading to a total coverage of 41% in April 1997 (Fig. 16-5A). The favorable weather conditions continued into the following year, and vegetation cover increased further to 70% until April 1998, predominantly in those areas that had been tilled only superficially. However, the prevalent erosion processes observed on the site could not be suppressed everywhere by the increasing vegetation cover. While sheet wash was stopped or diminished in many areas, one-fifth of the site experienced an intensification of sheet wash (Fig. 16-5D), and 45% of the unchanged area is actually still subject to moderate sheet wash. The spatial change patterns of both vegetation and erosion are remarkably detailed on this site, and contradictory developments often take place in close proximity, although from a perfunctory assessment of the site on a field trip or satellite image

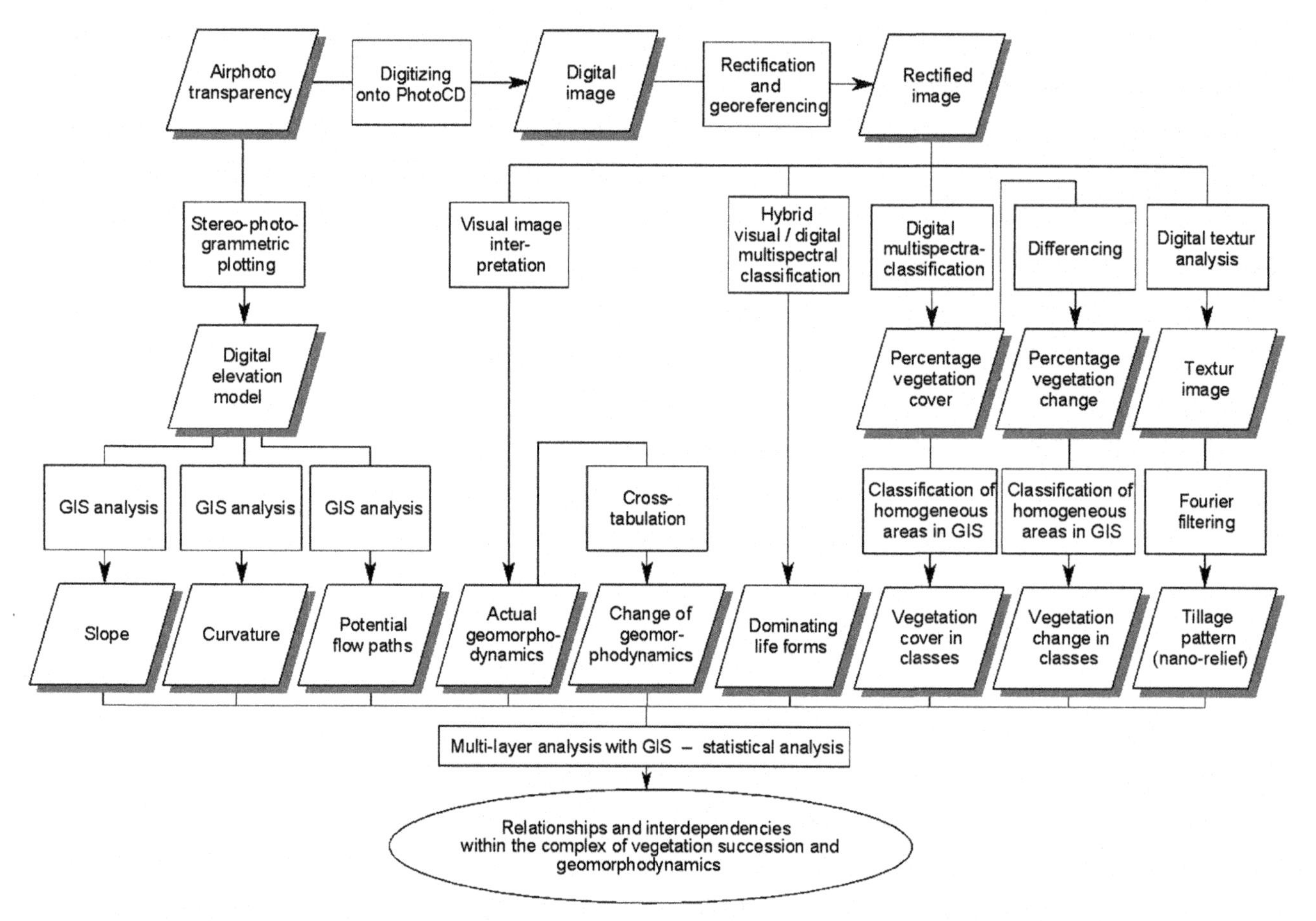

FIGURE 16-4 Flow chart with the main interpretation and processing steps involved in assembling the process-geomorphological information system for the EPRODESERT test sites. Taken from Marzolff (2003, fig. 5).

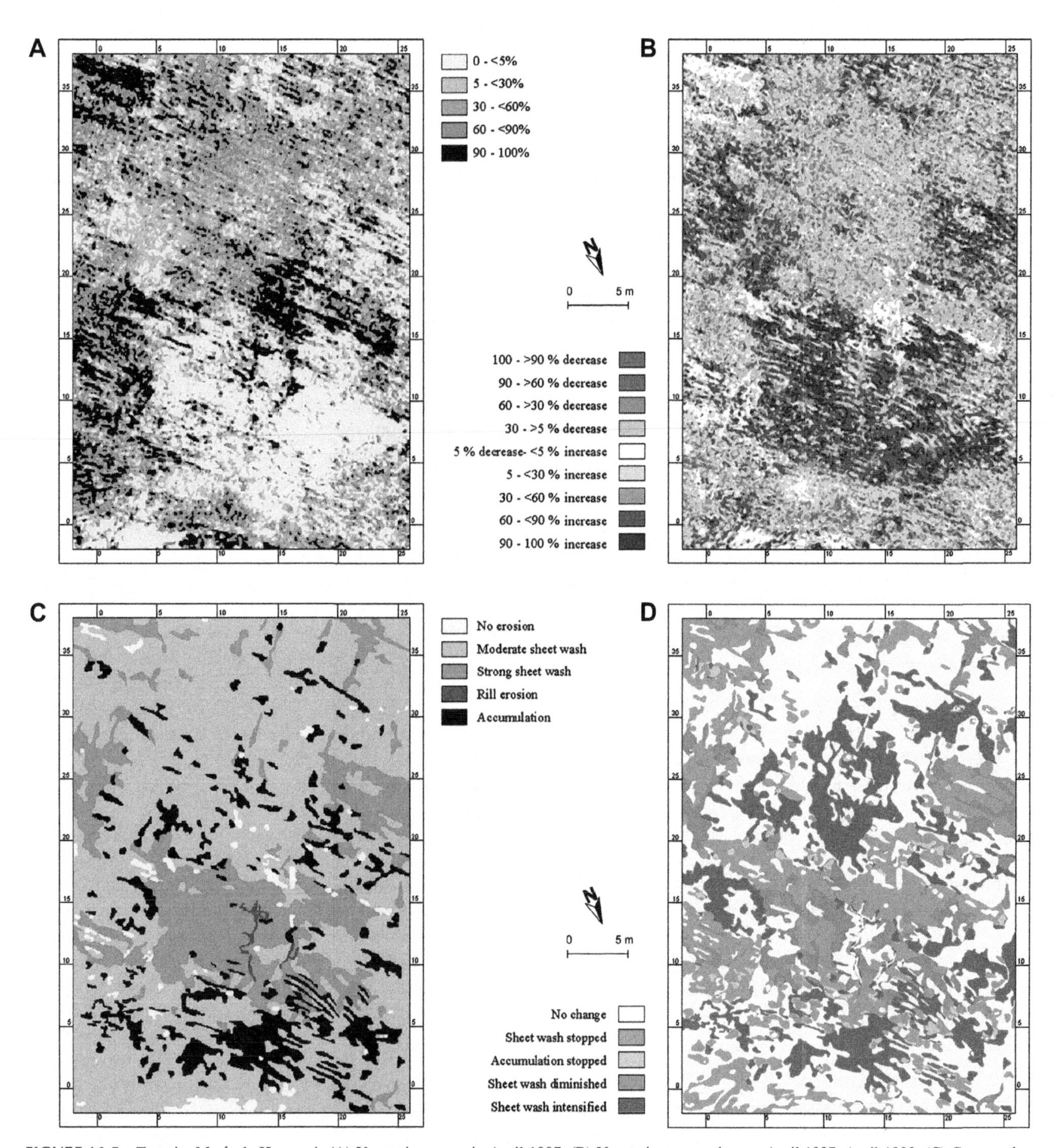

FIGURE 16-5 Test site María de Huerva 1. (A) Vegetation cover in April 1997. (B) Vegetation cover change, April 1997–April 1998. (C) Geomorphodynamics in April 1997. (D) Change of geomorphodynamics, April 1997–April 1998. Adapted from Marzolff (2003, color plates 1–4).

analysis, this would appear to be a typical rather homogeneous fallow field.

As expected, statistical analysis (predominantly by cross-tabulations and correspondence analyses, see Marzolff, 1999) showed a generally negative correlation of geomorphological process activity with vegetation cover density for all study sites. However, relationships are much more complex than anticipated and vary considerably between climatic regions and within those for different plant life forms. An influence of microrelief on geomorphological processes exists mainly for linear erosion forms. The influence of topography as an erosion-controlling factor outweighs the role of vegetation density only when vegetation cover is sparse; in this case, nanorelief (e.g. tillage pattern on set-aside field) considerably influences pattern and intensity of geomorphological processes, also.

As a most important result, it was concluded that on the observed test sites erosion processes are occurring in patterns of high spatial frequency at far higher percentages of vegetation cover than tends to be assumed by most investigations into land degradation (Marzolff, 1999). Generally, a vegetation cover of 30–40% is taken as a threshold beyond which runoff and soil erosion rates reach negligible amounts. In contrast, results of this study showed sheet erosion on fallow land with up to 70% overall vegetation cover. Looking at small-sized patterns, moderate sheet erosion can even be observed at up to 90% vegetation density, and process dynamics may during several observation periods intensify even with increasing vegetation cover.

In the semi-arid region in particular, where vegetation succession is limited by water stress, at least 60% vegetation cover is required in order to bring prevailing sheet erosion processes to a halt. A possible reason for the disparity between this threshold and the 30–40% value identified by numerous authors may be the improved assessment of percentage vegetation cover on the basis of SFAP. Terrestrial observations (field mapping) and conventional airphoto or satellite analysis tend to overestimate vegetation cover owing to the shadowing between plants and leaves.

16.3. INFLUENCE OF GRAZING ON VEGETATION COVER

Grazing by sheep and goats plays an important role in nearly all study areas investigated during the EPRODESERT project. Browsing as well as treading impede the regeneration of vegetation on many grazed areas in the Ebro Basin and the Pyrenees, and sheep trails encourage the development of erosion rills by exceptionally high runoff and erosion rates (Molinillo et al., 1997).

The Arnás catchment, located in the Upper Aragón River Basin of the Spanish Pyrenees, was cultivated totally with cereal until the middle of the twentieth century. Since its abandonment, large parts of the catchment were affected by a process of natural plant colonization with matorral composed of *Genista scorpius*, *Buxus sempervirens*, and *Rosa gr. canina* (García-Ruiz et al., 2005). The field covered by this image map (Fig. 16-6; see also Figs. 16-3D and 10-20) was abandoned in the late 1970s and subsequently used as sheep pasture. Close to the monitoring test site, a grazing control cage was installed in 1996. In the following years, plant species, vegetation cover and height were recorded in regular intervals both within the cage and on the neighboring reference sites (Ries et al., 2003, 2004). The test site as well as the grazing control sites were monitored with aerial photography, and vegetation cover maps were prepared for all sites. The small image ground sample distance (*GSD*), the presence of the cage grating and the high degree of shadowing between *G. scorpius* shrubs prevented the use of fully automatic image classification methods (see Chapter 11.5), and a hybrid method of multispectral thresholding and manual mapping had to be employed for the maps of the control sites (Fig. 16-7).

Results show that the exclusion of sheep from the area is able to boost vegetation cover significantly. While the reference-site cover is stable between 50% and 54%, vegetation cover within the cage increased from 60% to 92% between July 1996 and August 1998. The development of individual vegetation cover classes is equally interesting: 53% (cage) and 74% (reference site) of the areas initially show <60% cover by grass and herbs—the value shown to be an important threshold for reducing erosion processes from the aerial test-site monitoring. The percentage of these classes fell dramatically to 4% in the cage and at the same time increased slightly to 80% on the grazed reference site. At the time of writing, the control cage was completely filled with dense shrub growing out between the grating (Seeger 8/2009, pers. comm.).

Transferring the recovering rate of the vegetation from the control cage to the nearby monitoring test site showed that exclusion of sheep from the catchment would lead to nearly closed vegetation cover within only two years (Fig. 16-8; Ries et al., 2000). Interception by the protecting shrub canopy of *G. scorpius* matorral would reduce erosion rates even on the bare areas of the old trails beneath, where vegetation regeneration can be expected to be slower. In this study, high-resolution SFAP was able to capture patterns of vegetation exactly on the scale level where sheep have short-term influence on plant distribution.

16.4. COMBATING DESERTIFICATION AND SOIL DEGRADATION

The protecting role of vegetation is the base for many erosion control, soil conservation, and rangeland management measures. Vegetation cover reduces splash erosion due

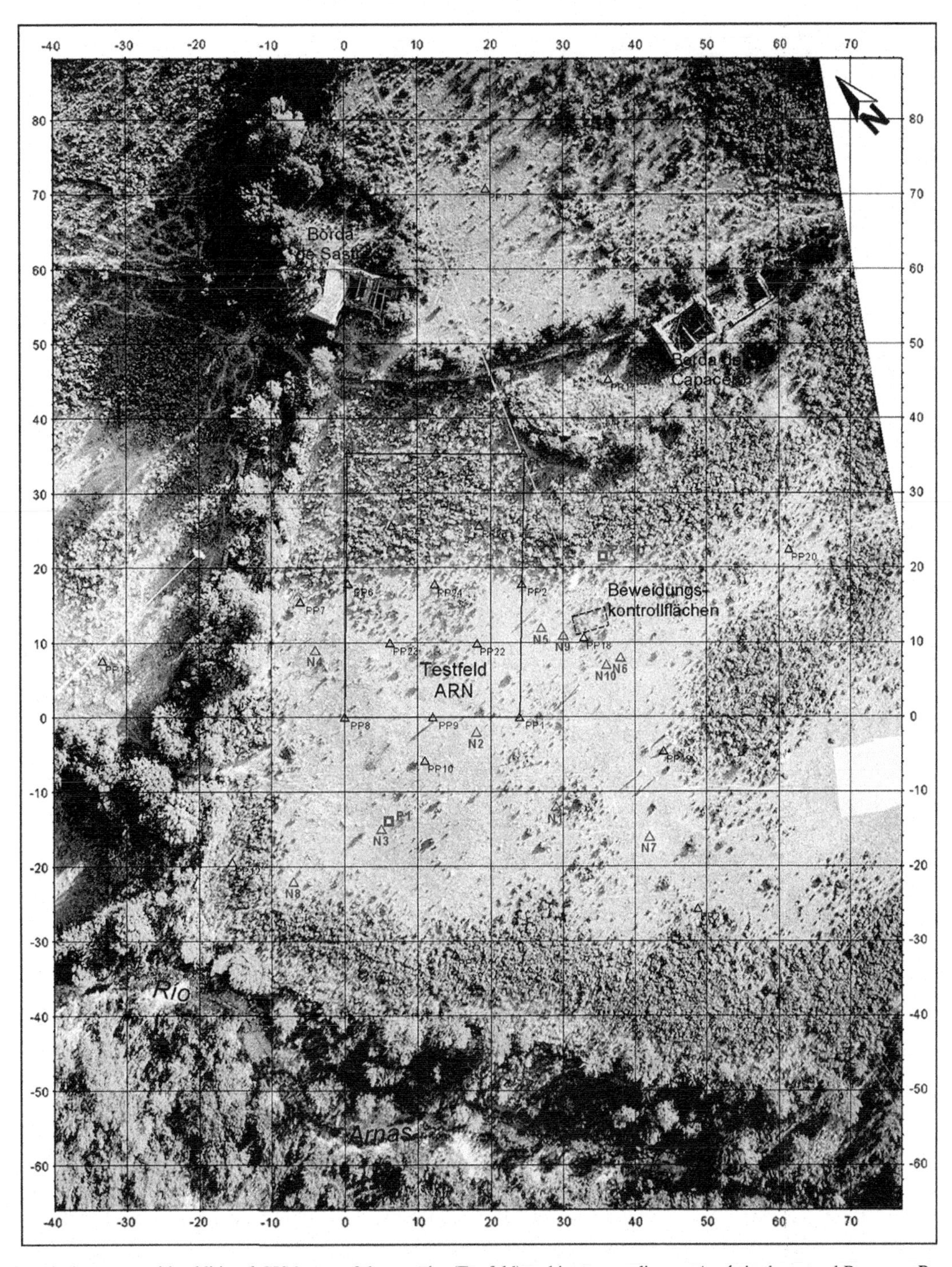

FIGURE 16-6 Airphoto map with additional GIS layers of the test site (Testfeld) and its surroundings at Arnás in the central Pyrenees, Province of Jaca, Spain. The grazing control site (Beweidungskontrollflächen) is to the right of the image center. Blue triangles indicate rainfall simulation microplots, brown squares soil profile pits. Hot-air blimp photograph taken by IM, JBR, and M. Seeger, August 1998. Image processing by IM; taken from Ries et al. (2000, fig. 3).

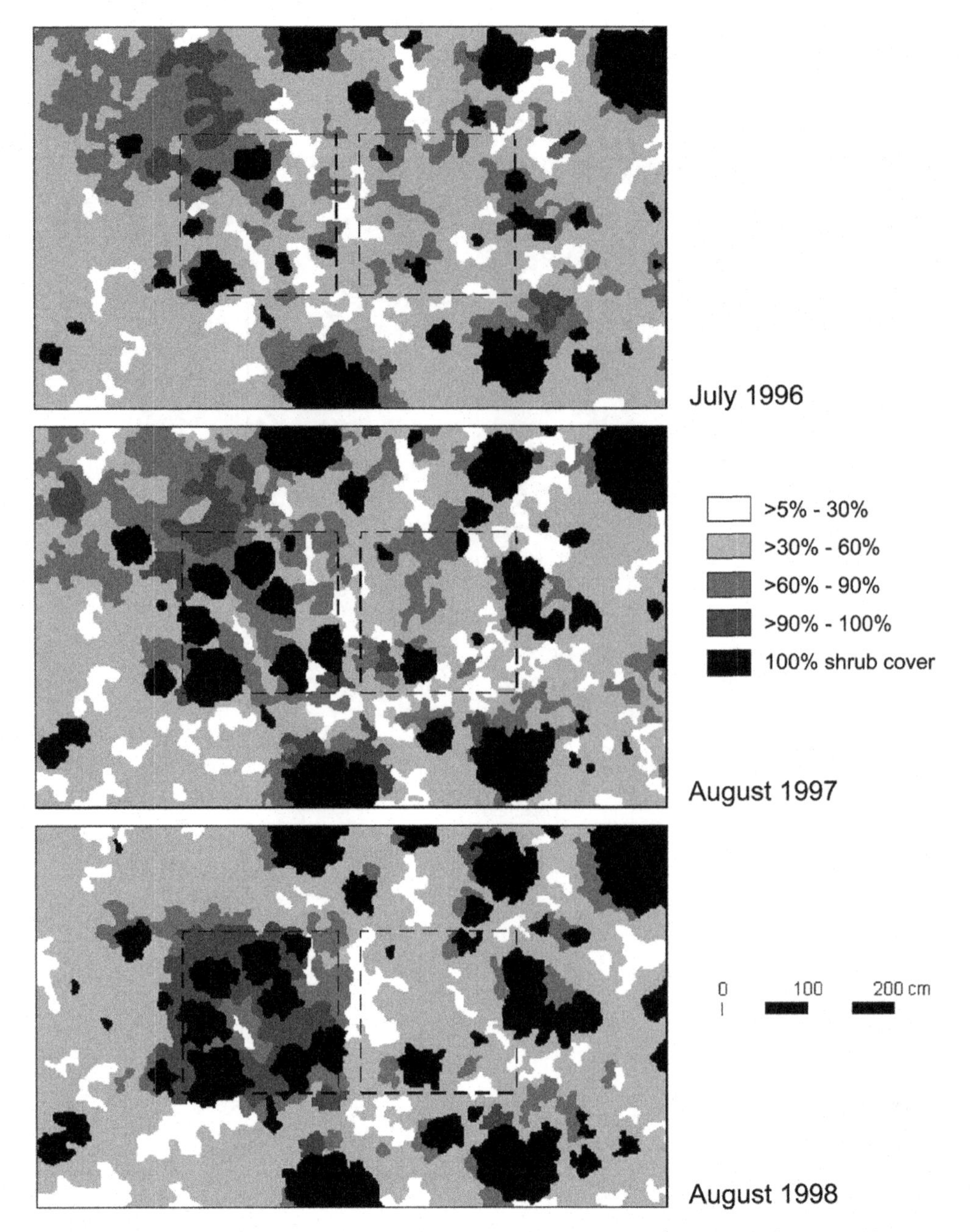

FIGURE 16-7 Vegetation cover of the grazing control cage (left) and reference site (right) at Arnás in July 1996, August 1997, and August 1998. White to dark gray are grass and herbs cover classes, black is full shrub cover. Image analysis by IM; adapted from Ries et al. (2000, fig. 10).

to interception of rainfall, decreases overland flow, and improves infiltration of precipitation and runoff water into the soil. One example for such vegetation regeneration measures was documented by SFAP on the wide and flat glacis areas of northern Burkina Faso (see Fig. 5-19), where the international non-governmental organization ADRA conducted tillage experiments in 2000 and 2001. A specially developed plow was used for carving deep furrows into the hard and bare clayey surface in order to reduce overland flow and encourage infiltration (Fig. 16-9). On the ridges piled up next to the furrows, various tree species were planted. It was expected that grasses would come up both in the furrows and the interlaced glacis strips, further increasing the erosion-control effects.

Experimental measurements of runoff and erosion rates on sites treated two rainy seasons apart could show that both factors were significantly reduced on earlier treated areas when compared to only recently treated

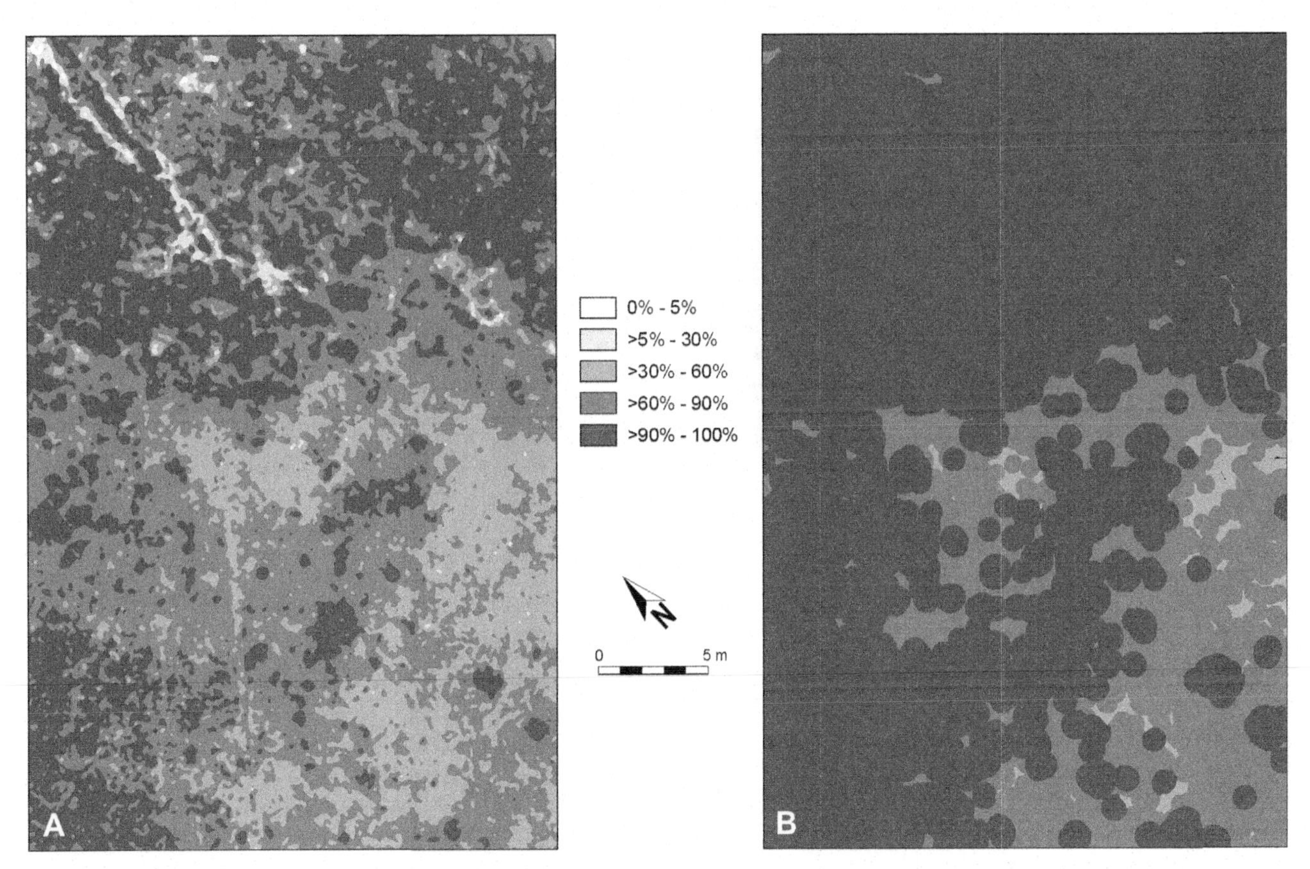

FIGURE 16-8 Vegetation cover at the test site Arnás. (A) Actual cover in July 1996, classification based on hot-air blimp photograph. Note the clearly visible open sheep trails between the shrubs where the sheep enter the test site through a wall opening in the upper left. (B) Simulated vegetation cover in summer 1998, assuming that grazing was discontinued two years previously. GIS analysis based on the results shown in Figure 16-7. Image processing by IM; adapted from Ries et al. (2000, figs. 7 and 11).

FIGURE 16-9 Tillage experiments for soil conservation near Gorom-Gorom, Province of Oudalan, Burkina Faso. Photo by JBR, July 2000.

sites, proving the short-term effectiveness of the measures. KAP was taken to document the vegetation development and assess its spatial extent (Fig. 16-10). The increase of grass cover, here shown in the dry season of December 2001, is stunning. In view of improving the resilience of this vulnerable area to climatic change, livestock pressure, and increasing food security, the experiments are promising. However, the success of the measures also depends on the future management of the improved sites. For a prevailing regeneration of the barren glacis areas and sustainable use as rangeland, grazing management strategies are of vital importance in a region where the number of cattle is directly correlated to fodder supply.

FIGURE 16-10 Tillage experiment sites seen from the air. (A) July 2000. (B) December 2001. Kite aerial photographs taken by IM, JBR, and K.-D. Albert.

16.5. SUMMARY

Small-format aerial photography offers great potential for analyzing vegetation patterns such as those examples presented here. Documenting and monitoring plant communities with heterogeneous distribution, for example the typically banded patterns of the low open shrubland of African tiger bush (*Brousse tigrée*), requires the use of corresponding image scales and resolutions that are most adequately provided by low-altitude photographs. Changes of land use and climate and their impact on erosion processes may be the greatest challenge that dryland research has to face in the next two decades. Improving the understanding of the relationships between vegetation and erosion on different scale levels continues to be an important task for which SFAP is able to make remarkable contributions.

Chapter 17

Soil Mapping and Soil Degradation

17.1. INTRODUCTION

Compiling soil maps is an essential objective of classic soil science. Together with relief maps or digital elevation models and land-use maps, they form the basis for many questions and problems in physical geography, e.g., concerning water balance or soil erosion, and have many other practical and scientific applications.

In central Europe, conventional aerial photographs are of limited use only for delineating soil units, because the fields are densely covered with vegetation or crop residue from May to September, the months in which most of the aerial photography surveys are carried out (Fig. 17-1). The soil surface is not visible during this period. Accordingly, soil mapping is done by fieldwork using soil profiles and hand augers, usually in autumn, winter (if the soil is not frozen), and early spring.

Such soil inventories with sampling by drilling and digging are quite time-consuming, so the distance between two sample points is rarely less than 20 m. Instead, the delineation of the soil units is usually aided by the interpretation of local topography, such as changes in slope, escarpments, and breaks of profile that may occur, for example, between hillslope and floodplain, between convex slope shoulder and concave slope foot, or associated with vegetation and land-use changes. Soil samples are taken in the center of homogeneous-looking areas as well as along the assumed soil map-unit boundaries. This method usually brings coherent results but may lead in some cases to circular reasoning in subsequent geomorphologic/soil-scientific studies, when the actual distribution and explanation of soil units in maps are analyzed on the basis of the topography and the geomorphologic process knowledge.

This might result in causal connections such as "summit position on the topographic map equates to intense erosion that results in a truncated soil profile." Soil-unit boundaries on a soil map would in most cases be verified by a digital

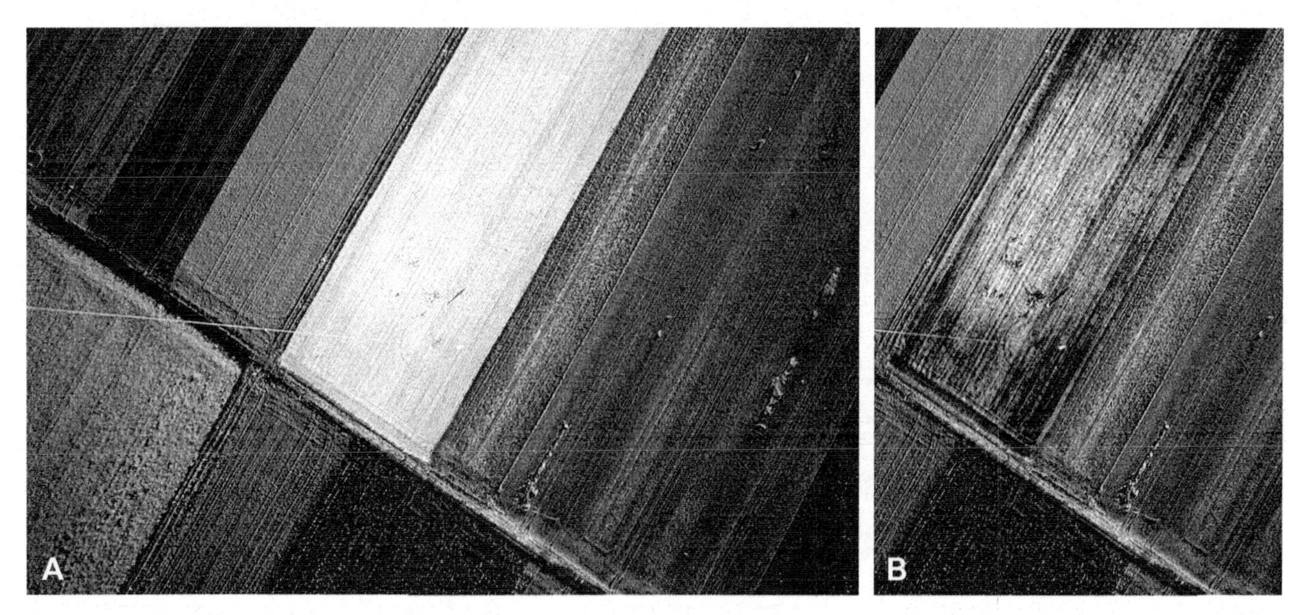

FIGURE 17-1 Agricultural landscape in the Wetterau (Hesse), Germany, in May. Nearly all fields are covered with crops and the soil surface is visible on solitary fallow fields only. As the automatically controlled image exposure is balanced for the complete scene (A), fine differences in soil colors are even less discernible; strong histogram adjustments (B) may help to enhance them. Hot-air blimp photography taken by JBR and A. Fengler, May 2000.

terrain model or contour map simply because of the intrinsical geomorphology background that the soil scientist relied on while compiling the map. Thus, two basic aspects have to be considered.

- Conventional soil mapping is usually based on expert knowledge from closely related disciplines. The map-unit boundaries are strongly correlated with changes in topography, vegetation, or land use.
- On areas that have been subject to land consolidation, which is particularly widespread in central and eastern Europe, this soil mapping technique is of limited use nowadays because such auxiliary topographic or structural features have been levelled out or cleared.

Recent studies on the current intensity and historic dimension of on-site soil erosion show highly differentiated results and sometimes surprising distributions of soil units.

17.2. SOILS AND LONG-TERM HUMAN LAND USE

In the periphery of a Neolithic settlement in the loess landscape in the Wetterau in Hesse, Germany, Fengler (2007) and JBR combined small-format aerial photography (SFAP) surveys with densely sampled soil mapping in order to document highly detailed patterns of soil units. The results were quite contradictory to initial expectations. Colluvium was found in ridge and other slope positions where geomorphologic process knowledge actually would rule out its existence. This soil distribution results from several millennia of human land use in this terrain that brought about artificial changes in relief that influenced erosion conditions at different times and under different land uses.

Soil units could be identified, interpreted, and delineated-on the basis of soil surface colors captured by aerial photographs that were taken between February and April as

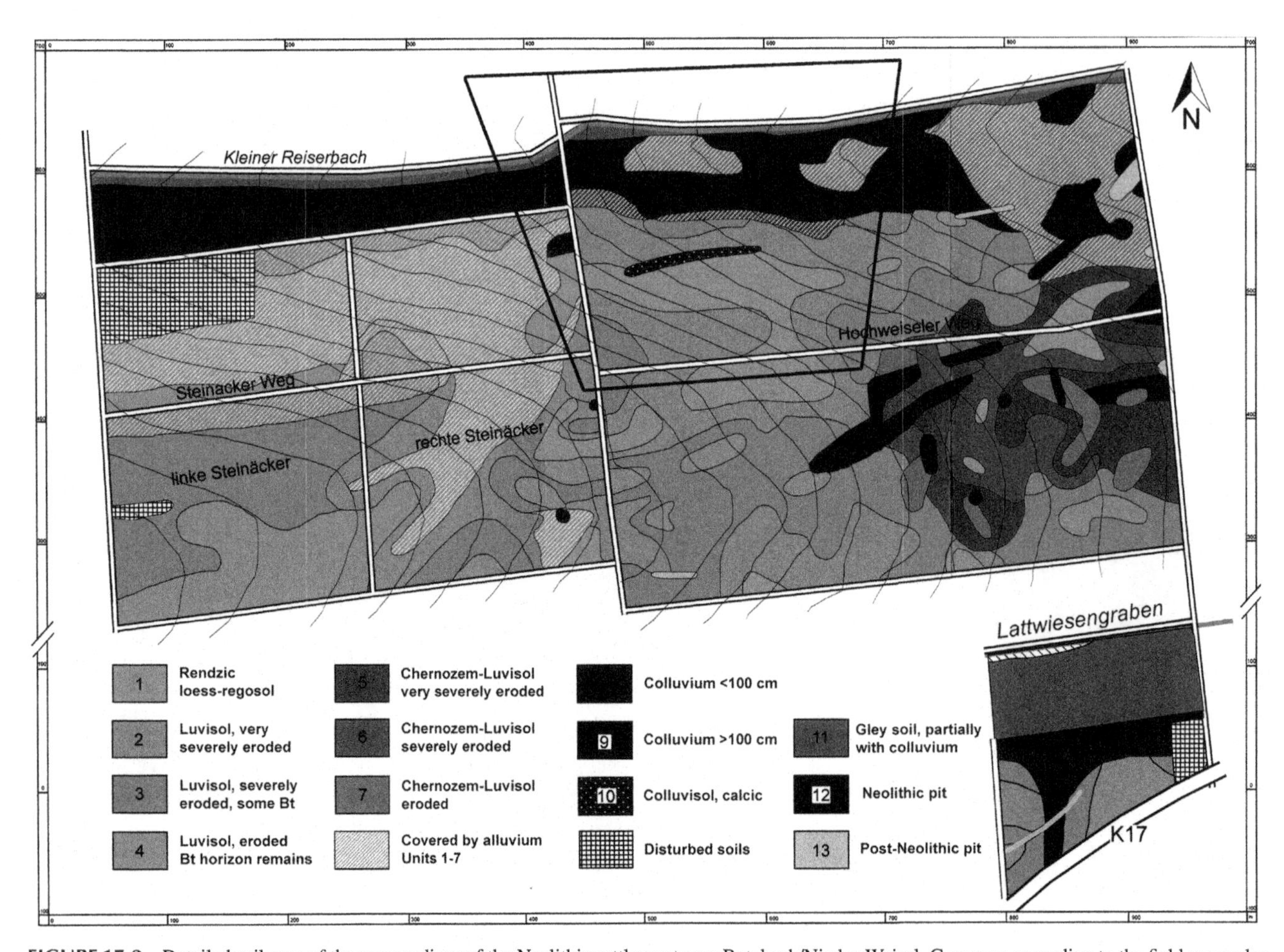

FIGURE 17-2 Detailed soil map of the surroundings of the Neolithic settlement near Butzbach/Nieder-Weisel, Germany, according to the field survey by Th. Hock and St. Mohr. Area in black frame corresponds to Figure 17-3. Adapted from Fengler (2007, Annex 1).

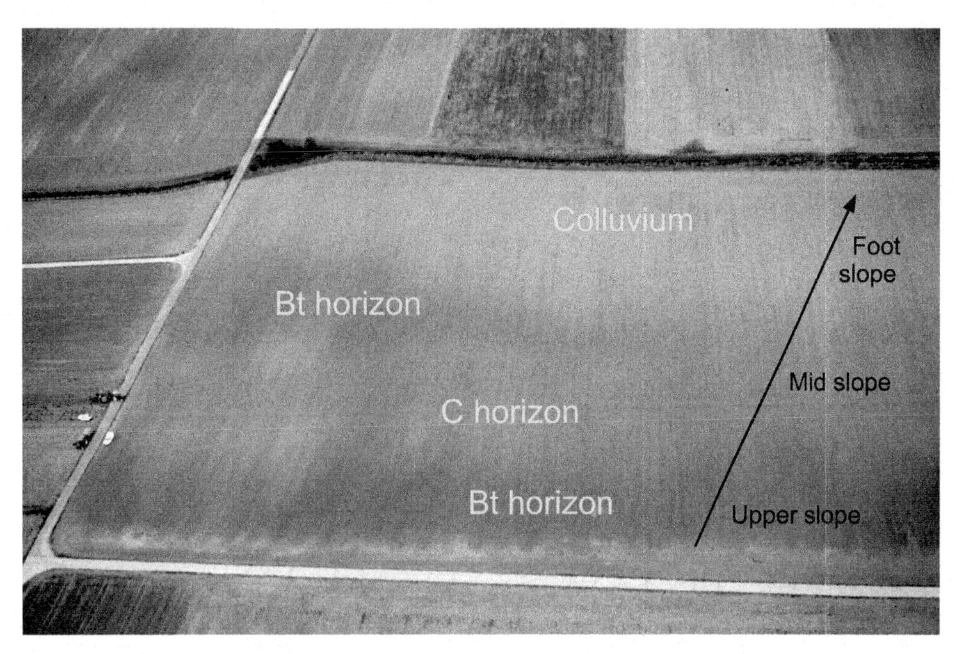

FIGURE 17-3 Detail of the area in a low-oblique small-format aerial photograph taken from a light airplane by A. Fengler, March 1998. Arrow indicates slope inclination. Note tractors and vehicles on left edge of scene.

well as in autumn. Similar to aerial archaeology (see Chapter 10.3.4), patterns became visible that could not have been detected from the ground perspective. At the same time, the limitations of airborne soil mapping became obvious. For example, soil moisture, treatment state, illumination conditions, and tillage directions have strong influences on soil color and thus on the interpretability of the aerial photographs. Without any fieldwork, an allocation of soil color to a specific soil unit is impossible, even for rather homogeneous sites, but made feasible with a small number of soil samples for reference. Mechanical tillage exerted for many decades has smudged the boundaries between formerly distinct soil units, thus rendering a clear delineation difficult. Image enhancement by histogram adjustments for increased color contrast may help to define such boundaries more clearly (see Fig. 11-9).

On the soil map of Butzbach/Nieder-Weisel (Fig. 17-2), the "Kleine Reiserbach" flows from the west to the east. Note the occurrence of Colluvisols south of the "Hochweiseler Weg" on top of the ridge near the Neolithic settlement. On the field "Rechte Steinäcker" in the left half of the map, even a completely eroded rendzic Regosol is covered by colluvium (as indicated by the lighter shade of pink).

In the oblique SFAP image printed here with only slight tonal enhancement (Fig. 17-3), the light C horizon of the mid slope and the equally light colluvium at the foot slope are clearly distinguished from the darker bands of the Bt horizon areas. Although the boundaries of the soil units are blurred due to mechanized tillage, they are much easier to trace on the basis of this photograph than to delineate in the field, where the soil mapper lacks the synoptic overview of the gently rolling hills.

17.3. SUMMARY

As these examples from central Europe demonstrate, small-format aerial photography has proven to be quite useful for detailed pedogeographic and geomorphologic studies that improve and correct the expert knowledge gained on the ground. When mapped in detail, soils and soil erosion may be much more complicated than commonly assumed. In fact, simple assumptions may prove to be quite contrary to actual soil conditions in places with long histories of human land use. Unfortunately, too little research up to now has endeavored to transfer the potential of SFAP for the loess landscapes shown here to other soilscapes.

Chapter 18

Architecture and Property Management

18.1. INTRODUCTION

Architecture refers to the design and construction of individual buildings, bridges, monuments, towers, and similar edifices as well as integrated building complexes. Holz et al. (1997) emphasized the need for large-scale, high-resolution aerial imagery for proper interpretation of such human structures. Small-format aerial photography (SFAP) has applications for all phases of architecture, construction, and property management from initial site selection to restoration of historic monuments. High spatial resolution and the ability to collect imagery from unconventional vantages are important attributes of SFAP. Aerial views depict the architectural structure or building site in relation to its surroundings and provide highly detailed site surveys or construction measurements (Figs. 18-1 to 18-3).

The fact that many architectural features are in urban settings introduces an added challenge to the small-format aerial photographer. Adjacent buildings, power lines, light poles, radio towers, and other obstacles may be numerous, along with many people, vehicles, and low-flying aircraft. In an age of increased security, SFAP may be restricted around major public buildings, such as sports arenas, museums, government offices, and the like. Thus, access to the airspace in close proximity to the target site may be difficult. Finally, it should be noted that cities give rise to increased turbulence and variations in near-surface wind compared with the surrounding countryside, which could affect SFAP platform operation.

For example, Wilson (2006) has undertaken kite aerial photography (KAP) of the golden statue, *Wisconsin*, atop the state capitol dome in Madison, Wisconsin, United States. The ability to maneuver the camera rig successfully within meters of the dome is truly amazing, given the difficulty of flying a kite in the middle of a city. However, scarcely any other means exist to place a camera in this aerial vantage. Only the most experienced kite flyer should attempt such a feat.

Another useful application of SFAP is to document before-and-after architect ural conditions, for example, during restoration or expansion of an existing building or

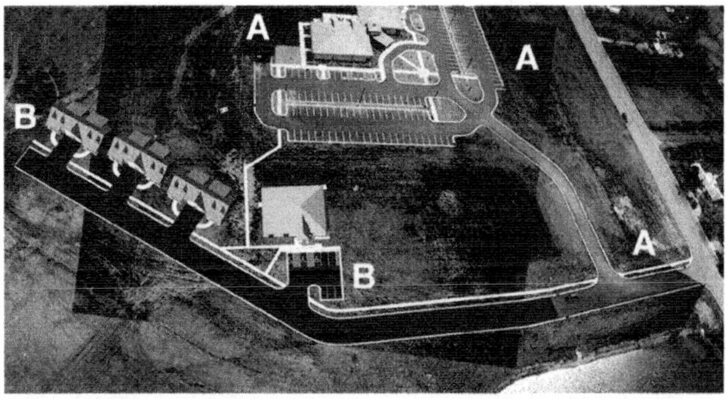

FIGURE 18-1 Mosaic of low-oblique photographs showing a development site in Bartlesville, Oklahoma, United States. (A) Existing buildings and roads, (B) architectural model of proposed new buildings and road superimposed on the landscape. Kite airphotos by SWA and JSA, April 2002; image courtesy of Mike Hughes Architects.

FIGURE 18-2 Basketball arena building site, Missouri State University. Note the large crane operating in the center of the construction site. Springfield, Missouri, United States. Kite airphoto by JSA and SWA, March 2008.

Small-Format Aerial Photography

FIGURE 18-3 View across dam construction site at Horse Thief Canyon, Kansas, United States. The partially completed dam extends across the scene and blocks Buckner Creek (right background). The outlet tower and channel are near the center of view, as heavy equipment moves across the dam structure. Kite aerial photo by JSA, May 2009.

FIGURE 18-4 Liberty Memorial, a national monument to World War I in Kansas City, Missouri, United States. (A) View of the original Liberty Memorial as constructed in the 1920s and closed for repairs. The center tower stands 217 feet (66 m) tall. View toward northeast; kite airphoto by JSA and J.T. Aber, November 1998. (B) Overview of renovated Liberty Memorial seen here just after its rededication in 2003. View northward with Kansas City skyline under clouds in the background. (C) Close-up detail of the tower and plaza of the renovated structure. (B) and (C) helium-blimp airphotos by JSA and SWA.

monument. Consider the Liberty Memorial in Kansas City, Missouri (Fig. 18-4), which is home to the United States official World War I museum. It was constructed in the 1920s, but by the 1990s it had fallen into such disrepair that it had to be closed to the public. Following extensive structural renovations and expansion, the memorial was rededicated in 2003, and a new underground museum was opened in 2006.

The examples below elaborate the expansion of a major art museum in Missouri and the use of SFAP for property and construction management at a small recreational lake in Kansas, United States.

18.2. NELSON–ATKINS MUSEUM OF ART, KANSAS CITY, MISSOURI

Origins of the Nelson–Atkins Museum of Art date from the late 1800s with William Rockhill Nelson (1841–1915), a newspaper publisher and philanthropist, and Mary McAfee Atkins (1836–1911), a widowed teacher. Nelson led the effort to beautify the city with parks, boulevards, and an art museum, and Atkins was dedicated to the acquisition and display of fine art. Both left well-endowed trusts for this purpose. In 1927, by mutual consent of their trustees, the Nelson and Atkins funds were combined to create the financial basis for building a single world-class art museum (Ward and Fidler, 1993; Nelson-Atkins Museum, undated).

Construction of the original Nelson–Atkins building began in 1930 and was completed in 1933. Kansas City architects Wight and Wight designed a neoclassical beaux-arts structure built of Indiana limestone featuring massive ionic columns with interior marble and bronze finishes. The east wing was named the *Atkins Museum of Fine Arts*, and the rest of the building was called the *William Rockhill Nelson Gallery of Art*. The museum was situated on the Nelson's baronial estate covering 22 acres (~9 ha). The building sits on a hilltop facing south toward the valley of Brush Creek, and the surrounding grounds were developed into a public park. By the mid-twentieth century, it already had developed a renowned reputation and was particularly noteworthy for its Asian collections of Chinese, Indian, and Japanese art as well as European and American paintings. It was, in addition, an iconic architectural symbol of Kansas City, Missouri (Figs. 18-5 and 18-6). The name of the whole was shortened to *Nelson–Atkins Museum of Art* on its 50th anniversary in 1983.

During the 1990s, planning for a major addition began. This would be the first fundamental change to the museum in more than six decades, and it was intended to be a significant contribution to architecture as well as art display. The expansion dealt with every aspect of museum function—galleries, offices, library, receiving, storage, preparation, parking, cafe, and shop (Wood and Slegman, 2007). Original ideas for the expansion focused mainly on the northern side of the existing building, which was the main public entrance, so that the magnificent southerly view of the existing building and park would not be disturbed.

Six internationally famous architects were invited to submit design concepts in 1999. Most envisioned plans to develop the northern side of the existing building. The winning proposal from Steven Holl was radically different, however. Holl was inspired by a Chinese silk handscroll in the museum's collection, *The North Sea* by Zhou Chen (1455–1536), which depicts a series of buildings tumbling down a secluded hillside next to the sea. Holl's design features a long, narrow, low building that extends southward from the eastern end of the existing Nelson–Atkins building. In Holl's description, the original building is stone, heavy, and inward, whereas the new building is feather, light, and outward (Wood and Slegman, 2007). Together they complement each other—stone and feather. This concept utilizes the park atmosphere and sculpture garden of the

FIGURE 18-5 View across the south lawn of the Nelson–Atkins Museum of Art looking north toward the southern façade of the original building with the downtown Kansas City skyline in the background. The shuttlecock sculptures suggest a giant badminton court in which the building represents the net. Right side of the scene is the site of the future Bloch building. Kite airphoto by JSA and J.T. Aber, November 1998.

FIGURE 18-6 Close-up view of the southern façade of the Nelson–Atkins building showing the classic ionic columns of the south entrance. Kite airphoto by SWA and JSA, June 2008; Kansas City, Missouri, United States.

FIGURE 18-7 New Bloch building under construction on the eastern side of the Nelson–Atkins Museum of Art campus. (A) Superwide-angle view northward with the downtown Kansas City skyline in the background. (B) Closer view of the new Bloch building. Five raised "lenses" are identified simply by number (1–5); these portions extend above ground in the completed structure. Helium-blimp airphotos by JSA, April 2005.

museum campus without compromising the grand northern and southern façades of the original building.

Considered daring and innovative, Holl's design immediately generated substantial controversy because of its ultra-modern style that contrasts sharply with the traditional neoclassical appearance of the Nelson–Atkins building. As with other recent museum architectural disputes, such as the *Pyramide du Louvre* in the 1980s, the new building design sparked heated debate within the local community. Construction of the new building, named for Henry W. and Marion Bloch, moved forward in 2001, in spite of local controversy. Indeed debate intensified during construction, as it was quite difficult to appreciate on the ground how the new building would look when finished (Fig. 18-7).

The Bloch building was completed and opened to the public in 2007. The new building is 840 feet (256 m) long, and much of it lies below ground. Rising above ground are five so-called lenses clad in frosted glass panels (Fig. 18-8). The primary purpose of these lenses is to channel diffuse natural light into the interior display spaces via vaulted ceilings (Fig. 18-9). At night, the lenses glow with interior lighting. The Bloch building is set next to the outdoor sculpture garden, which overlaps onto the grassy roof of the building between lenses (Fig. 18-10).

In addition to the Bloch building, the expansion project includes a new underground parking garage on the northern side of the Nelson–Atkins building. The roof of the parking garage supports a large, shallow, water-filled reflecting pool designed by Walter de Maria called *One Sun / 34 Moons* (Fig. 18-11). The pool contains a gold-covered, bronze-and-steel slab that represents the sun and 34 circular lenses that

FIGURE 18-8 Overview of the Nelson–Atkins Museum of Art with most of the finished Bloch building visible on the right side. Note the odd shape of each lens in the Bloch building. View toward north; kite airphoto by SWA and JSA, June 2008.

FIGURE 18-9 (A) Low-oblique view of Bloch building lenses 2–5. Asterisk (*) marks door in lens 4 that leads out to the sculpture park. (B) Interior view in lens 4 of the Bloch building showing the frosted glass wall and vaulted ceiling designed to pass diffuse light into the galleries. Door in background opens to sculpture park walkway shown in (A). Photos by JSA and SWA, June 2008.

FIGURE 18-10 Close-up view of the sculpture park in vicinity of lenses 4 and 5. Sculptures: a, *Turbo* by Tony Cragg (bronze, 2001); b, *Ferryman* by Tony Cragg (bronze, 1997); c, *Three Bowls* by Ursula von Rydingsvard (cedar and graphite, 1990); and d, patio extension of the indoor Noguchi sculpture court. Kite airphoto by SWA and JSA, June 2008.

pass daylight through water into the garage underneath. At night, light from below illuminates the reflecting pool and roof plaza. While construction of the new parking garage and Bloch building went on, considerable renovation also took place throughout the interior and exterior of the original Nelson–Atkins building (Wood and Slegman, 2007). The exterior received cleaning and tuck-pointing of the limestone, a new roof was installed, and sculptures were rearranged (Fig. 18-12).

Holl envisioned the Bloch building as a feather, but others have likened its lenses as fingers of a hand rising out of the ground—lens 1 is the thumb and lenses 2–5 are the long fingers. This theme is seen in the painting, *Art Part*, by Elizabeth Murray displayed in lens 5 (Fig. 18-13). Much of the initial controversy about the museum expansion has diminished now, but lingering uncertainty continues. Many people remain skeptical. Upon leaving the new Bloch building and entering the original Nelson–Atkins building,

FIGURE 18-11 Northern side of the museum complex. *One Sun / 34 Moons* by Walter de Maria is the black square reflecting pool above the parking garage, upper left portion of view. Lens 1 of the Bloch building is under construction in right background. Roof of the Nelson–Atkins building is lower right, and a shuttlecock rests on the lawn in lower left corner of scene. Helium-blimp airphoto by JSA, April 2005.

FIGURE 18-12 Renovated east sculpture terrace. Selected sculptures: (a) *Peace on Earth* by Jacques Lipchitz (bronze, 1969) and (b) *Storage* by Judith Shea (bronze, 1999). Kite airphoto by SWA and JSA, June 2008.

FIGURE 18-13 *Art Part* by Elizabeth Murray (oil on 22 canvases, ~3 m wide, 1981). The theme of a long paintbrush resting on a stylized hand reflects the overall design and appearance of the Bloch building. Photo by JSA, June 2008.

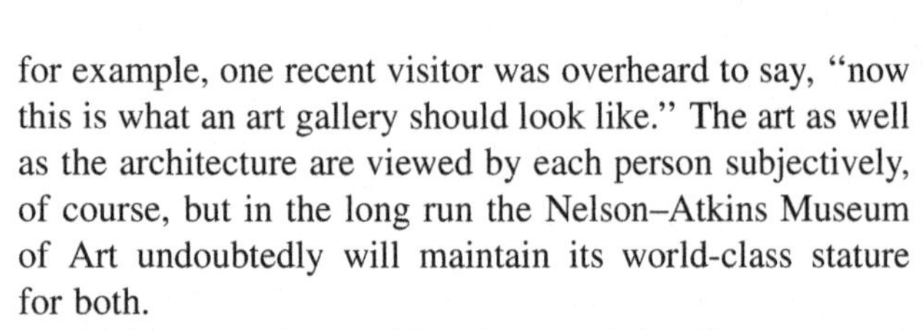

for example, one recent visitor was overheard to say, "now this is what an art gallery should look like." The art as well as the architecture are viewed by each person subjectively, of course, but in the long run the Nelson–Atkins Museum of Art undoubtedly will maintain its world-class stature for both.

Architecture is considered one of the fine arts, and architectural photography is essential for demonstrating buildings, structures, and monuments (Dubois, 2007). During the process of museum expansion, SFAP provided a means to document the unfolding architectural drama from a low-height vantage that would be difficult to achieve by other methods. Tethered kites and a small helium blimp were utilized as lifting platforms depending on wind conditions. In this urban setting, manned aircraft are not allowed to fly less than 1000 feet (300 m) above the ground, whereas all the photographs presented here were acquired from heights of less than 500 feet (150 m). This allows for both panoramic overviews (see Figs. 18-5, 18-7, and 18-8) as well as intimate close-up shots from unusual angles (see Figs. 18-6, 18-10 to 18-12). Such SFAP imagery uniquely illustrates and elaborates Holl's stone-and-feather concept for the Nelson–Atkins Museum of Art.

FIGURE 18-14 Digital orthophotograph of Lake Kahola vicinity, east-central Kansas, United States. The lake is approximately 2.5 km long and 0.5 km wide. At the resolution of this image, individual buildings and docks are barely visible. Derived from a conventional panchromatic airphoto; orthophoto dataset obtained from the Kansas Geological Survey; taken from Aber and Aber (2003, fig. 1).

18.3. PROPERTY MANAGEMENT, LAKE KAHOLA, KANSAS

Lake Kahola is a relatively small, man-made reservoir located in the Flint Hills of east-central Kansas (Fig. 18-14). It was built in the 1930s as a water-supply lake for the nearby city of Emporia, and the lake surroundings were developed as a park with recreational cabins and homes. The building lots were leased from the city under a long-term arrangement with the Kahola Park Cabin Owners Association. Early in the twenty-first century, however, the city began the complex legal process of selling the land to the Cabin Owners Association. This sale was completed in 2007, when the renamed Kahola Homeowners Association (KHA) took ownership of the lake and surrounding property.

Over the years, development of cabin sites at Lake Kahola had proceeded in a somewhat piecemeal manner, which was aggravated by the lack of original survey markers. The existing plat of lease lots was a schematic blueprint chart of unknown age, which was not a legally valid survey (Fig. 18-15). This problem had become exacerbated in recent years by the construction of increasingly large recreational homes and numerous garages, boathouses, docks, and other structures. Two issues were of primary concern: (1) wise management for future development, and (2) arbitration of disputes between adjacent lot owners. In such densely developed situations, a few inches

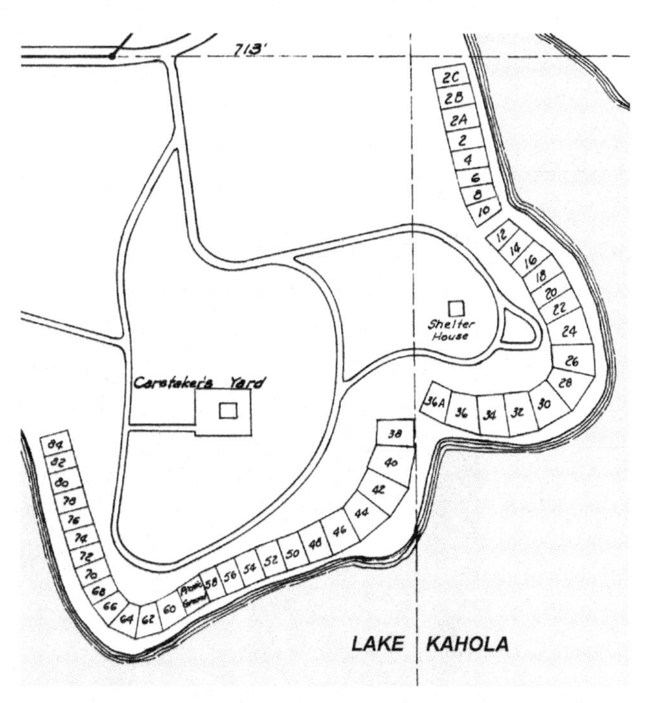

FIGURE 18-15 Portion of existing property chart for northeastern part of Lake Kahola, Kansas. Size and shape of building lots are idealized and do not correspond to actual lot dimensions. Adapted from a blueprint of unknown age (Aber and Aber, 2003, fig. 2).

(cm) are sometimes critical for accurate planning and construction purposes.

Recognizing these issues, the Cabin Owners Association began an effort in 2000 to place permanent survey markers at the corners of all building lots in the park. To supplement the ground-based survey markers and lot measurements, large-scale airphotos were selected as the tool for documenting lakeshore development and lot boundaries. Vertical KAP was conducted during the winter, leaf-off period in order to obtain views with minimal obstruction from trees (Aber and Aber, 2003). This approach proved to be the most cost-effective means to acquire imagery with sufficiently high spatial resolution for property management applications at Lake Kahola.

Conducting vertical photography around an irregular shoreline proved challenging, especially in regard to numerous power lines, roads, fences, and trees in vicinity of the buildings. The kite flyer and camera operator were usually out of visual sight of each other and had to communicate via personal radios to keep the camera on target. Because each section of the shore could be photographed only with certain wind direction, multiple field sessions were necessary. Furthermore, images had to be acquired without snow cover, so that survey markers could be located. These requirements limited the number of suitable days in which fieldwork could be conducted. Collection of kite aerial photographs spanned a total of four months in 2002 (January–March, December) in order to complete all building sites around the lake perimeter.

Lot boundary markers were identified with additional ground survey, and these markers were annotated on the images as well (Fig. 18-16). Relating ground survey markers to the airphotos was straightforward in most cases, except where long winter shadows fell across features in the images. Individual images are arranged in sequence to provide complete coverage of all buildings and lot boundaries around the lake, and sets of overlapping images were mosaicked together. However, the ground survey markers were not georeferenced, so a controlled mosaic could not be produced. Nonetheless, approximate spatial dimensions of images could be calculated, based on field measurements of selected ground objects (Fig. 18-17).

The images were assembled in web page format on a compact disk (CD), so that users can easily select images for visual display or paper printout. The KHA is able to access the images quickly to evaluate architectural plans, building permits, and property changes. The images in this database could be updated periodically to add new annotation as needed, for example, locations of water wells, propane tanks, and sanitary holding tanks. Already in the short time since the airphotos were acquired in 2002, many changes have appeared in connection with construction of new houses, retaining walls, parking areas, sheds, garages, decks, etc. This trend accelerated after the KHA acquired private ownership of the lake in 2007. New SFAP may be necessary by 2012 to reflect property changes around Lake Kahola.

FIGURE 18-16 Vertical kite aerial photograph showing a portion of northeastern Lake Kahola, February 2002. The image is annotated with lot numbers and survey markers (red). Features as small as 15 cm, the size of survey markers, can be identified in the original image. Note long tree shadows. North toward top; compare with right side of chart in Figure 18-15. Adapted from Aber and Aber (2003, fig. 3).

FIGURE 18-17 Mosaic assembled from two images using *D Joiner* software showing a portion of northeastern Lake Kahola, February 2002. Scale bar applied on the basis of ground measurements. North toward top; compare with upper-right portion of chart in Figure 18-15.

18.4. SUMMARY

Small-format aerial photography has worthwhile applications for many aspects of architectural planning, construction, and renovation as well as general property survey and management. Primary advantages include low-height, large-scale images that depict architectural features in detail. Oblique shots provide unusual viewing angles and portray building sites in relation to surrounding features. In some cases, unmanned tethered platforms—kite or blimp—may be the only feasible way to position a camera close to a high monument or tall building for aerial inspection. Vertical views may be acquired for accurate survey and mapping purposes where high spatial resolution (5–10 cm) is necessary. The ability to rephotograph sites periodically during construction or renovation adds an important temporal dimension often lacking from conventional aerial photography.

Chapter 19

Golf Course Management

19.1. OVERVIEW

Golf course management is a sizable and rapidly growing industry worldwide. In the United States, 12,000 public and private golf courses generate $18 million annual revenue (First Research, 2009). Demand depends upon demographics and population growth tempered by economic conditions that influence disposable income and leisure time. In many developing countries, building golf courses is done to increase tourism and enhance local economies, which is the case in central and eastern Europe (Fig. 19-1; Royal & Ancient Golf, 2008).

FIGURE 19-1 Golf course near the Tatra Mountains in Slovakia at Stará Lesná. Part of a touristic complex of hotels, ski lifts, hiking and bicycle trails, sport camps, and other recreational activities based on the scenic beauty and clean environment of the Tatra Mountains. (A) View toward northeast. (B) View toward east. Kite airphotos by SWA and JSA, July 2007.

Golf courses represent a type of landscape architecture in which the topography, soils, drainage, and vegetation are altered greatly from natural conditions. In many situations, the maintenance of turf requires substantial use of fertilizers, herbicides, and pesticides, as well as frequent

FIGURE 19-2 Point Pinos lighthouse (lower right) surrounded by the Pacific Grove Municipal Golf Links, Pacific Grove, California, United States. Rocky shore of the Pacific coast appears beyond the golf course. (A) Color-visible image. (B) Color-infrared view. Golf course turf is light pink. Sand dunes are stabilized by exotic ice plants (*Carpobrotus edulis*), which appear bright red. Conifers around the lighthouse are dark reddish-brown in color. Taken from Aber et al. (2003, figs. 2 and 3).

irrigation, mowing, and soil treatments. These practices may have deleterious side effects, and many golf courses are attempting to minimize their environmental impacts nowadays. According to best management practices, ecological and conservation issues are important for golf to exist in an environmentally sustainable manner (Royal & Ancient Golf, 2006).

Low-height, high-resolution aerial photography is a tool for managers to visualize and evaluate the condition of the golf course and consequences of management practices. For example, managers could use these photographs to monitor fairway species encroachment, measure the degree of shading from trees, evaluate irrigation systems, visualize golf-cart traffic patterns, or track changes to green dimensions caused by mowing patterns.

Unmanned small-format aerial photography (SFAP) is a means to acquire such airphotos at relatively low cost compared with manned (airplane or helicopter) methods. Such photographs may provide general views of golf courses and their surrounding land use and also may depict vegetation conditions in both normal-color and color-infrared formats (Fig. 19-2). The following example demonstrates the potential of kite aerial photography (KAP) for analysis of turf and irrigation conditions in the semi-arid climate of southwestern Kansas, United States (Aber et al., 2003).

19.2. GARDEN CITY, KANSAS

In dry regions, irrigation is a major issue facing golf courses. Such is the case at the Southwind Country Club, located about 3 km (2 miles) south of Garden City, Kansas, where the supply of irrigation water is limited by water-rights appropriation. The Southwind Country Club includes suburban housing around the golf course (Fig. 19-3). The housing division and golf course each have a high-capacity water well, tapping the High Plains aquifer. In recent years, the golf course has "borrowed" excess water from the housing division. However, the housing division will have little excess water in the near future, as new houses are constructed and water use increases. In an attempt to evaluate the effectiveness of irrigation on the golf course, color-infrared KAP was conducted in late May, 2002 to show cool-season turf at its peak growth stage.

The golf course is constructed on rolling sand hills terrain, and the native vegetation is sand-sage prairie. The irrigated bentgrass (*Agrostis palustris*) fairways and bluegrass (*Poa pratensis*) roughs contrast sharply with dry sand-sage prairie in color-infrared photographs (Fig. 19-4). The irrigation plan consists of overlapping water circles along the fairways. Color-infrared images demonstrated clearly

FIGURE 19-3 Overview of the Southwind Country Club, near Garden City, Kansas, United States. Suburban housing borders the golf course, upper left and right. View toward north; kite aerial photo by JSA and SWA, May 2002.

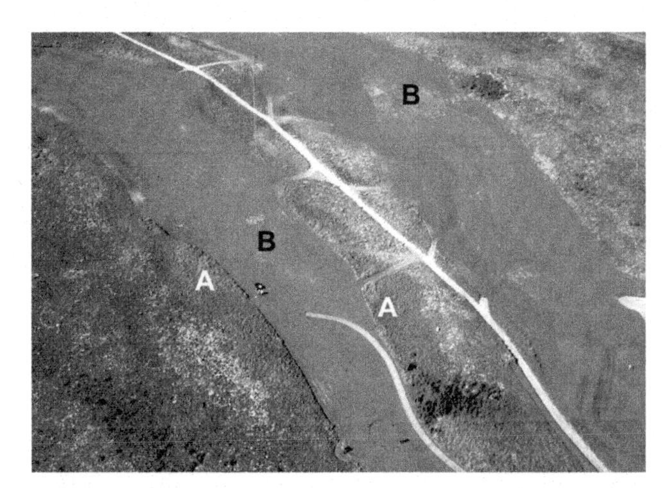

FIGURE 19-4 Color-infrared airphoto of the 15th and 17th fairways at Southwind Country Club, near Garden City, Kansas, United States. Bright pink and red indicate photosynthetically active, irrigated vegetation of the golf course. Some irrigation water has enhanced growth of the sand-sage prairie adjacent to the fairways (A). Blue patches in the fairways (B) indicate zones with weak grass. Kite flyers are standing left of scene center. Taken from Aber et al. (2003, fig. 4).

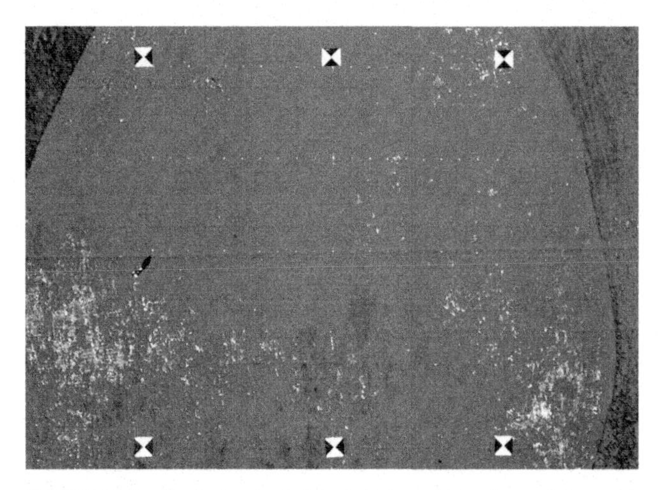

FIGURE 19-5 Vertical, color-infrared airphoto of test plot on fairway 14, Southwind Country Club, Garden City, Kansas, United States. Large aerial targets are 5 ft × 5 ft (1.5 m × 1.5 m). Small dots indicate grid cells. Pink turf is healthy bentgrass; darker red is encroaching blue/ryegrass mixture (bottom center); gray-white indicates weak or dead turf. Person standing to left. Taken from Aber et al. (2003, fig. 5).

that water circles extend beyond the fairways into the adjacent sand-sage prairie in many places. Such knowledge may help golf course managers to situate and operate the irrigation system better in order to conserve water usage.

The bentgrass fairways are susceptible to a condition called "localized dry spot" wherein the soil surface becomes water repellent. These dry spots leave the turf vulnerable to desiccation in dry winters which often leads to winterkill. A test plot (30 m × 30 m) was established to examine the effectiveness of different soil surfactants designed to alleviate localized dry spots.

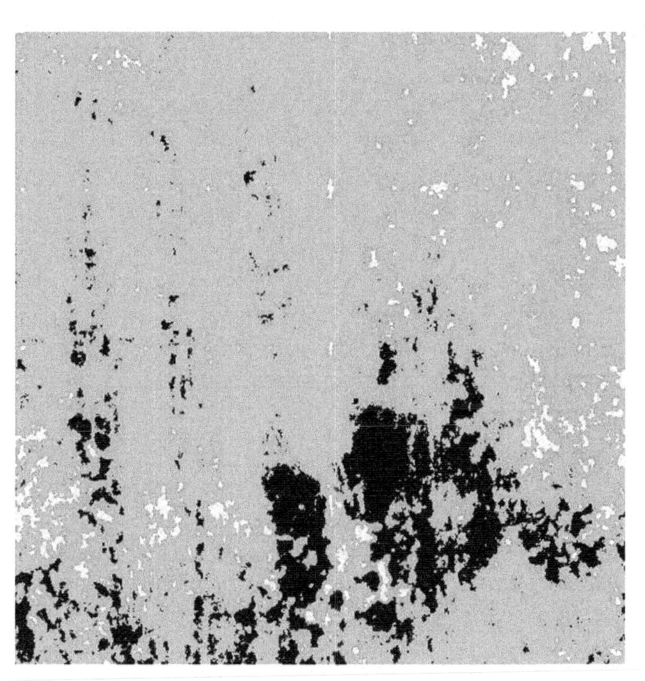

FIGURE 19-6 Reclassified image of the vertical color-infrared airphoto showing the 30 m × 30 m test plot, in which bentgrass (gray), blue/ryegrass (black) encroachment, and winterkilled turf (white) are delineated and quantified. Compare with previous figure; taken from Aber et al. (2003, fig. 6).

A vertical color-infrared photograph (Fig. 19-5) of the test plot early in the growing season revealed the degree of winterkill that occurred in the previous winter. It also showed darker-colored turf that is related to the encroachment of mixed bluegrass and ryegrass (*Lolium perenne*) species. Various combinations of these grasses were overseeded in past years to provide quick cover following winterkill. Geospatial analysis techniques were applied to quantify the degree of blue/ryegrass encroachment and winterkill (Fig. 19-6). The analysis of the study plot showed that 86% was bentgrass, 11% was a blue/ryegrass combination, and 3% was winterkilled.

19.3. SUMMARY

Golf is growing in popularity around the world, and golf courses represent substantial human alterations of the natural environment. Small-format aerial photography is especially useful for golf course managers to visualize and evaluate the effectiveness and impacts of their practices. Low-height, large-scale images, particularly color-infrared photographs, depict vegetation conditions related to irrigation, mowing, and applications of fertilizer, herbicides, pesticides, and other treatments. This type of detailed information is valuable for managers to improve their methods in order to maintain viable golf courses and to achieve minimal environmental impacts.

Combined References

AAAS/NSF 2005. Science and Engineering Visualization Challenge. *Autumn color, Estonian bog,* first place in the photography category. Science 309, p. 1991. Available from <http://www.sciencemag.org/cgi/content/full/309/5743/1991b>; (accessed December 2009).

Aaviksoo, K. 1988. Interpretation of Estonian mire sites from aerial photos. Acta et Commentationes Universitatis Tartuensis 812, Estonia, p. 193-208 [in Russian].

Aaviksoo, K. 1995. Vegetation of Endla Nature Reserve classified on the basis of Landsat TM data. In Aaviksoo, K., Kull, K., Paal, J. and Trass H. (eds.), Consortium Masingii. Tartu University, Tartu, Estonia, p. 27-36.

Aaviksoo, K., Kadarik, H. and Masing, V. 1997. Kaug- ja lähivõtteid 30 Eesti soost. Aerial views and close-up pictures of 30 Estonian mires. Tallinn, EEIC, Ministry of the Environment, Tallinn, Estonia, 96 p.

Aaviksoo, K., Paal, J. and Dislis, T. 2000. Mapping of wetland habitat diversity using satellite data and GIS: An example from the Alam-Pedja Nature Reserve, Estonia. Proceedings Estonian Academy of Science, Biology/Ecology 49/2, p. 177-193.

Aber, J.S. 2004. Lighter-than-air platforms for small-format aerial photography. Kansas Academy of Science, Transactions 107, p. 39-44.

Aber, J.S., Aaviksoo, K., Karofeld, E. and Aber, S.W. 2002. Patterns in Estonian bogs depicted in color kite aerial photographs. Suo 53/1, p. 1-15.

Aber, J.S. and Aber, S.W. 2001. Potential of kite aerial photography for peatland investigations with examples from Estonia. Suo 52/2, p. 45-56.

Aber, J.S. and Aber, S.W. 2003. Applications of kite aerial photography: Property survey. Kansas Academy of Science, Transactions 106, p. 107-110.

Aber, J.S. and Aber, S.W. 2009. Kansas physiographic regions: Bird's-eye views. Kansas Geological Survey, Educational Series 17, 76 p.

Aber, J.S., Aber, S.W., Buster, L., Jensen, W.E. and Sleezer, R.L. 2009. Challenge of infrared kite aerial photography: A digital update. Kansas Academy of Science, Transactions 112, p. 31-39.

Aber, J.S., Aber, S.W., Janoèko, J., Zabielski, R. and Górska-Zabielska, M. 2008. High-altitude kite aerial photography. Kansas Academy of Science, Transactions 111, p. 49-60.

Aber, J.S., Aber, S.W. and Leffler, B. 2001. Challenge of infrared kite aerial photography. Kansas Academy of Science, Transactions 104, p. 18-27.

Aber, J.S., Aber, S.W., Pavri, F., Volkova, E. and Penner, R.L. 2006. Small-format aerial photography for assessing change in wetland vegetation, Cheyenne Bottoms, Kansas. Kansas Academy of Science, Transactions 109, p. 47-57.

Aber, J.S. and Ber, A. 2007. Glaciotectonism. Developments in Quaternary Science 6, Elsevier, Amsterdam, 246 p.

Aber, J.S., Eberts, D. and Aber, S.W. 2005. Applications of kite aerial photography: Biocontrol of salt cedar (*Tamarix)* in the western United States. Kansas Academy of Science, Transactions 108, p. 63-66.

Aber, J.S., Kalm, V. and Lewicki, M. 2001. Geomorphic interpretation of Landsat imagery for western Estonia. Slovak Geological Magazine 7/3, p. 237-242.

Aber, J.S., Sobieski, R., Distler, D.A. and M.C. Nowak, M.C. 1999. Kite aerial photography for environmental site investigations in Kansas. Kansas Academy of Science, Transactions 102, p. 57-67.

Aber, J.S., Zupancic, J. and Aber, S.W. 2003. Applications of kite aerial photography: Golf course management. Kansas Academy of Science, Transactions 106, p. 211-214.

Achtelik, M. 2008. Octocopter im Anflug. TUM-Studenten bauen autonom fliegende Kameraträger. TUMcampus 2/08, p. 11-12.

Ahvenniemi, M., Ojala, K. and Raitala, J. 1998. A 15 channel TM classification of bog types in Finnish Lapland. Available from <http://www.oulu.fi/astronomy/planetology/Kaukok/Vuotos/aVuotosShort.html>; (accessed January 2010).

Albert, K.-D. 2002. Die Altdünenlandschaft im Sahel NE Burkina Fasos: Geomorphogenese und Geomorphodynamik einer semiariden Kulturlandschaft. Ph.D. thesis, Department of Physical Geography, Frankfurt University. Frankfurt am Main, Germany, 246 p. + app.

Albertz, J. 2007. Einführung in die Fernerkundung: Grundlagen der Interpretation von Luft- und Satellitenbildern. 3rd edition. Wissenschaftliche Buchgesellschaft, Darmstadt, Germany, 254 p.

Allikvee, H. and Masing, V. 1988. Põhja-Eesti kõrgustiku suurte mosaiiksoode valdkond. In Valk, U. (ed.), Eesti Sood (Estonian peatlands). Valgus, Tallinn, Estonia, p. 264-270.

Altan, M.O., Celikoyan, T.M., Kemper, G. and Toz, G. 2004. Balloon photogrammetry for cultural heritage. International Archives of the Photogrammetry, Remote Sensing and Spatial Information Sciences 35 (B5), p. 964–968.

Amsbury, D.L., Evans, C. and Ackleson, S. 1994. Earth observations during space shuttle flight STS-49: Endeavor's mission to planet Earth. Geocarto International 9/2, p. 67-80.

Andres, W. 1972. Beobachtungen zur jungquartären Formungsdynamik am Südrand des Anti-Atlas (Marokko). In Büdel, J. and Rathjens, C. (eds.), Neue Wege der Geomorphologie. Zur Differenzierung der Abtragungsprozesse in verschiedenen Klimaten. Zeitschrift für Geomorphologie N.F. (Suppl. 14), p. 66-80.

Anonymous 2008. On assignment: Crash landing. National Geographic 213/5, p. 189.

Archibold, O.W., Lévesque, L.M. J., de Boer, D.H., Aitken, A.E. and Delanoy, L. 2003. Gully retreat in a semi-urban catchment in Saskatoon, Saskatchewan. Applied Geography 23, p. 261–279.

Arthus-Bertrand, Y. 2002. Earth from above. Revised and expanded edition. Abrams, New York, 462 p.

Ascending Technologies 2009 (anonymous). Innovative Multirotor-Flugsysteme - Innovative multi-rotor air-vehicles. Available from <http://www.asctec.de>; (accessed February 2010).

Asner, G.P., Braswell, B.H., Schimel, D.S. and Wessman, C.A. 1998. Ecological research needs from multiangle remote sensing data. Remote Sensing of Environment 63, p. 155-165.

Avery, T.E. and Berlin, G.L. 1992. Fundamentals of remote sensing and airphoto interpretation 5th edition. Macmillan, New York, 474 p.

Avni, Y. 2005. Gully incision as a key factor in desertification in an arid environment, the Negev highlands, Israel. Catena 63, p. 185–220.

Bähr, H.-P. 2005. Digitale Bildverarbeitung: Anwendung in Photogrammetrie, Fernerkundung und GIS. 4th edition. Wichmann, Heidelberg, 325 p.

Baker, A.K.M., Fitzpatrick, R.W. and Koehne, S.R. 2004. High resolution low altitude aerial photography for recording temporal changes in dynamic surficial environments. In Roach, I.C. (ed.), Regolith 2004. CRC LEME, Kensington, Australia, p. 21–25.

Baker, S. 1994. San Francisco in Ruins: The 1906 aerial photographs of George R. Lawrence. Landscape 10/10, p. 9-14.

Baker, S. 1997. Controversy: Was it kites or a balloon? KiteLines 12/3, p. 46-51.

Banerjee, I. and McDonald, B.C. 1975. Nature of esker sedimentation. In Jopling, A.V. and McDonald, B.C. (eds.), Glaciofluvial and glaciolacustrine sedimentation. Society Economic Paleontologists and Mineralogists, Special Publication 23, p. 132-154.

Barnard, F. R. 1927. One picture is worth ten thousand words. Advertisement in the Journal *Printers' Ink*, 10 March 1927.

Barrette, J., August, P. and Golet, F. 2000. Accuracy assessment of wetland boundary delineation using aerial photography and digital orthophotography. Photogrammetric Engineering & Remote Sensing 66/4, p. 409-416.

Bárta, V. and Barta, V. 2006 Nad Slovenskom: Over Slovakia. AB Art Press, Slovenská L'upèa, Slovakia, 96 p.

Bartels, M. and Wei, H. 2007. Pattern recognition in photogrammetry and remote sensing: PRRS 2006 in conjunction with ICPR 2006. Photogrammetric Record 22/117, p. 97-99.

Batut, A. 1890. La photographie aérienne par cerf-volant. Gauthier-Villars et fils, Paris, 74 p.

Bauer, M., Befort, W., Coppin, Ir. Pol R. and Huberty, B. 1997. Proceedings of the first North American symposium on small format aerial photography. American Society of Photogrammetry and Remote Sensing, Bethesda, Maryland, United States, 218 p.

Beauffort, G. de and M. Dusariez, M. 1995. Aerial photographs taken from a kite: Yesterday and today. KAPWA-Foundation Publishing, 142 p.

Benton, C. C. 2009: Notes on kite aerial photography. Available from <http://steel.ced.berkeley.edu/cris/kap/wind/>; (accessed December 2009).

Berni, J.A.J., Zarco-Tejada, P.J., Suárez, L. and Fereres, E. 2009. Thermal and narrowband multispectral remote sensing for vegetation monitoring from an unmanned aerial vehicle. IEEE Transactions on Geoscience and Remote Sensing 47/3, p. 722–738.

Betts, H.D., Trustrum, N.A. and De Rose, R.C. 2003. Geomorphic changes in a complex gully system measured from sequential digital elevation models, and implications for management. Earth Surface Processes and Landforms 28, p. 1043–1058.

Beutnagel, R. 2009. DOPERO. Available from <http://www.dopero.de/>; (accessed December 2009).

Beutnagel, R., Bieck, W. and Bohnke, O. 1995. Picavet - past and present. Aerial Eye, 1/4.

Bigras, C. 1997. Kite aerial photography of the Axel Heiberg Island fossil forest. In Bauer, M., Befort, W., Coppin, Ir. Pol R. and Huberty, B. (eds.), Proceedings of the first North American symposium on small format aerial photography. American Society of Photogrammetry and Remote Sensing, Bethesda, Maryland, United States, p. 147-153.

Bitelli, G., Girelli, V.A., Tini, M.A. and Vittuari, L. 2004. Low-height aerial imagery and digital photogrammetrical processing for archaeological mapping. International Archives of the Photogrammetry, Remote Sensing and Spatial Information Sciences 35 (B5), p. 498–504.

Bitelli, G., Unguendoli, M. and Vittuari, L. 2001. Photographic and photogrammetric archaeological surveying by a kite system. Proceedings of the 3rd International Congress on Science and Technology for the Safeguard of Cultural Heritage in the Mediterranean Basin, Elsevier, Paris, p. 538–543.

Boike, J. and Yoshikawa, K. 2003. Mapping of periglacial geomorphology using kite/balloon aerial photography. Permafrost and Periglacial Processes 14, p. 81–85.

Bollacker, K.D. 2010. Avoiding a digital dark age. American Scientist 98/2, p. 106-110.

Braasch, O. and Planck, D. 2005. Vom heiteren Himmel: Luftbildarchäologie. Gesellschaft für Archäologie in Württemberg und Hohenzollern e.V., Esslingen am Neckar, Germany, 68 p.

Bruegge, C.J., Schaepman, M., Strub, G., Beisl, U., Demircan, A., Geiger, B., Painter, T.H., Paden, B.E. and Dozier, J. 2004. Field measurements of bi-directional reflectance. In Schönermark, M. von, Geiger, B. and Röser, H.P. (eds.), Reflection properties of vegetation and soil—with a BRDF data base. Wissenschaft und Technik Verlag, Berlin, p. 195-224.

Budworth, G. 1999. The complete book of knots. Barnes and Noble, New York, 160 p.

Buerkert, A., Mahler, F. and Marschner, H. 1996. Soil productivity management and plant growth in the Sahel: Potential of an aerial monitoring technique. Plant and Soil 180/1, p. 29–38.

Burge, D. 2007. 100 years from now ... In Peres, M.R. (ed.), Focal encyclopedia of photography. 4th edition. Elsevier, Amsterdam, p. 456-460.

Burkard, M.B. and Kostaschuk, R.A. 1997. Patterns and controls of gully growth along the shoreline of Lake Huron. Earth Surface Processes and Landforms 22/10, p. 901–911.

Busemeyer, K.-L. 1987. Zur Konzeption eines ferngelenkten, gefesselten Heißluftkammersystems. In International Society for Photogrammetry and Remote Sensing (ISPRS) and Deutsches Bergbau-Museum (DBM) (eds.), Luftaufnahmen aus geringer Flughöhe. Veröffentlichungen aus dem Deutschen Bergbaumuseum 41. Bochum, Germany, p. 49-56.

Busemeyer, K. L. 1994. Luftschiffeinsatz in Pakistan. Modell, 1994 /1, p. 4–7.

Caulfield, P. 1987. Capturing the landscape with your camera: Techniques for photographing vistas and closeups in nature. Amphoto, Watson-Guptill Publ., New York, 160 p.

Chandler, J. H. 1999. Effective application of automated digital photogrammetry for geomorphological research. Earth Surface Processes and Landforms 24/1, p. 51–63.

Chandler, J.H., Fryer, J.G. and Jack, A. 2005. Metric capabilities of low-cost digital cameras for close range surface measurement. Photogrammetric Record 20, p. 12-26.

Charman, D. 2002. Peatlands and environmental change. J. Wiley & Sons, London & New York, 301 p.

Chorier, N. and Mehta, Z. 2007. Kite's eye view: India between earth and sky. Lustre Press, Roli Books, Roli & Janssen BV, Netherlands, 192 p.

Christlein, R. and Braasch, O. 1998. Das unterirdische Bayern. 7000 Jahre Geschichte und Archäologie im Luftbild. 3rd edition. Theiss, Stuttgart, Germany, 275 p.

Clark, R.N. 2008a. Digital Camera Sensor Performance Summary. Available from <http://www.clarkvision.com/imagedetail/digital.sensor.performance.summary/>; (accessed December 2009).

Clark, R.N. 2008b. Digital Cameras: Does Pixel Size Matter? Factors in Choosing a Digital Camera. Available from <http://www.clarkvision.com/imagedetail/does.pixel.size.matter/>; (accessed December 2009).

Claussen, C., Niesen, M., Möller, M. and Born, J. 2008. MAVinci – and your visions fly. Available from <http://www.mavinci.de>; (accessed December 2009).

Cohen, J.E. and Small, C. 1998. Hypsographic demography: The distribution of human population by altitude. Proceedings of the National Academy of Sciences, Applied Physical Sciences, Social Sciences 95, p. 14009-14014.

Cohen, J.-L. 2006. Above Paris: The aerial survey of Roger Henrard. Princeton Architectural Press, New York, 320 p.

Colby, J. D. 1991. Topographic normalization in rugged terrain. Photogrammetric Engineering & Remote Sensing 57/5, p. 531-537.

Colwell, J.E. 1974. Vegetation canopy reflectance. Remote Sensing of Environment 3, p. 175-183.

Colwell, R.N. 1956. Determining the prevalence of certain cereal crop diseases by means of aerial photography. Hilgardia 26, p. 223-286.

Colwell, R.N. 1997. History and place of photographic interpretation. In, Philipson, W.R. (ed.), Manual of photographic interpretation. 2nd edition. American Society for Photogrammetry and Remote Sensing, Bethesda, Maryland, United States, p. 3-47.

Combrink H.J., Delgado-Cartay, M.D., Higgins S.I., February, E. and Müller, M. 2009. Spatial and seasonal patterns of NDVI along a rainfall gradient in an African savanna. Presentation at the 7th Annual Savanna Science Network Meeting 2009, Skukuza, Mpumalanga, South Africa. Available from <http://www.sanparks.org/parks/kruger/conservation/scientific/noticeboard/science_network_meeting_2009/Presentations/combrink.pdf>; (accessed December 2009).

Comer, R.P., Kinn, G., Light, D. and Mondello, C. 1998. Talking digital. Photogrammetric Engineering & Remote Sensing 64, p. 1139-1142.

Cowardin, L.M., Carter, V., Golet, F.C. and La Roe, E.T. 1979. Classification of wetlands and deepwater habitats in the United States. U.S. Dept. Interior, Fish & Wildlife Service, FWS/OBS-79/31.

Defibaugh, D. 2007. Landscape photography. In Peres, M.R. (ed.), Focal encyclopedia of photography. 4th edition. Elsevier, Amsterdam, p. 333-334.

DeLoach, C.J. 2004. *Tamarix ramosissima*. Crop Protection Compendium Database [CD-ROM]. CAB International, Wallingford, United Kingdom.

DeLoach, C.J. and Gould, J. 1998. Biological control of exotic, invading saltcedar (*Tamarix* spp.) by the introduction of *Tamarix*-specific control insects from Eurasia. Proposal to USDI Fish and Wildlife Service. USDA Agricultural Research Service, Temple, TX and USDA Animal and Plant Health Inspection Service, Plant Protection and Quarantine, Phoenix, Arizona, United States.

DeLoach, C.J., Knutson, A.E., Moran, P.J., Michels, G.J., Thompson, D.C., Carruthers, R.I., Nibling, F., Fain, T.G. 2007. Biological control of Saltcedar (Cedro salado) (*Tamarix* spp.) in the United States, with implications for Mexico. In Lira-Saldivar, R.H. (ed.), Bioplaguicidas y Control Biologico. International Symposium of Sustainable Agriculture, Mexico, 24-26 October 2007, p. 142-172.

Deneault, D. 2007. Tracking ground targets with measurements obtained from a single monocular camera mounted on an unmanned aerial vehicle. M.S. Thesis, Department of Mechanical and Nuclear Engineering, Kansas State University. Manhattan, Kansas, United States, 120 p.

Dewey, J. 2004. Geological Society of America, GeoTales, vol. 1.

Dickenson, W.R. 2009. Pacific atoll living: How long already and until when? Geology 19/3, cover and p. 4-10.

Dippie, B.W., Heyman, T.T., Mulvey, C. and Troccoli, J.C. 2002. George Catlin and his Indian gallery. Edited by Gurney, G. and Heyman, T.T., Smithsonian American Museum of Art, W.W. Norton & Co., 294 p.

Drury, S.A. 1987. Image interpretation in geology. Allen & Unwin, London, 243 p.

Dubois, W.W. 2007. Architectural photography. In Peres, M.R. (ed.), Focal encyclopedia of photography. 4th edition. Elsevier, Amsterdam, p. 325.

Dunford, R., Michel, K., Gagnage, M., Piégay, H. and Trémelo, M.-L. 2009. Potential and constraints of unmanned aerial vehicle technology for the characterization of Mediterranean riparian forest. International Journal of Remote Sensing 30/19, p. 4915–4935.

Eastaway, P. 2007. Nature photography. In Peres, M.R. (ed.), Focal encyclopedia of photography. 4th edition. Elsevier, Amsterdam, p. 576-577.

Eberts, D., White, L., Broderick, S., Nelson, S.M., Wynn, S. and Wydoski, R. 2003. Biological control of saltcedar at Pueblo, Colorado: Summary of research and insect, vegetation and wildlife monitoring - 1997-2002. Technical Memorandum No. 8220-03-06. U.S. Bureau of Reclamation, Technical Service Center, Denver, Colorado, United States.

Eisenbeiss, H. 2008. The autonomous mini helicopter: a powerful platform for mobile mapping. International Archives of the Photogrammetry, Remote Sensing and Spatial Information Sciences 37 (B1), p. 977–984.

Elwell, H. and Stocking, M.A. 1974. Rainfall parameters and a cover model to predict runoff and soil loss from grazing trials in the Rhodesian sandveld. Proceedings of the Grassland Society of South Africa 9, p. 157-164.

Eriksen, P. and Olesen, L.H. 2002. Fortiden set fra himlen: Luftfotoarkæologi i Vestjylland. Holstebro Museum, Denmark, 160 p.

Espinar, V. and Wiese, D. 2006. Guided to gather. Toy plane upgraded with telemetry. GPS World 2006/2, p. 32-38.

Everaerts, J. 2008. The use of unmanned aerial vehicles (UAVs) for remote sensing and mapping. International Archives of the Photogrammetry, Remote Sensing and Spatial Information Sciences 37 (B1), p. 1187–1191.

Everitt, J.H., Yang, C. and Davis, M.R. 2006. Remote mapping of saltcedar in the Rio Grande System of Texas. Texas Journal of Science 58/1, p. 13-22.

Eyton, J.R. 1990. Color stereoscopic effect cartography. Cartographica 27, p. 20-29.

Federal Aviation Administration 2007. Balloon Flying Handbook. Skyhorse Publishing, New York, 128 p.

Feil, C. and Rose, E. 2005. The Finger Lakes region of New York: A view from above. VFA Publishing, Scarborough, Maine, United States, 160 p.

Fengler, A. 2007. Flächen- und standortbezogene geomorphologische und bodenkundliche Untersuchungen zur Bodentypenverteilung in einem Lössgebiet der Wetterau im Umfeld eines neolithischen Siedlungsplatzes. Habelt, Bonn, Germany, 141 p.

Finney, A. 2007. Infrared photography. In Peres, M.R. (ed.), Focal encyclopedia of photography. 4th edition. Elsevier, Amsterdam, p. 556-562.

First Research 2009 (anonymous). Golf courses: Industry profile excerpt. Available from <http://www.firstresearch.com/Industry-Research/Golf-Courses.html>; (accessed February 2010).

Fouché, P.S. and Booysen, N.W. 1994. Assessment of crop stress conditions using low altitude aerial color-infrared photography and computer image processing. Geocarto International 9/2, p. 25–31.

Fraser, C.S. 1997. Digital camera self-calibration. Journal of Photogrammetry & Remote Sensing 52/4, p. 149–159.

Fryer, J.G., Mitchell, H.L. and Chandler, J.H. (eds.), 2007. Applications of 3D measurement from images. Whittles, Dunbeath, United Kingdom, 304 p.

Gambino, M. 2008. Danger zones. Smithsonian 38/10, p. 52-57.

García-Ruiz, J.M., Arnáez, J., Beguería, S., Seeger, M., Martí-Bono, C., Regués, D., Lana-Renault, N. and White, S. 2005. Runoff generation in an intensively disturbed, abandoned farmland catchment, Central Spanish Pyrenees. Catena 59/1, p. 79-92.

García-Ruiz, J.M., Lasanta, T., Ruiz-Flano, P., Ortigosa, L., White, S., González, C. and Martí, C. 1996. Land-use changes and sustainable development in mountain areas: A case study in the Spanish Pyrenees. Landscape Ecology 11/5, p. 267-277.

Geißler, C. 2007. Einfluss der Landbedeckung auf die aktuelle Geomorphodynamik in *gully*-Einzugsgebieten im semi-ariden Spanien. M.Sc. thesis, Department of Physical Geography, Frankfurt University. Frankfurt am Main, Germany, 124 p. + app.

Geodis 2006. Vysoké Tatry: Atlas ortofotomáp, 1:15 000 (2. vyd.). Geodis Slovakia, Bratislava, Slovakia, 128 p.

Gérard, B., Buerkert, A., Hiernaux, P. and Marschner, H. 1997. Non-destructive measurement of plant growth and nitrogen status of pearl millet with low-altitude aerial photography. In Ando, T., Fujita, K., Mae, T., Matsumoto, H., Mori, S. and Sekiya, J. (eds.), Plant nutrition - for sustainable food production and environment. Kluwer Academic Pub., Netherlands, p. 373-378.

Gerster, G. 2004. Weltbilder: 70 Flugbilder aus sechs Kontinenten. Schirmer/Mosel, München, 144 p.

Giménez, R., Marzolff, I., Campo, M.A., Seeger, M., Ries, J. B., Casalí, J. and Álvarez-Mozos, J. 2009. High-resolution photogrammetric and field measurements of gullies with contrasting morphology. Earth Surface Processes and Landforms 34, p. 1915-1926.

Gómez Lahoz, J. and González Aguilera, D. 2009. Recovering traditions in the digital era: The use of blimps for modelling the archaeological cultural heritage. Journal of Archaeological Science 36/1, p. 100–109.

Gravenor, C.P., Green, R. and Godfrey, J.D. 1960. Air photographs of Alberta. Research Council of Alberta, Bulletin 5, 79 p.

Graves, B. 2007. Remote control, small-format aerial photography: The Easy Star way. Unpub. research report, Earth Science Department, Emporia State University, Kansas, United States, 36 p.

Grenzdörffer, G. 2004. Digital low-cost remote sensing with PFIFF, the integrated digital remote sensing system. International Archives of the Photogrammetry, Remote Sensing and Spatial Information Sciences 35 (B1) p. 235 – 239.

Grenzdörffer, G. J., Engel, A. and Teichert, B. 2008. The photogrammetric potential of low-cost UAVs in forestry and agriculture. International Archives of the Photogrammetry, Remote Sensing and Spatial Information Sciences 37 (B1), p. 1207–1213.

Gruber, M., Leberl, F. and Perko, R. 2003. Paradigmenwechsel in der Photogrammetrie durch digitale Luftbildaufnahme? Photogrammetrie - Fernerkundung - Geoinformation 2003/4, p. 285–297.

Grün, A. 2008. Scientific-technological developments in photogrammetry and remote sensing between 2004 and 2008. In Li, Z., Chen, J.and Baltsavias, E.P. (eds.), Advances in photogrammetry, remote sensing, and spatial information. 2008 ISPRS Congress book. CRC Press, Boca Raton, United States, p. 21–25.

GSA Rock-Color Chart 1991. Geological Society of America, 8th printing, 1995.

Haack, B.N., Guptill, S.C., Holz, R.K., Jampoler, S.M., Jensen, J.R. and Welch, R.A. 1997. Urban analysis and planning. In Philipson, W.R. (ed.), Manual of photographic interpretation. 2nd edition. American Society for Photogrammetry and Remote Sensing, Bethesda, Maryland, United States, p. 517-553.

Hake, G., Grünreich, D. and Meng, L. 2002. Kartographie: Visualisierung raum-zeitlicher Informationen. 4th edition. De Gruyter, Berlin, 604 p.

Hall, R.C. 1997. Post war strategic reconnaissance and the genesis of project Corona. In R.A. McDonald (ed.), Corona: Between the Sun & the Earth, The first NRO reconnaissance eye in space. American Society of Photogrammetry and Remote Sensing, p. 25-58.

Hall, S.S. 1992. Mapping the next millenium: The discovery of new geographies. Random House, New York, 477 p.

Ham, W.E. and Curtis, N.M. Jr. 1960. Common minerals, rocks, and fossils of Oklahoma. Oklahoma Geological Survey, Guide Book X, 28 p.

Hamblin, W.K. 2004. Beyond the visible landscape: Aerial panoramas of Utah's Geology. BYU Geology, Provo, Utah, United States, 300 p.

Hapke, B., DiMucci, D., Nelson, R. and Smythe, W. 1996. The cause of the hot spot in vegetation canopies and soils: Shadow-hiding versus coherent backscatter. Remote Sensing of Environment 58, p. 63-68.

Hapke, C. and Richmond, B. 2000. Monitoring beach morphology changes using small-format aerial photography and digital softcopy photogrammetry. Environmental Geosciences 7/1, p. 32–37.

Hardin, P. J. and Jackson, M. W. 2005. An unmanned aerial vehicle for rangeland photography. Rangeland Ecology & Management 58/4, p. 439–442.

Hart, C. 1982. Kites: An historical survey. 2nd edition. Appel Publ., Mt. Vernon, New York, United States, 210 p.

Heckes, J. 1987. Überblick über Kammerträger für Luftaufnahmen im Nahbereich. In International Society for Photogrammetry and Remote Sensing (ISPRS) and Deutsches Bergbau-Museum (DBM) (eds.), Luftaufnahmen aus geringer Flughöhe. Veröffentlichungen aus dem Deutschen Bergbaumuseum 41. Bochum, Germany, p. 25-32.

Heisey, A. 2007. Archaeology of the skies. American Archaeology 11/4, p. 20-27.

Heisey, A. and Kawano, K. 2001. In the Fifth World: Portrait of the Navajo Nation. Rio Nuevo Publishers, Tucson, Arizona, United States, 88 p.

Henke de Oliveira, C. 2001. Análise de padrões e processos no uso do solo, vegetação, crescimento e adensamento urbano. Estudo de caso: Município de Luiz Antônio (SP). Ph.D. thesis, University of São Carlos, Brazil, 101 p.

Henry, J.-B., Malet, J.-P., Maquaire, O. and Grussenmeyer, P. 2002. The use of small-format and low-altitude aerial photos for the realization of high-resolution DEMs in mountainous areas: Application to the Super-Sauze earthflow (Alpes-de-Haute-Provence, France). Earth Surface Processes and Landforms 27, p. 1339–1350.

Holz, R.K., Baker, R.D., Baker, S., Haack, B.N., Lindgren, D.T. and Stow, D.A. 1997. Structures and cultural features. In Philipson, W.R. (ed.), Manual of photographic interpretation. 2nd edition. American Society for Photogrammetry and Remote Sensing, Bethesda, Maryland, United States, p. 269-308.

Hornschuch, A. and Lechtenbörger C. 2004. Fernerkundungsdaten und topographische Karten zur Dokumentation des sabäischen Kulturerbes in der Republik Jemen. Kartographische Nachrichten 3/2004, p. 112-117.

Hughes, T.J., Denton, G.H., Andersen, B.G., Schilling, D.H., Fastook, J.L. and Lingle, C.S. 1981. The last great ice sheets: A global view. In Denton, G.H. and Hughes, T.J. (eds.), The last great ice sheets. J. Wiley & Sons, New York, p. 263-317.

Hunt, D. 2002. Kite aerial photography electronic resources – Picavet basics. Available from <http://www.kaper.us/basics/BASICS_picavet.html>; (accessed December 2009).

Hunt, E.R., Cavigelli, M., Daughtry, C.S.T., McMurtrey III, J. and Walthall, C.L. 2005. Evaluation of digital photography from model aircraft for remote sensing of crop biomass and nitrogen status. Precision Agriculture 6, p. 359–378.

Hunt, E.R. Jr., Everitt, J.H., Ritchie, J.C., Moran, M.S., Booth, D.T., Anderson, G.L., Clark, P.E. and Seyfried, M.S. 2003. Applications and research using remote sensing for rangeland management. Photogrammetric Engineering & Remote Sensing 69, p. 675-693.

Ilomets, M. 1982. The productivity of *Sphagnum* communities and the rate of peat accumulation in Estonian bogs. In Masing, V. (ed.), Peatland ecosystems. Valgus, Tallinn, Estonia, p. 102-116.

Imeson, A.C. and Prinsen, H.A.M. 2004. Vegetation patterns as biological indicators for identifying runoff and sediment source and sink areas for semi-arid landscapes in Spain. Agriculture, Ecosystems and Environment 104/2, p. 333–342.

Inoue, Y., Morinaga, S. and Tomita, A. 2000. A blimp-based remote sensing system for low-altitude monitoring of plant variables: A preliminary experiment for agricultural and ecological applications. International Journal of Remote Sensing 21/2, p. 379–385.

Jacobson, C. 1999. Knots for the outdoors. 2nd edition. Globe Pequot Press, Old Saybrook, Connecticut, United States, 58 p.

Jahnke, K.-D. 1993. Ferngesteuerte Heißluftballone. Geschichte, Bau und Betrieb. FMT-Fachbuch, Baden-Baden, Germany, 111 p.

Jensen, J.R. 2005. Introductory digital image processing: A remote sensing perspective. 3rd edition. Pearson/Prentice Hall, Upper Saddle River, New Jersey, United States, 526 p.

Jensen, J.R. 2007. Remote sensing of the environment: An Earth resource perspective. 2nd edition. Prentice Hall Series in Geographic Information Science, Upper Saddle River, New Jersey, United States, 592 p.

Jensen, J.R., Narumalani, S., Weatherbee, O., Morris, K.S.and Mackey, H.E. 1993. Predictive modeling of cattail and waterlily distribution in a South Carolina reservoir using GIS. Photogrammetric Engineering & Remote Sensing 58/11, p. 1561-1568.

Jensen, T., Apan, A., Young, F. and Zeller, L. 2007. Detecting the attributes of a wheat crop using digital imagery acquired from a low-altitude platform. Computers and Electronics in Agriculture 1, p. 66-77.

Jia, L., Buerkert, A., Chen, X., Roemheld, V. and Zhang, F. 2004. Low-altitude aerial photography for optimum N fertilization of winter wheat on the North China Plain. Field Crops Research 89/2, p. 389–395.

John, S. and Klein, A. 2004. Hydrogeomorphic effects of beaver dams on floodplain morphology: Avulsion processes and sediment fluxes in upland valley floors (Spessart, Germany). Quaternaire 15/1-2, p. 219-231.

Jones, G. P., Pearlstine, L. G. and Percival, H. F. 2006. An assessment of small unmanned aerial vehicles for wildlife research. Wildlife Society Bulletin 34/3, p. 750–758.

Juvonen, T.-P., Ojanen, M., Tanttu, J.T., and Rosnell, J. 1997. Kaukokartoituksen ympäristösovellukset: Suotyyppien erottaminen pintalämpötilojen perusteella (Environmental applications of remote sensing methods: Discriminating mire site types by surface temperatures). Suo 48/1, p. 9-19.

Kalisch, A. 2009. Ableitung und Analyse von Erosionsrinnen-Netzwerken aus digitalen Geländemodellen mittels großmaßstäbiger Photogrammetrie und GIS – Südwest-Marokko. M.Sc. thesis, Department of Physical Geography, Frankfurt University. Frankfurt am Main, Germany, 111 p.

Karofeld, E. 1998. The dynamics of the formation and development of hollows in raised bogs in Estonia. The Holocene 8/6, p. 697-704.

Kasser, M. and Egels, Y. 2002. Digital photogrammetry. Taylor & Francis, London, 351 p.

Keränen, R. 1980. Vetypallon ja radio-ohjattavan kameran avulla tapahtuvasta ilmakuvauksesta. Terra 92/1, p. 34-37.

Kessler, T. 2007. Photographic optics. In Peres, M.R. (ed.), Focal encyclopedia of photography. 4th edition. Elsevier, Amsterdam, p. 711-724.

Klint, K.E.S. and Pedersen, S.A.S. 1995. The Hanklit glaciotectonic thrust fault complex, Mors, Denmark. Geological Survey of Denmark, Series A, no. 35, 30 p.

Koff, T. 1997. Der Einfluss der Entwicklung eines Hochmoores auf die Ausbildung der Pollenspektren am Beispiel des Nigula-Hochmoores (SW Estland). Telma 27, p. 75-90.

Konecny, G. 2003. Geoinformation: Remote sensing, photogrammetry and geographic information systems. Taylor & Francis, London, 248 p.

Koo, T. K. 1993. Low altitude, small format aerial photogrammetry. The Australian Surveyor 38/4, p. 294-297.

Kraus, K. 2004. Photogrammetrie. Band 1: Geometrische Informationen aus Photographien und Laserscanneraufnahmen. De Gruyter, Berlin, 516 p.

Kraus, K., Harley, I. and Kyle, S. 2007. Photogrammetry: Geometry from images and laser scans. De Gruyter, Berlin, 459 p.

Kriss, M. 2007. Solid state imaging sensors. In Peres, M.R. (ed.), Focal encyclopedia of photography. 4th edition. Elsevier, Amsterdam, p. 370-371.

Kronberg, P. 1995. Tektonische Strukturen in Luftbildern und Satellitenaufnahmen: ein Bildatlas. Emke, Stuttgart, Germany, 204 p.

Laliberte, A.S., Rango, A., Herrick, J.E., Fredrickson, E.L. and Burkett, L. 2007. An object-based image analysis approach for determining fractional cover of senescent and green vegetation with digital plot photography. Journal of Arid Environments 69/1, p. 1–14.

Lambers, K., Eisenbeiss, H., Sauerbier, M., Kupferschmidt, D., Gaisecker, T., Sotoodeh, S. and Hanusch, T. 2007. Combining photogrammetry and laser scanning for the recording and modelling of the Late Intermediate Period site of Pinchango Alto, Palpa, Peru. Journal of Archaeological Science 34/10, p. 1702–1712.

Lane, S.N. 2000. The measurement of river channel morphology using digital photogrammetry. Photogrammetric Record 16/96, p. 937–961.

Lane, S.N., Chandler, J.H. and Richards, K.S. 1998. Landform monitoring, modelling and analysis: Land form in geomorphological research. In Lane, S.N. (ed.), Landform monitoring, modelling and analysis. Wiley, Chichester, United Kingdom, p. 1–17.

Langford, M. and Bilissi, E. 2007. Langford's Advanced Photography. 7th revised edition. Elsevier, Boston, 432 p.

Lauer, D.T., Morain, S.A. and Salomonson, V.V. 1997. The Landsat program: Its origins, evolution, and impacts. Photogrammetric Engineering & Remote Sensing 63, p. 831-838.

Leachtenauer, J.C., Daniel, K. and Vogl, T.P. 1997. Digitizing Corona imagery: Quality vs. cost. In R.A. McDonald (ed.), Corona: Between the Sun & the Earth, The first NRO reconnaissance eye in space. American Society Photogrammetry and Remote Sensing, Bethesda, Maryland, United States, p. 189-203.

Lee, W. T. 1922. The face of the earth as seen from the air: A study in the application of airplane photography to geography. American Geographical Society, New York, 110 p.

Leroy, M. and Bréon, F.-M. 1996. Angular signatures of surface reflectances from airborne POLDER data. Remote Sensing of Environment 57/2, p. 97-107.

Lesschen, J.P., Cammeraat, L.H., Kooijman, A.M. and van Wesemael, B. 2008. Development of spatial heterogeneity in vegetation and soil properties after land abandonment in a semi-arid ecosystem. Journal of Arid Environments 72/11, p. 2082–2092.

Li, J., Li, Y., Chapman, M.A. and Rüther, H. 2005. Small format digital imaging for informal settlement mapping. Photogrammetric Engineering & Remote Sensing 71/4, p. 435-442.

Li, Z., Zhu, Q. and Gold, C. 2005. Digital terrain modeling: Principles and methodology. CRC Press, Boca Raton, United States, 323 p.

Lillesand, T.M. and Kiefer, R.W. 1994. Remote sensing and image interpretation. J. Wiley & Sons, New York, 750 p.

Lillesand, T.M., Kiefer, R.W. and Chipman, J.W. 2008. Remote sensing and image interpretation. Wiley, Hoboken, New Jersey, United States, 756 p.

Longley, P.A., Goodchild, M.A., Maguire D. J. and Rhind D. W. 2006. Geographical information systems and science. 2nd edition. Wiley, Chichester, United Kingdom, 517 p.

Loopmann, A., Pirrus, R. and Ilomets, M. 1988. Nigula Riiklik Looduskaitseala. In Valk, U. (ed.), Eesti sood. Valgus, Tallinn, Estonia, p. 227-233.

Lowman, P.D. Jr. 1999. Landsat and Apollo: The forgotten legacy. Photogrammetric Engineering & Remote Sensing 65, p. 1143-1147.

Lucht, W. 2004. Viewing the Earth from multiple angles: Global change and the science of multiangular reflectance. In Schönermark, M. von, Geiger, B. and Röser, H.P. (eds.), Reflection properties of vegetation and soil—with a BDRF data base. Wissenschaft und Technik Verlag, Berlin, p. 9-29.

Lueder, D.R. 1959. Aerial photographic interpretation—Principles and applications. McGraw-Hill, New York, 462 p.

Luftfahrtbundesamt 2009. Das Luftfahrt-Bundesamt - mit Sicherheit zum Ziel. Available from <http://www.lba.de>; (accessed December 2009).

Luhmann, T. 2003. Nahbereichsphotogrammetrie: Grundlagen, Methoden und Anwendungen. 2nd revised edition. Wichmann, Heidelberg, 586 p.

Luhmann, T., Robson, S., Kyle, S. and Harley, I. 2007. Close range photogrammetry: Principles, techniques and applications. Wiley, Chichester, United Kingdom, 528 p.

Lynch, D.K. and Livingston, W. 1995. Color and light in nature. Cambridge University Press, Cambridge, United Kingdom, 254 p.

Mack, J. 2007. The art of small things. Harvard University Press, Cambridge, Massachusetts, United States, 218 p.

Malin, D. and Light, D.L. 2007. Aerial photography. In Peres, M.R. (ed.), Focal encyclopedia of photography. 4th edition. Elsevier, Amsterdam, p. 501-504.

Marani, M., Belluco, E., Ferrari, S., Silvestri, S., D'Alpaos, A., Lanzoni, S., Feola, A. and Rinaldo, A. 2006. Analysis, synthesis and modelling of high-resolution observations of salt-marsh eco-geomorphological patterns in the Venice lagoon. Estuarine, Coastal and Shelf Science 69/3, p. 414–426.

Marinello, F., Bariani, P., Savio, E., Horsewell, A. and Chiffre, L. de 2008. Critical factors in SEM 3D stereo microscopy. Measurement Science and Technology 6, p. 65705.

Markowski, S. 1993. Nad zamkami Polski: Above Poland's castles. Wydawnictwo Postscriptum, Kraków, Poland, 144 p.

Martínez-Casasnovas, J.A., Ramos, M.C. and Poesen, J. 2004. Assessment of sidewall erosion in large gullies using multi-temporal DEMs and logistic regression analysis. Geomorphology 58/1-4, p. 305–321.

Marzolff, I. 1999. Großmaßstäbige Fernerkundung mit einem unbemannten Heißluftzeppelin für GIS-gestütztes Monitoring von Vegetationsentwicklung und Geomorphodynamik in Aragón (Spanien). Doctoral dissertation, Department of Physical Geography, Albert-Ludwigs University Freiburg. Freiburg, Germany, 227 p + app.

Marzolff, I. 2003. Aplicación de la fotografía aérea de alta resolución tomada a partir de un zepelín aerostático teledirigado al seguimiento de la dinámica vegetal y geomorfológica en campos abandonados en Aragón (España). – Hochauflösendes Luftbild-Monitoring von Vegetationsentwicklung und Geomorphodynamik in Aragón (Spanien) mittels ferngesteuerter Heißluftzeppeline. In Marzolff, I., Ries, J. B., La Riva, J. de and Seeger, M. (eds.), Landnutzungswandel und Landdegradation in Spanien – El cambio en el uso del suelo y la degradación del territorio en España. Ergebnisse des Workshops vom 18.-21.10.2001 in Frankfurt am Main. Frankfurt am Main, Germnay/Zaragoza, Spain (Sonderband Frankfurter Geowissenschaftliche Arbeiten/Monografías de la Universidad de Zaragoza), p. 143–162.

Marzolff, I., Albert, K.-D. and Ries, J.B. 2002. Fernerkundung vom Fesseldrachen. Luftbild-Monitoring gibt Aufschluss über Schluchterosion in der Sahelzone. Forschung Frankfurt 3/2002, p. 16-22.

Marzolff, I. and Poesen, J. 2009. The potential of 3D gully monitoring with GIS using high-resolution aerial photography and a digital photogrammetry system. Geomorphology 111/1-2, p. 48–60.

Marzolff, I. and Ries J. B. 1997. 35-mm photography taken from a hot-air blimp. In Bauer, M., Befort, W., Coppin, Ir. Pol R. and Huberty, B. (eds.), Proceedings of the First North American Symposium on Small Format Aerial Photography. American Society of Photogrammetry and Remote Sensing, Bethesda, Maryland, United States, p. 91-101.

Marzolff, I. and Ries, J. B. 2007. Gully monitoring in semi-arid landscapes. Zeitschrift für Geomorphologie 51/4, p. 405–425.

Marzolff, I., Ries, J. B. and Albert, K.-D. 2003. Kite aerial photography for gully monitoring in Sahelian landscapes. Proceedings of the Second Workshop of the EARSeL Special Interest Group on Remote Sensing for Developing Countries, 18-20 September 2002, Bonn, Germany [CD-ROM].

Masing, V. 1997. Ancient mires as nature monuments. Monumenta Estonica, Estonian Encyclopaedia Publ., Tallinn, Estonia, 96 p.

Masing, V. 1998. Multilevel approach in mire mapping, research, and classification. International Mire Conservation Group, Classification Workshop, Greifswald Meeting (1998). Available from <http://www.imcg.net/docum/greifswa/masing.htm>; (accessed December 2009).

Mather, P.M. 2004. Computer processing of remotely-sensed images: An introduction. 3rd edition. Wiley, Chichester, United Kingdom, 324 p.

McGlone, J.C. (ed.) 2004. Manual of photogrammetry. 5th edition. American Society for Photogrammetry and Remote Sensing, Bethesda, Maryland, United States, 1151 p.

Meehan, L. 2003. Digital photography basics. Collins and Brown, United Kingdom, 96 p.

Menzies, J. and Rose, J. 1987. Drumlins—Trends and perspectives. Episodes 10, p. 29-30.

Mills, J.P., Newton, I. and Graham, R.W. 1996. Aerial photography for survey purposes with a high resolution, small format, digital camera. Photogrammetric Record 15/88, p. 575-587.

Mitsch, W.J. and Gosselink, J.G. 2007. Wetlands. 4th edition. J. Wiley & Sons, Hoboken, New Jersey, 582 p.

Miyamoto, M., Yoshino, K., Nagano, T., Ishida, T. and Sata, Y. 2004. Use of balloon aerial photography for classification of Kushiro wetland vegetation, northeastern Japan. Wetlands 24/3, p. 701-710.

Molinillo, M., Lasanta, T. and García-Ruiz, J.M. 1997. Managing mountainous degraded landscapes after farmland abandonment in the Central Spanish Pyrenees. Environmental Management 21/4, p. 587–598.

Mount, R. 2005. Acquisition of through-water aerial survey images: Surface effects and the prediction of sun glitter and subsurface illumination. Photogrammetric Engineering & Remote Sensing 71/12, p. 1407-1415.

Müller, E. and Wöhlecke, W. 2002. Modellflug und Luftrecht. RC-Network Magazin. Available from <http://www.rc-network.de/magazin/artikel_02/art_02-0001/art_02-0001-00.html>; (accessed December 2009).

Mullins, H.T. and Hinchey, E.J. 1989. Erosion and infill of New York Finger Lakes: Implications for Laurentide ice sheet deglaciation. Geology 17, p. 622-625.

Murtha, P.A., Deering, D.W., Olson, C.E. Jr. and Bracher, G.A. 1997. Vegetation. In Philipson, W.R. (ed.), Manual of photographic interpretation. 2nd edition. American Society for Photogrammetry and Remote Sensing, Bethesda, Maryland, United States, p. 225-255.

Nachtergaele, J. and Poesen, J. 1999. Assessment of soil losses by ephemeral gully erosion using high-altitude (stereo) aerial photographs. Earth Surface Processes and Landforms 24/8, p. 693–706.

National Academy of Sciences 2000. The impact of selling the Federal Helium Reserve. National Academy Press. Available from <http://www.nap.edu/catalog.php?record_id=9860>; (accessed February 2010).

Nelson-Atkins Museum of Art, Architecture & History (undated, anonymous). Available from <http://www.nelson-atkins.org/art/HistTreasuredHist.cfm>; (accessed December 2009).

Nilson, T. and Kuusk, A. 1989. A reflectance model for the homogeneous plant canopy and its inversion. Remote Sensing of Environment 27, p. 157-167.

Ogleby, C. 2007. Photogrammetry. In Peres, M.R. (ed.), Focal encyclopedia of photography. 4th edition. Elsevier, Amsterdam, p. 583-584.

Oostwoud Wijdenes, D. and Bryan, R.B. 2001. Gully-Head erosion processes on a semi-arid valley floor in Kenya. Earth Surface Processes and Landforms 26/9, p. 911–933.

Orru, M., Širokova, M. and Veldre, M. 1993. Eesti soo (Estonian mires). Geological Survey of Estonia, map scale = 1:400,000.

Osterman, M. 2007. Selected photographs from the 19th century. In Peres, M.R. (ed.), Focal encyclopedia of photography. 4th edition. Elsevier, Amsterdam, p. 135.

Padeste, R. 2007. Imaging systems. In Peres, M.R. (ed.), Focal encyclopedia of photography. 4th edition. Elsevier, Amsterdam, p. 364-370.

Pawson, D. 1998. The handbook of knots: A step-by-step guide to tying and using more than 100 knots. DK Publishing, New York, 160 p.

Pedersen, S.A.S. 1996. Progressive glaciotectonic deformation in Weichselian and Paleogene deposits at Feggeklit, northern Denmark. Geological Society of Denmark, Bulletin 42, p. 153-174.

Pedersen, S.A.S. 2000. Superimposed deformation in glaciotectonics. Geological Society of Denmark, Bulletin 46, p. 125-144.

Pedersen, S.A.S. 2005. Structural analysis of the Rubjerg Knude glaciotectonic complex, Vendsyssel, northern Denmark. Geological Survey of Denmark and Greenland, Bulletin 8, 192 p.

Perko, R., Fürnstahl, P., Bauer, J. and Klaus, A. 2005. Geometrical accuracy of Bayer pattern images. In Skala, V. (ed.), Proceedings of the 13th International Conference in Central Europe on Computer Graphics, Visualization and Computer Vision (WSCG 2005), Plzeò, Czech Republic, January 31 - February 4, 2005: Short papers. Univ. of West Bohemia, Plzeò, p. 117-120.

Persoz, F., Larsen, E. and Singer, K. 1972. Helium in the thermal springs of Unartoq, South Greenland. Grønlands Geologiske Undersøgelse, Rapport Nr. 44, 21 p.

Peterson, U. and Aunap, R. 1998. Changes in agricultural land use in Estonia in the 1990s detected with multitemporal Landsat MSS imagery. Landscape and Urban Planning 41, p. 193-201.

Pfeiffer, B. and Weimann, G. 1991. Geometrische Grundlagen der Luftbildinterpretation: Einfachverfahren der Luftbildauswertung. Wichmann, Karlsruhe, Germany, 132 p.

Philipson, W.R. (ed.) 1997. Manual of photographic interpretation. 2nd edition. American Society for Photogrammetry and Remote Sensing, Bethesda, Maryland, United States, 689 p.

Pitt, D.G. and Glover, G.R. 1993. Large-scale 35-mm aerial photographs for assessment of vegetation-management research plots in eastern Canada. Canadian Journal of Forest Research 23/10, p. 2159-2169.

Plegnière, S. 2009. Gully-Monitoring in den Bardenas Reales (Nordspanien) mithilfe der Analyse großmaßstäbiger Luftbilder. M.Sc. Thesis, Departments of Physical Geography and Remote Sensing, Trier University. Trier, Germany, 154 p.

Poesen, J., Nachtergaele, J., Verstraeten, G. and Valentin, C. 2003. Gully erosion and environmental change: importance and research needs. Catena 50/2-4, p. 91–133.

Poesen, J., Vandekerckhove, L., Nachtergaele, J., Oostwoud Wijdenes, D., Verstraeten, G. and van Wesemael, B. 2002. Gully erosion in dryland environments. In Bull, L.J. and Kirkby, M.J. (eds.), Dryland rivers: Hydrology and geomorphology of semi-arid channels. Wiley, Chichester, United Kingdom, p. 229-262.

Polyak, L., Gataullin, V., Okuneva, O. and Stelle, V. 2000. New constraints on the limits of the Barents-Kara ice sheet during the last glacial maximum based on borehole stratigraphy from the Pechora Sea. Geology 28/7, p. 611-614.

Prest, V.K. 1983. Canada's heritage of glacial features. Geological Survey of Canada, Miscellaneous Report 28, 119 p.

Preu, C., Naschold, E. and Weerakkody, U. 1987. Der Einsatz einer Ballon-Fotoeinrichtung für küstenmorphologische Fragestellungen. In Hofmeister, B. and Voss, F. (eds.), Beiträge zur Geographie der Küsten und Meere. Berliner geographische Studien 25, p. 377-388.

Prosilica 2009. Prosilica cameras go airborne. Prosilica Camera News 11, May 2009, p. 2-4.

Pu, R., Gong, P., Tians, Y., Miao, X., Carruthers, R.I., Anderson, G.L. 2008. Using classification and NDVI differencing methods for monitoring sparse vegetation coverage: A case study of saltcedar in Nevada, USA. International Journal of Remote Sensing 29/14, p. 3987-4011.

Quilter, M.C. and Anderson, V.J. 2000. Low altitude/large scale aerial photographs: A tool for range and resource managers. Rangelands 22/2, p. 13-17.

Rango, A., Laliberte, A.S., Steele, C., Herrick, J.E., Bestelmeyer, B., Schmugge, T., Roanhorse, A. and Jenkins, V. 2006. Using unmanned aerial vehicles for rangelands: Current applications and future potentials. Environmental Practice 8/3, p. 159–168.

Ranson, K.J., Irons, J.R. and Williams, D.L. 1994. Multispectral bidirectional reflectance of northern forest canopies with the Advanced Solid-State Array Spectroradiometer (ASAS). Remote Sensing of Environment 47, p. 276-289.

Ray, S.F. 2002. Applied photographic optics: Lenses and optical systems for photography, film, video and digital imaging. Focal Press, Oxford, United Kingdom, 656 p.

Richards, J.A. and Jia, X. 2006. Remote sensing digital image analysis: An introduction. 4th edition. Springer, Berlin, 439 p.

Rieke-Zapp, D., Tecklenburg, W., Peipe, J., Hastedt, H. and Haig, C. 2009. Evaluation of the geometric stability and the accuracy potential of digital cameras—Comparing mechanical stabilisation versus parameterisation. ISPRS Journal of Photogrammetry & Remote Sensing 64/3, p. 248-258.

Ries, J.B. 2000. Geomorphodynamik und Landdegradation auf Brachflächen zwischen Ebrobecken und Pyrenäen. Großmaßstäbiges Monitoring zur Erfassung und Prognose des Prozessgeschehens im Landnutzungswandel als Beitrag zur Methodenentwicklung. Habilitation thesis, Faculty of Geosciences/Geography, Frankfurt University. Frankfurt am Main, Germany, 598 p. + app.

Ries, J. B. 2002. Geomorphodynamics on fallow land and abandoned fields in the Ebro Basin and the Pyrenees - Monitoring of processes and development. In Schmidt, K.-H. and Vetter, T., (eds.), Late Quaternary Geomorphodynamics. Zeitschrift für Geomorphologie N.F., Suppl.-Bd. 127. Borntraeger, Berlin, p. 21-45.

Ries, J. 2003. Landnutzungswandel und Landdegradation in Spanien - eine Einführung. Cambios de uso del suelo y degradación de territorio en España - una breve introducción. In Marzolff, I., Ries, J. B., La Riva, J. de and Seeger, M. (eds.), Landnutzungswandel und Landdegradation in Spanien – El cambio en el uso del suelo y la degradación del territorio en España. Ergebnisse des Workshops vom 18.-21.10.2001 in Frankfurt am Main, Germany. Sonderband Frankfurter Geowissenschaftliche Arbeiten/Monografías de la Universidad de Zaragoza, Frankfurt am Main/Zaragoza, p. 11–29.

Ries, J.B. and Hirt, U. 2008. Permanence of soil surface crusts on abandoned farmland in the Central Ebro Basin/Spain. Catena 72/2, p. 282–296.

Ries, J.B. and Marzolff, I. 2003. Monitoring of gully erosion in the Central Ebro Basin by large scale aerial photography taken from a remotely controlled blimp. Catena 50/2-4, p. 309-328.

Ries, J.B. and Marzolff, I. 2007. Großmaßstäbiges Gully-Monitoring in semiariden Landschaften (MoGul – DFG-Projekt RI 835/2-1 bzw. MA 2549/1-1): Abschlussbericht/Final report. Frankfurt University, Germany, 19 p. Available from <http://www.geo.uni-frankfurt.de/fb/fb11/ipg/ag/ma/downloads/Abschlussbericht_MoGul-Projekt_2007.pdf>; (accessed December 2009).

Ries, J.B., Marzolff, I., Seeger, M., 2000. Der Beweidungseinfluß auf Vegetationsbedeckung und Bodenerosion in der Flysch-Zone der spanischen Pyrenäen. In Zollinger, G. (ed.), Aktuelle Beiträge zur angewandten physischen Geographie der Tropen, Subtropen und der Regio Trirhenia., Festschrift zum 60, p. 167-194. Geburtstag von Prof. Dr. Rüdiger Mäckel., Freiburg, Germany.

Ries, J. B., Marzolff, I. and Seeger, M. 2003. Einfluss der Beweidung auf Vegetationsbedeckung und Geomorphodynamik zwischen Ebrobecken und Pyrenäen. Geographische Rundschau 55/5, p. 52-59.

Ries, J. B., Seeger, M. and Marzolff, I. 1998. El proyecto EPRODESERT. Cambios de uso del suelo y morfodinámica en el Nordeste de España. Geograficalia 35, p. 205-225.

Ries, J. B., Seeger, M. and Marzolff, I. 2004. Influencia del pastoreo en la cubierta vegetal y la geomorfodinámica en el transecto Depresión del Ebro-Pirineos. Geograficalia 45, p. 5-19.

Robinson, M. 2003a. The genius of Rogallo's Wing. Kiting 25/2, p. 27, 32.

Robinson, M. 2003b. The flying cowboy. Kiting 25/3, p. 27, 34-35.

Romer, G.B. 2007. Introduction to the biographies of selected innovators of photographic technology. In Peres, M.R. (ed.), Focal encyclopedia of photography. 4th edition. Elsevier, Amsterdam, p. 123-134.

Rosenthaler, L. 2007. Digital archiving. In Peres, M.R. (ed.), Focal encyclopedia of photography. 4th edition. Elsevier, Amsterdam, p. 359-364.

Rotomotion 2009. Complete UAS and UAV Helicopters. Available from <http://www.rotomotion.com>; (accessed December 2009).

ROTROB 2009. MARVIN - Multipurpose Aerial Robot Vehicle with Intelligent Navigation. Available from <http://rotrob.com>; (accessed December 2009).

Royal & Ancient Golf Club of St. Andrews 2006 (anonymous). Course management best practice guidelines. Available from <http://www.randa.org/>; (accessed December 2009).

Royal & Ancient Golf Club of St. Andrews 2008 (anonymous). Eastern European golf set to grow sustainably. Available from <http://www.randa.org/>; (accessed December 2009).

Ryerson, R.A., Curran, P.J. and Stephans, P.R. 1997. Agriculture. In Philipson, W.R. (ed.), Manual of photographic interpretation. 2nd edition. American Society for Photogrammetry and Remote Sensing, Bethesda, Maryland, United States, p. 365-397.

Sandmeier, S. 2004. Spectral variability of BRDF-data. In Schönermark, M. von, Geiger, B. and Röser, H.P. (eds.), Reflection properties of vegetation and soil—with a BRDF data base. Wissenschaft und Technik Verlag, Berlin, p. 131-146.

Schaaf, C.B. and Strahler, A.H. 1994. Validation of bidirectional and hemispherical reflectances from a geometric-optical model using ASAS imagery and pyranometer measurements of a spruce forest. Remote Sensing of Environment 49, p. 138-144.

Scheritz, M., Dietrich, R., Scheller, S., Schneider, W. and Boike, J. 2008. High resolution digital elevation model of polygonal patterned ground on Samoylov Island, Siberia, using smallformat photography. In Kane, D.L., Hinkel, K.M. (eds.), Proceedings of the Ninth International Conference on Permafrost. June 29–July 3, 2008, University of Alaska, Fairbanks, United States, p. 1589-1594.

Schneider, S. 1974. Luftbild und Luftbildinterpretation. De Gruyter, Berlin, 530 p.

Schot, P.P. 1999. Wetlands. In Nath, B. et al. (eds.), Environmental Management in Practice 3. Routledge, London & New York, p. 62-85.

Schwarm, L. and Adams, R. 2003. On fire / Larry Schwarm. Duke University Press, Durham, North Carolina, United States (pages not numbered).

Seang, T.P. and Mund, J.-P. 2006. Balloon-based geo-referenced digital photo technique - a low cost high-resolution option for developing countries. Proceedings of the XXIII International FIG Congress, 8-13 October 2006, Munich, Germany: Shaping the Change, p. 1-12.

Seang, T.P., Mund, J.P. and Symann, R. 2008. Practitioner's guide: Low cost amateur aerial pictures with balloon and digital camera. MethodFinder, Usingen, Germany, 14 p.

Seeger, M., Marzolff, I. and Ries, J.B. 2009. Identification of gully-development processes in semi-arid NE Spain. Zeitschrift für Geomorphologie 53/4, p. 417-431.

Shackelford, A. 2004. Development of urban area geospatial information products from high resolution satellite imagery using advanced image analysis techniques. Doctoral dissertation, University of Missouri-Columbia, United States, 185 p.

Shan, J., Fu, C.-S., Li, B., Bethel, J., Kretsch, J. and Mikhail, E. 2006. Principles and evaluation of autostereoscopic photogrammetric measurement. Photogrammetric Engineering & Remote Sensing 72/4, p. 365–372.

Shaw, J. 1994. Landscape photography: Professional techniques for shooting spectacular scenes. Amphoto, Watson-Guptill Publ., New York, 144 p.

Shortis, M.R., Bellman, C.J., Robson, S., Johnston, G.J. and Johnson, G.W. 2006. Stability of zoom and fixed lenses used with digital SLR cameras. International Archives of the Photogrammetry, Remote Sensing and Spatial Information Sciences 36/5, p. 285-290.

Shortis, M.R., Seager, J.W., Harvey, E.S. and Robson, S. 2005. The influence of Bayer filters on the quality of photogrammetric measurement. In Beraldin, J.A., El-Hakim, S.F., Gruen, A. and Walton, J.S. (eds.), Videometrics VIII. Proceedings of the Electronic Imaging, Science and Technology Symposium, 18 - 20 January 2005, San Jose, California, USA. SPIE 5665. Bellingham, Washington, United States, p. 164-171.

Smith, M.J., Chandler, J.H. and Rose, J. 2009. High spatial resolution data acquisition for the geosciences: Kite aerial photography. Earth Surface Processes and Landforms 34/1, p. 155–161.

Steenblock, G. 2006. Hauptsache gut versichert. RC-Network Magazin. Available from <http://www.rc-network.de/magazin/artikel_06/art_06-021/art_021-01.html>; (accessed December 2009).

Steiert, J. and Meinzer, W. 1995. Playas: Jewels of the plains. Texas Tech University Press, Lubbock, Texas, United States, 134 p.

Steinbeck, E.A. (ed.) 1975. Steinbeck: A life in letters. Viking Press, New York, 906 p.

Stroebel, L. 2007. Perspective. In Peres, M.R. (ed.), Focal encyclopedia of photography. 4th edition. Elsevier, Amsterdam, p. 728-733.

Sutton, K. 1999. From 'chute to kite: How the classic flow form came to be. Kitelines 13/1, p. 39-41.

Svendsen, J.I. et al. 1999. Maximum extent of the Eurasian ice sheets in the Barents and Kara Sea region during the Weichselian. Boreas 28/1, p. 234-242.

Tatem, A.J., Goetz, S.J. and Hay, S.I. 2008. Fifty years of Earth-observation satellites. American Scientist 96, p. 390-398.

Teng, W.L., Loew, E.R., Ross, D.I., Zsilinszky, V.G., Lo, C.P., Philipson, W.R., Philpot, W.D. and Morain, S.A. 1997. Fundamentals of photographic interpretation. In Philipson, W.R. (ed.), Manual of photographic interpretation. 2nd edition. American Society for Photogrammetry and Remote Sensing, Bethesda, Maryland, United States, p. 49-113.

Thamm, H.P. and Judex, M. 2006. The "low cost drone" – An interesting tool for process monitoring in a high spatial and temporal resolution. International Archives of the Photogrammetry, Remote Sensing and Spatial Information Sciences 36/7, p. 140-144.

Thornes, J. (ed.) 1990. Vegetation and erosion. Processes and environments. Wiley, Chichester, New York, Brisbane, 518 p.

Thornes, J. 2008. Vegetation and erosion. Introduction to session SSS2 of the European Geophysical Union General Assembly 2009, Vienna, Austria. Available from <http://meetingorganizer.copernicus.org/EGU2009/session/878>; (accessed December 2009).

Tielkes, E. 2003. L'œil du cerf-volant: Evaluation et suivi des états de surface par photographie aérienne sous cerf-volant. Margraf Publ., Weikersheim, Germany, 113 p.

Tiner, R.W. 1997. Wetlands. In Philipson, W.R. (ed.), Manual of photographic interpretation. 2nd edition. American Society for Photogrammetry and Remote Sensing, Bethesda, Maryland, United States, p. 475-494.

Trémeau, A., Tominaga, S. and Plataniotis, K.N. 2008. Color in image and video processing: Most recent trends and future research directions. EURASIP Journal on Image and Video Processing, Article ID 581371, p. 1-26.

Troll, C. 1939. Luftbildplan und ökologische Bodenforschung: ihr zweckmäßiger Einsatz für die wissenschaftliche Erforschung und praktische Erschließung wenig bekannter Länder. Zeitschrift der Gesellschaft für Erdkunde zu Berlin 7/8, p. 241-298.

Tschudi, M.A., Maslanik, J.A. and Perovich, D.K. 2008. Derivation of melt pond coverage on Arctic sea ice using MODIS observations. Remote Sensing of Environment 112/5, p. 2605–2614.

Tucker, A. 2009. Winging it. Smithsonian 39/10, p. 46-53.

Tucker, C.J. 1979. Red and photographic infrared linear combinations for monitoring vegetation. Remote Sensing of Environment 8, p. 127-150.

Ullmann, H. 1971. Hochmoor-Luftbilder mit Hilfe eines Kunststoffballons. Österreich. Bot. Zeitschrift 119, p. 548-556.

U.S. Library of Congress 2007 (anonymous). Selected photographers & examples of their work: George R. Lawrence. Available from <http://lcweb2.loc.gov/ammem/collections/panoramic_photo/pnphtgs.html>; (accessed December 2009).

Vandaele, K., Poesen, J., Marques de Silva, J.R., Govers, G. and Desmet, P.J. 1997. Assessment of factors controlling ephemeral gully erosion in southern Portugal and central Belgium using aerial photographs. Zeitschrift für Geomorphologie 41, p. 273–287.

Vandekerckhove, L., Poesen, J. and Govers, G. 2003. Medium-term gully headcut retreat rates in Southeast Spain determined from aerial photographs and ground measurements. Catena 50/2-4, p. 329–352.

Vandekerckhove, L., Poesen, J., Oostwoud Wijdenes, D. and Gyssels, G. 2001. Short-term gully retreat rates in Mediterranean environments. Catena 44/2, p. 133–161.

Verhoeven, G. 2008. Imaging the invisible using modified digital still cameras for straightforward and low-cost archaeological near-infrared photography. Journal of Archaeological Science 35/12, p. 3087-3100.

Verhoeven, G., Loenders, J., Vermeulen, F. and Docter, R. 2009. Helikite aerial photography or HAP – A versatile means of unmanned, radio controlled low altitude aerial archaeology. Archaeological Prospection 16/2, p. 125-138.

Vericat, D., Brasington, J., Wheaton, J.M. and Cowie, M. 2009. Accuracy assessment of aerial photographs acquired using lighter-than-air blimps: Low-cost tools for monitoring fluvial systems. River Research and Applications 25/8, p. 985-1000.

Vierling, L.A., Fersdahl, M., Chen, X., Li, Z. and Zimmerman, P. 2006. The short wave aerostat-mounted imager (SWAMI): A novel platform for acquiring remotely sensed data from a tethered balloon. Remote Sensing of Environment 103/3, p. 255–264.

Wackrow, R. 2008. Spatial measurement with consumer grade digital cameras. Doctoral dissertation, Faculty of Engineering, Loughborough University. Loughborough, United Kingdom, 187 p.

Wackrow, R., Chandler, J.H. and Bryan, P. 2007. Geometric consistency and stability of consumer-grade digital cameras for accurate spatial measurement. Photogrammetric Record 22/118, p. 121-134.

Walker, J. W. and de Vore, S. L. 1995. Low altitude large scale reconnaissance: A method of obtaining high resolution vertical photographs for small areas. Interagency Archeological Services, Denver, United States, 161 p.

Wanzke, H. 1984. The employment of a hot-air ship for the stereophotogrammetric documentation of antique ruins. International Archives of Photogrammetry and Remote Sensing 25 (A5), p. 746-756.

Ward, R. and Fidler, P.J. 1993. The Nelson-Atkins Museum of Art: A handbook of the collection. Hudson Hills Press, New York, 414 p.

Warner, W.S., Graham, R.W. and Read, R.E. 1996. Small format aerial photography. American Society for Photogrammetry and Remote Sensing, Bethesda, Maryland, United States, 348 p.

Wiesnet, D.R., Wagner, C.R. and Philpot, W.D. 1997. Water, snow, and ice. In Philipson, W.R. (ed.), Manual of photographic interpretation. 2nd edition. American Society for Photogrammetry and Remote Sensing, Bethesda, Maryland, United States, p. 257-267.

Wildi, E. 2006. Master composition guide for digital photographers. Amherst Media, Buffalo, New York, 123 p.

Williams, J. 2005. Standard atmosphere tables. USA Today. Available from <http://www.usatoday.com/weather/wstdatmo.htm>; (accessed December 2009).

Williams, R.S. Jr. 1986. Glaciers and glacial landforms. In Short, N.M. and Blair, R.W. Jr. (eds.), Geomorphology from space. NASA Special Publication 486, p. 521-596.

Williams, R.S. Jr. and Carter, W.D. (eds.) 1976. ERTS-1: A new window on our planet. U.S. Geological Survey, Professional Paper 929, 362 p.

Wilmarth, V.R., Kaltenbach, J.L. and Lenoir, W.B. 1977. Skylab explores the Earth. NASA Special Publication 380, 517 p.

Wilson, C. 2006. Hanging by a thread: A kite's view of Wisconsin. Itchy Cat Press, Blue Mounds, Wisconsin, United States, 131 p.

Wolf, P.R. and Dewitt, B.A. 2000. Elements of photogrammetry with applications in GIS. McGraw Hill, Boston, 608 p.

Wood, T. and Slegman, A. 2007. Bold expansion: The Nelson-Atkins Museum of Art Block Building. Scala Publishers, London, 64 p.

Woodhouse, N. 2009. LPS eATE Preview. Available from <http://labs.erdas.com>; (accessed December 2009).

Wu, Y., Zhang, Q. and Liu, S. 2008. A contrast among experiments in three low-altitude unmanned aerial vehicles photography: Security, quality & efficiency. International Archives of the Photogrammetry, Remote Sensing and Spatial Information Sciences 37 (B1), p. 1223-1227.

Wu, Y., Zheng, Q., Zhang, Y., Liu, B., Cheng, H. and Wang, Y. 2008. Development of gullies and sediment production in the black soil region of northeastern China. Geomorphology 101/4, p. 683–691.

Zahorcak, M. 2007. Evolution of the photographic lens in the 19th century. In Peres, M.R. (ed.), Focal encyclopedia of photography. 4th edition. Elsevier, Amsterdam, p. 157-176.

Zavalata, E. 2000. The economic value of controlling an invasive shrub. Ambio 29/8, p. 462-467.

Zhan, Q., Shi, W. and Xiao, Y. 2005. Quantitative analysis of shadow effects in high-resoluton images of urban areas. International Archives of the Photogrammetry, Remote Sensing and Spatial Information Sciences, Vol. XXXVI, Part 8/W27, 6 p. [CD-ROM]. Available from <http://www.isprs.org/commission8/workshop_urban/zhan.pdf>; (accessed December 2009).

Zimmerman, J.L. 1990. Cheyenne Bottoms: Wetland in jeopardy. University Press of Kansas, Lawerence, Kansas, United States, 197 p.

Zuckerman, J. 1996. Techniques of natural light photography. Writer's Digest Books, Cincinnati, Ohio, United States, 134 p.

Index

Page numbers marked with f refer to figure captions, page numbers marked with t refer to table captions.

C

J

K

L

M

T

Printed in the United States
By Bookmasters